Sven Helmer, Alexandra Poulovassilis, and Fatos Xhafa

Reasoning in Event-Based Distributed Systems

T0205436

Studies in Computational Intelligence, Volume 347

Editor-in-Chief

Prof. Janusz Kacprzyk
Systems Research Institute
Polish Academy of Sciences
ul. Newelska 6
01-447 Warsaw
Poland
E-mail: kacprzyk@ibspan.waw.pl

Further volumes of this series can be found on our homepage:
springer.com

Vol. 324. Alessandro Soro, Vargiu Eloisa, Giuliano Armano, and Gavino Paddeu (Eds.)
Information Retrieval and Mining in Distributed Environments,
2010
ISBN 978-3-642-16088-2

Vol. 325. Quan Bai and Naoki Fukuta (Eds.)
Advances in Practical Multi-Agent Systems, 2010
ISBN 978-3-642-16097-4

Vol. 326. Sheryl Brahnam and Lakhmi C. Jain (Eds.)
Advanced Computational Intelligence Paradigms in Healthcare
5, 2010
ISBN 978-3-642-16094-3

Vol. 327. Slawomir Wiak and
Ewa Napieralska-Juszczak (Eds.)
Computational Methods for the Innovative Design of Electrical
Devices, 2010
ISBN 978-3-642-16224-4

Vol. 328. Raoul Huys and Viktor K. Jirsa (Eds.)
Nonlinear Dynamics in Human Behavior, 2010
ISBN 978-3-642-16261-9

Vol. 329. Santi Caballé, Fatos Xhafa, and Ajith Abraham (Eds.)
Intelligent Networking, Collaborative Systems and Applications,
2010
ISBN 978-3-642-16792-8

Vol. 330. Steffen Rendle
Context-Aware Ranking with Factorization Models, 2010
ISBN 978-3-642-16897-0

Vol. 331. Athena Vakali and Lakhmi C. Jain (Eds.)
New Directions in Web Data Management 1, 2011
ISBN 978-3-642-17550-3

Vol. 332. Jianguo Zhang, Ling Shao, Lei Zhang, and
Graeme A. Jones (Eds.)
Intelligent Video Event Analysis and Understanding, 2011
ISBN 978-3-642-17553-4

Vol. 333. Fedja Hadzic, Henry Tan, and Tharam S. Dillon
Mining of Data with Complex Structures, 2011
ISBN 978-3-642-17556-5

Vol. 334. Álvaro Herrero and Emilio Corchado (Eds.)
Mobile Hybrid Intrusion Detection, 2011
ISBN 978-3-642-18298-3

Vol. 335. Radomir S. Stankovic and Radomir S. Stankovic
From Boolean Logic to Switching Circuits and Automata, 2011
ISBN 978-3-642-11681-0

Vol. 336. Paolo Remagnino, Dorothy N. Monekosso, and Lakhmi
C. Jain (Eds.)
Innovations in Defence Support Systems – 3, 2011
ISBN 978-3-642-18277-8

Vol. 337. Sheryl Brahnam and Lakhmi C. Jain (Eds.)
Advanced Computational Intelligence Paradigms in
Healthcare 6, 2011
ISBN 978-3-642-17823-8

Vol. 338. Lakhmi C. Jain, Eugene V. Aidman, and
Canicious Abeynayake (Eds.)
Innovations in Defence Support Systems – 2, 2011
ISBN 978-3-642-17763-7

Vol. 339. Halina Kwasnicka, Lakhmi C. Jain (Eds.)
Innovations in Intelligent Image Analysis, 2010
ISBN 978-3-642-17933-4

Vol. 340. Heinrich Hussmann, Gerrit Meixner, and
Detlef Zuehlke (Eds.)
Model-Driven Development of Advanced User Interfaces, 2011
ISBN 978-3-642-14561-2

Vol. 341. Stéphane Doncieux, Nicolas Bredeche, and
Jean-Baptiste Mouret(Eds.)
New Horizons in Evolutionary Robotics, 2011
ISBN 978-3-642-18271-6

Vol. 342. Federico Montesino Pouzols, Diego R. Lopez, and
Angel Barriga Barros
Mining and Control of Network Traffic by Computational
Intelligence, 2011
ISBN 978-3-642-18083-5

Vol. 343. Kurosh Madani, António Dourado Correia,
Agostinho Rosa, and Joaquim Filipe (Eds.)
Computational Intelligence, 2011
ISBN 978-3-642-20205-6

Vol. 344. Atilla Elçi, Mamadou Tadiou Koné, and
Mehmet A. Orgun (Eds.)
Semantic Agent Systems, 2011
ISBN 978-3-642-18307-2

Vol. 345. Shi Yu, Léon-Charles Tranchevent,
Bart De Moor, and Yves Moreau
Kernel-based Data Fusion for Machine Learning, 2011
ISBN 978-3-642-19405-4

Vol. 346. Weisi Lin, Dacheng Tao, Janusz Kacprzyk, Zhu Li,
Ebroul Izquierdo, and Haohong Wang (Eds.)
Multimedia Analysis, Processing and Communications, 2011
ISBN 978-3-642-19550-1

Vol. 347. Sven Helmer, Alexandra Poulovassilis, and Fatos Xhafa
Reasoning in Event-Based Distributed Systems, 2011
ISBN 978-3-642-19723-9

Sven Helmer, Alexandra Poulovassilis, and Fatos Xhafa

Reasoning in Event-Based Distributed Systems

 Springer

Dr. Sven Helmer
University of London
Dept. of Computer Science and Inf. Systems,
Birkbeck
Malet Street
London WC1E 7HX
UK
E-mail: sven@dcs.bbk.ac.uk

Dr. Fatos Xhafa
University of London
London Knowledge Lab, Birkbeck
23-29 Emerald Street
London WC1N 3QS
UK
E-mail: fatos@lsi.upc.edu

Prof. Alexandra Poulovassilis
University of London
London Knowledge Lab, Birkbeck
23-29 Emerald Street
London WC1N 3QS
UK
E-mail: ap@dcs.bbk.ac.uk

ISBN 978-3-642-26786-4   ISBN 978-3-642-19724-6 (eBook)

DOI 10.1007/978-3-642-19724-6

Studies in Computational Intelligence   ISSN 1860-949X

Typeset & Cover Design: Scientific Publishing Services Pvt. Ltd., Chennai, India.

Printed on acid-free paper

9 8 7 6 5 4 3 2 1

springer.com

Fatos Xhafa dedicates this book to the memory of his father.

Preface

Event-based distributed systems are playing an ever increasing role in areas such as enterprise management, information dissemination, finance, environmental monitoring and geo-spatial systems. Event-based processing originated with the introduction of Event-Condition-Action (ECA) rules to database systems in the 1980s. Since then, the use of ECA rules and the reactive computing paradigm has spread widely into middleware, stream processing, wireless sensor networks and radio frequency identification (RFID).

The wide propagation of event-based processing spanning numerous application domains explains why many different distributed architectures are being used for event-based systems, including publish-subscribe, Peer-to-Peer, Grid, event-stream processing and message queues. As such systems become more complex and more pervasive, intelligent techniques are needed for detecting and processing events that are of interest to users from the possibly huge volumes of low-level event occurrences. Complex Event Processing aims to correlate simple event occurrences into more meaningful derived events and is the topic of several chapters of this book. Other research issues include detection of new or unusual events, optimisation of event processing, event consumption policies, privacy and security, system dynamicity and responsiveness, and quality of service guarantees.

Intelligent and logic-based approaches provide sound foundations for addressing many of the research challenges, and this book covers a broad range of recent advances contributed by leading experts in the field. Reasoning about the properties and behaviour of event-based distributed systems presents significant challenges beyond those of centralised systems due to their greater complexity and dynamicity, and their temporal, spatial and context-aware characteristics. Nevertheless, this also opens up opportunities for building highly scalable and adaptable systems. The fundamental concepts presented in this book are illustrated with examples drawn from applications in areas such as supply chain management, environmental and traffic monitoring, patient monitoring, data centre and network monitoring, fraud detection, smart homes, role-based access control, spacecraft and satellite data monitoring, online collaboration in virtual organisations, monitoring market data, and monitoring business processes.

The target audience of the book are senior year undergraduate and graduate students, as well as instructors, researchers and industry professionals. The book covers theoretical approaches, architectural frameworks, system implementations and applications. The first three chapters provide foundational material which gives the necessary background for reading the other chapters for those who are unfamiliar with the subject. The chapters have been contributed by many leading experts in the field and we hope that the book will be become a useful reference and resource for those who are already working in this exciting and rapidly evolving field or are moving into it.

Acknowledgment

The editors are grateful to the authors of this volume for their contributions. We would like to thank Professor Janusz Kacprzyk (Editor-in-Chief, Springer Studies in Computational Intelligence Series) for the the support and encouragement and to Dr. Thomas Ditzinger (Springer Engineering Inhouse Editor, Studies in Computational Intelligence Series), and Ms. Heather King (Editorial Assistant, Springer Verlag, Heidelberg) for their support and excellent collaboration during the edition of the book.

Fatos Xhafa's work was undertaken during his stay at Birkbeck, University of London, UK (on leave from Technical University of Catalonia, Spain) and was supported by the General Secretariat of Universities of the Ministry of Education, Spain.

December 2010

Sven Helmer
Alex Poulovassilis
Fatos Xhafa

Contents

List of Contributors

Raman Adaikkalavan
Computer Science & Informatics,
Indiana University South Bend
e-mail: raman@cs.iusb.edu

Darko Anicic
FZI Forschungszentrum Informatik,
Haid-und-Neu-Str. 10-14, 76131
Karlsruhe, Germany
e-mail: darko.anicic@fzi.de

Jean Bacon
Computer Laboratory, University of
Cambridge, JJ Thomson Avenue,
Cambridge CB3 0FD, UK
e-mail: jean.bacon@cl.cam.ac.uk

Pedro Bizarro
University of Coimbra, Portugal
e-mail: bizarro@dei.uc.pt

Simon Brodt
Institute for Informatics, University
of Munich, Oettingenstr. 67, 80538
Munich, Germany
e-mail: brodt@pms.ifi.lmu.de

Francois Bry
Institute for Informatics, University
of Munich, Oettingenstr. 67, 80538
Munich, Germany
e-mail: bry@pms.ifi.lmu.de

Stefano A. Cerri
CNRS, Centre National de Recherche Scientifique, LIRMM, Laboratoire d'Informatique de Robotique et

de Microelectronique de Montpellier
161 rue Ada, 34392 Montpellier
Cedex 5 France
e-mail: cerri@lirrm.fr

Sharma Chakravarthy
Computer Science and Engineering,
The University of Texas At
Arlington
e-mail: sharma@cse.uta.edu

Alexandru Costan
University Politehnica of Bucharest,
313 Splaiul Independentei, 060042
Bucharest, Romania
e-mail: alexandru.costan@cs.pub.ro

Valentin Cristea
University Politehnica of Bucharest,
313 Splaiul Independentei, 060042
Bucharest, Romania
e-mail: valentin.cristea@cs.pub.ro

Ciprian Dobre
University Politehnica of Bucharest,
313 Splaiul Independentei, 060042
Bucharest, Romania
e-mail: ciprian.dobre@cs.pub.ro

Pascal Dugenie
CNRS, Centre National de Recherche Scientifique, LIRMM, Laboratoire d'Informatique de Robotique et
de Microelectronique de Montpellier
161 rue Ada, 34392 Montpellier
Cedex 5 France
e-mail: dugenie@lirmm.fr

Michael Eckert
TIBCO Software, Balanstr. 49,
81669 Munich, Germany
e-mail: meckert@tibco.com

Opher Etzion
IBM Haifa Research Lab, Haifa,
Israel
e-mail: opher@il.ibm.com

David M. Eyers
Computer Laboratory, University of
Cambridge, JJ Thomson Avenue,
Cambridge CB3 0FD, UK
e-mail: david.eyers@cl.cam.ac.uk

Avigdor Gal
Technion –Israel Institute of Technology, Faculty of Industrial Engineering & Management, Technion
City, 32000 Haifa, Israel
e-mail: avigal@ie.technion.ac.il

Dieter Gawlick
Oracle, California
e-mail: dieter.gawlick@oracle.com

Diogo Guerra
FeedZai, Portugal
e-mail:
diabetesogo.guerra@feedzai.com

Paul Fodor
Stony Brook University, Stony
Brook, NY 11794, U.S.A.
e-mail: pfodor@cs.sunysb.edu

Gregor Hackenbroich
SAP Research Dresden, Chemnitzer
Straße 48, 01187 Dresden, Germany
e-mail:
gregor.hackenbroich@sap.com

Steffen Hausmann
Institute for Informatics, University
of Munich, Oettingenstr. 67, 80538

Munich, Germany
e-mail: hausmann@pms.ifi.lmu.de

Sven Helmer
Department of Computer Science
and Information Systems, Birkbeck,
University of London, UK
e-mail: sven@dcs.bbk.ac.uk

Zbigniew Jerzak
SAP Research Dresden, Chemnitzer
Straße 48, 01187 Dresden, Germany
e-mail: zbigniew.jerzak@sap.com

Anja Klein
SAP Research Dresden, Chemnitzer
Straße 48, 01187 Dresden, Germany
e-mail: anja.klein@sap.com

Yonit Magid
IBM Haifa Research Lab, Haifa,
Israel
email: yonit@il.ibm.com

Matteo Migliavacca
Department of Computing, Imperial
College London, 180 Queen's Gate,
London SW7 2AZ, UK
e-mail: migliava@doc.ic.ac.uk

Ken Moody
Computer Laboratory, University of
Cambridge, JJ Thomson Avenue,
Cambridge CB3 0FD, UK
e-mail: ken.moody@cl.cam.ac.uk

Ioannis Papagiannis
Department of Computing, Imperial
College London, 180 Queen's Gate,
London SW7 2AZ, UK
e-mail: ip108@doc.ic.ac.uk

Peter Pietzuch
Department of Computing, Imperial
College London, 180 Queen's Gate,

London SW7 2AZ, UK
e-mail: prp@doc.ic.ac.uk

Florin Pop
University Politehnica of Bucharest,
313 Splaiul Independentei, 060042
Bucharest, Romania
e-mail: florin.pop@cs.pub.ro

Olga Poppe
Institute for Informatics, University
of Munich, Oettingenstr. 67, 80538
Munich, Germany
e-mail: poppe@pms.ifi.lmu.de

Alex Poulovassilis
London Knowledge Lab, Birkbeck,
University of London
e-mail: ap@dcs.bbk.ac.uk

Ella Rabinovich
IBM Haifa Research Lab, Haifa,
Israel
e-mail: ellak@il.ibm.com

George Roussos
Birkbeck College, University of
London, WC1E 7HX
e-mail: gr@dcs.bbk.ac.uk

Sebastian Rudolph
Karlsruhe Institute of Technology,
Germany
e-mail: rudolph@kit.edu

Brian Shand
CBCU / Eastern Cancer Registry
and Information Centre, National
Health Service, Unit C –Magog
Court, Shelford Bottom, Hinton
Way, Cambridge CB22 3AD, UK
e-mail: Brian.Shand@cbcu.nhs.uk

Inna Skarbovsky
IBM Haifa Research Lab, Haifa,
Israel
e-mail: inna@il.ibm.com

Roland Stühmer
FZI Forschungszentrum Informatik,
Germany
e-mail: stuehmer@fzi.de

Nenad Stojanovic
FZI Forschungszentrum Informatik,
Germany
e-mail: nenad.stojanovic@fzi.de

Rudi Studer
FZI Forschungszentrum Informatik,
Germany
e-mail: studer@fzi.de

Segev Wasserkrug
IBM Haifa Research Lab, Haifa,
Israel
e-mail: segevw@il.ibm.com

Fatos Xhafa
Department of Languages and
Informatics Systems, Technical
University of Catalonia, Spain
e-mail: fatos@lsi.upc.edu

Nir Zolotorevsky
IBM Haifa Research Lab, Haifa,
Israel
e-mail: nirz@il.ibm.com

Michael Zoumboulakis
Birkbeck College, University of
London, WC1N 3QS
e-mail: mz@dcs.bbk.ac.uk

Introduction to Reasoning in Event-Based Distributed Systems

Sven Helmer, Alex Poulovassilis, and Fatos Xhafa

1 Event-Based Distributed Systems

Event-based distributed systems have played an important role in distributed computing in the recent past and their influence is steadily increasing. With the rapid development of Internet technologies, such systems are gaining in importance in a broad range of application domains, including enterprise management, environmental monitoring, information dissemination, finance, pervasive systems, autonomic computing, geo-spatial systems, collaborative working and learning, and online games. Event-based computing is becoming a central aspect of emerging large-scale distributed computing paradigms such as Grid, Peer-to-Peer and Cloud computing, wireless networking systems, and mobile information systems.

The general motivation for distributed processing (which applies also to event-based distributed systems) is that it allows for more scalable and more reliable systems. Moreover, in the context of event-based distributed systems specifically, events occurring over a wide area can be partially filtered and aggregated locally before being shipped to their destination, i.e. to the component or components of

Sven Helmer
Department of Computer Science and Information Systems, Birkbeck,
University of London, UK
Tel.: +44 (0)20 7631 6718; Fax: +44 (0)20 7631 6727
e-mail: sven@dcs.bbk.ac.uk

Alex Poulovassilis
London Knowledge Lab, Birkbeck, University of London
Tel.: +44 (0)20 7631 6705 / (0)20 7763 2120; Fax: +44 (0)20 7631 6727
e-mail: ap@dcs.bbk.ac.uk

Fatos Xhafa
Department of Languages and Informatic Systems, Technical University of Catalonia, Spain
Tel.: +34 93 4137880; Fax: +34 93 4137833
e-mail: fatos@lsi.upc.edu

S. Helmer et al.: Reasoning in Event-Based Distributed Systems, SCI 347, pp. 1–10.
springerlink.com   © Springer-Verlag Berlin Heidelberg 2011

the system responsible for processing information about event occurrences and reacting appropriately.

In event-based systems, an 'event' can take many forms. However, common to all events is that they are abstractions of observations made in a certain environment. Event-based processing originated in the area of active databases in the 1980s with the introduction of triggers to database systems (we refer readers to [9, 11] for an overview of active databases). A trigger is an Event-Condition-Action (ECA) rule which is checked by the DBMS whenever an event occurs that matches the Event part of the rule. The Condition part of the rule acts as a sentinel filtering out events that are not relevant, and only those events satisfying the condition cause the Action part of the rule to be executed. Since these early beginnings, ECA rules have become a reactive computing paradigm in their own right, independent of their use in database systems. For example, the ECA concept has been applied in the context of middleware architectures such as CORBA [6], J2EE [2] and TIBCO [4], in stream processing [5], reactive web-based applications [8], wireless sensor networks [1] and radio frequency identification [10]. We refer readers to [3] for an overview of the development of the ECA paradigm in databases during the 1980s and 1990s, and a discussion of its more recent use in areas such as distributed event specification and detection, information filtering, web page change monitoring, stream data processing, role-based access control and autonomic computing.

What are some of the reasons for the popularity of event-based processing? Firstly, it offers a lot of flexibility to applications that need to monitor an environment and react to the occurrence of significant events. Event-based systems rely mainly on asynchronous communication to do this. In contrast to synchronous communication, processes do not have to stop working while waiting for data. Generally this means that event sources 'push' data about event occurrences to event consumers. Moreover, developers are able to build more loosely-coupled systems, as system components only need minimum knowledge about each other. Different levels of coupling are possible, ranging from fixed point-to-point communications to a network of brokers.

Event-based processing began with systems that handled simple events, such as the insertion of a tuple into a database table or the reading of a temperature value by a sensor. However, as systems are processing increasing volumes of events, users are not necessarily interested in keeping track of thousands or even millions of simple events, but rather in seeing and handling the 'big picture'. This can be achieved by employing intelligent techniques for deriving events at higher levels of abstraction. For example, the area of Complex Event Processing (CEP), which is the focus of several of the chapters in this book, aims to correlate simple events temporally and/or spatially in order to infer more meaningful composite events. CEP comes with a trade-off, though, in that a balance needs to be struck between the expressiveness of CEP languages on the one hand and the performance of CEP systems processing them on the other: the more expressive a language, the more powerful the event processing engine needs to be.

Many different architectures, languages and technologies have been and are being used for implementing event-based distributed systems — we refer readers to [7] for an overview of the range of technologies used and an analysis of common research issues arising across different application domains. This large variability is due to the fact that much of the development of such systems has been undertaken independently by different communities. While there is some overlap in the different approaches, often the main driving force during development has been the needs and requirements of specific application domains. Although a merging and generalisation of concepts is under way, c.f. [7], this will still take some time to accomplish. In the meantime, Chapters 2 and 3 of this book present an overview of distributed architectures for event-based systems (Chapter 2) and languages for specifying and detecting complex events (Chapter 3), aiming to allow the reader to gain an understanding of the state-of-the-art in these two fundamental areas of the field.

Many research challenges are being addressed by the event-based distributed computing research community, ranging from the detection of events of interest, to filtering, aggregating, dissemination and querying of event data, and more advanced topics relating to the semantics of event languages and issues such as performance and optimisation of event processing, specification and implementation of event consumption policies (i.e. policies for consuming simple events during the derivation of composite events), handling event data arising from heterogeneous sources, resource management, ensuring the privacy and security of information, and quality of service guarantees. Intelligent techniques are playing an important role in addressing these challenges. In particular, reasoning and logic-based approaches provide fundamental principles for tackling these research issues and developing solutions on the basis of formal, yet powerful, foundations. Chapters 4 to 9 of the book describe a broad range of current research in this direction. Chapters 12 and 13 focus on two specialist topics, namely the role of context in event-based distributed systems, and the handling of uncertainty in the detection of simple events and the inference of higher-level derived events from simple events.

Event-based systems can serve as a basis for the development of advanced applications in diverse domains, ranging from traditional areas of logistics and production to more recent mobile and pervasive applications. As event-based processing is often application-driven, the concepts presented in the book are illustrated using examples from a broad range of application scenarios in areas such as supply chain management, environmental and traffic monitoring, patient monitoring, data centre and network monitoring, fraud detection, smart homes, access control, spacecraft and satellite data monitoring, online collaboration, monitoring market data, and monitoring business processes. Chapters 9, 10 and 11 focus specifically on applications in event-based pervasive computing, patient care, and collaboration in social organisations.

2 Reasoning in Event-Based Distributed Systems

Reasoning about the properties and behaviour of event-based distributed systems differs in many ways from reasoning about traditional event management systems due to their greater complexity and dynamicity as well as their temporal, spatial, domain-specific and context-sensitive characteristics. A variety of different event types may arise in event-based distributed systems, including signal, textual, visual and other kinds of events. The time associated with an event may be a single time point or a time interval. Reasoning in event-based distributed systems thus needs to address new issues and research challenges including:

- correlating simple distributed events and reasoning about them in real-time;
- spatio-temporal reasoning over events, with both point and interval semantics;
- reasoning about uncertain events and under real-time constraints;
- reasoning about events in continuous time;
- context-aware reasoning.

Depending on the application supported by an event-based system, different architectures and models are being used, including publish-subscribe, Peer-to-Peer (P2P), Grid computing, event-stream processing, and message queues. A common factor across different application domains is an ever-increasing complexity. Users and developers expect that modern systems are able to cope with not only simple events reporting a change in a single data item but also composite events, i.e. the detection of complex patterns of event occurrences that are possibly spatially distributed and may span significant periods of time. This gives rise to the need for development of

- semantic foundations for event-based distributed systems;
- formal models and languages for expressing composite events;
- reasoning techniques for such models and languages;
- formal foundations for event consumption policies;
- data mining and machine learning techniques for detecting new or unusual event patterns;
- scalable techniques for processing efficiently large volumes of complex distributed events.

The event-based approach is becoming a central aspect of new Grid and P2P systems, ubiquitous and pervasive systems, wireless networking, and mobile information systems. The architectures and middleware of such systems use event-based approaches not only to support efficiency, scalability and reliability but also to support a new class of demanding applications in a variety of domains. Event-based approaches are showing their usefulness not only for stand-alone platforms and applications but also for achieving interoperability between, and integration of, different event-based applications, particularly in the business domain. Research challenges here include:

- development of distributed architectures for supporting intelligent event processing;
- development of event-driven Service Oriented Architectures to support interoperability and integration of distributed applications;
- handling the heterogeneity that arises when event data is produced by different sources, e.g. by enriching the data with additional semantic metadata or through ontology-based mediation services;

- developing techniques for specifying, analysing and enforcing policies for re-source management, security and privacy;
- monitoring and delivering quality of service requirements.

Further research issues are identified in Chapter 2 of this book, which gives an overview of the architectural aspects of event-based distributed systems, focusing particularly on the intelligent and reasoning techniques incorporated within such systems. Some of these are explored in more detail in later chapters of the book, including:

- optimisation of event processing in the face of large numbers of distributed users, high volumes of distributed event data, and the need for timely response with low resource consumption;
- dynamic adaptation to new situations as applications are executing, e.g. changes in the availability of resources, the types of event data being produced, the complex events that need to be detected, or the quality of service requirements;
- specifying, implementing and reasoning with policies for security and access control to physical and virtual environments and resources;
- responsiveness to the requirements of diverse users, for example through rule-based mechanisms for capturing different users needs and preferences and making recommendations to users.

3 Overview of the Book

This book aims to give a comprehensive overview of recent advances in intelligent techniques and reasoning in event-based distributed systems. It divides roughly into five sections:

(i) Chapters 2 and 3 survey different architectures for event-based distributed systems and different languages for complex event processing, covering foundational material that is developed further in subsequent chapters of the book.
(ii) Chapters 4, 5 and 6 explore in more detail some of the event language paradigms identified in Chapter 3, addressing issues such as semantics, performance and optimisation of event-based processing.
(iii) Chapters 7 and 8 focus on security and access control in event-based distributed systems.
(iv) Chapters 9, 10 and 11 discuss the requirements for intelligent processing in several application domains and the development of techniques targeting these domains: event detection in pervasive computing environments, emergency patient care, and online collaboration in social organisations.
(v) Chapters 12 and 13 conclude by addressing two specialist aspects of reasoning in event-based distributed systems: handling of context and handling of uncertainty.

Overview of the Chapters

Chapter 2, *"Distributed architectures for event-based systems"*, gives an overview of the architectural aspects of event-based distributed systems. The authors present a logical architecture that comprises the main components involved in generating, transmitting, processing and consuming event data, based on the ECA paradigm. They discuss the nature of primitive, composite and derived events, and the ways in which these are processed in this logical architecture. They then describe how to realise the archetypal logical architecture in a variety of settings. The alternative system architectures they present are discussed within five themes — Complex Event Processing, Service Oriented Architectures, Grid Architectures, P2P Architectures and Agent-based Architectures. For each of these themes, the authors discuss previous work, recent advances, and future research trends.

Chapter 3, *"A CEP Babelfish: Languages for Complex Event Processing and Querying Surveyed"*, describes five commonly used approaches for specifying and detecting event patterns: (i) event query languages based on a set of event composition operators, (ii) data stream query languages, (iii) production rules, (iv) state machines, and (v) logic languages. The authors illustrate each of these approaches by means of a use case in Sensor Networks and discuss appropriate application areas for each approach. They note that there is no single best approach that fits all application requirements and settings, and that therefore commercial CEP products tend to adopt multiple approaches within one system or even combined within one language.

The next three chapters explore some of the language approaches (i)-(v) above in more detail. Chapter 4, *"Two Semantics for CEP, no Double Talk: Complex Event Relational Algebra (CERA) and its Application to XChange^{EQ}"*, discusses the semantics of Event Query Languages, from both declarative and operational perspectives. The authors note that for such languages the declarative semantics serve as a reference for a correct operational semantics, and also as the basis for developing optimisation techniques for event query evaluation. The chapter focuses particularly on the XChange^{EQ} event query language, which is based on the event composition operators approach. The authors adopt a model-theoretic approach for the declarative semantics, and an event relational algebra as the basis for the operational semantics, and hence for event query evaluation and optimisation. They confirm the relationship between the two versions of the semantics by sketching a proof of the correctness of the operational semantics with respect to the declarative semantics.

Chapter 5, *"ETALIS: Rule-Based Reasoning in Event Processing"*, presents a logic language for specifying complex events. Similarly to Chapter 4, the authors give both a model-theoretic declarative semantics for their language and a rule-based operational semantics. There is a detailed discussion of event consumption policies, defined over both time points and time intervals. Their language is compiled into Prolog for execution, and the authors compare the event detection performance of two Prolog implementations of their approach with a state machine-based implementation. A brief discussion follows of how a logic-based approach can support reasoning over complex events, for example as relating to their relative order and other more complex relationships between events.

Chapter 6, *"GINSENG Data Processing Framework"*, describes a hybrid approach to distributed complex event processing, combining the capabilities and benefits of publish/subscribe, event stream processing (ESP) and business rule processing (BRP). The authors describe the GINSENG modular middleware architecture for distributed event processing, which supports the integration of a variety of event producing and event consuming external components through its extensible content-based publish/subscribe mechanism. The middleware handles the stateless parts of business rules close to the data sources through the publish/subscribe mechanism, while the stateful parts of rules are handled by the ESP and BRP engines. The authors describe the implementation of these two engines, and focuses particularly on performance monitoring and control. They discuss data quality-driven optimisation of event stream processing, using an evolutionary approach. They also mention the possibility of using temporal reasoning to identify event occurrences that can no longer match any rule and can therefore be garbage-collected. The chapter concludes with a review of related work in middleware technology, publish/subscribe, ESP and BRP.

Chapter 7, *"Security Policy and Information Sharing in Distributed Event-Based Systems"*, discusses security policies and information sharing in event-based distributed systems, using health care service provision as the motivating application. The authors assume a publish/subscribe mechanism for event-based communication, and they discuss how role-based access control can be used to enforce authorisation policies at the client level, while taking context into account as well. They discuss additional requirements for providing secure information flow between distributed system components, both within and between administrative domains, and how these requirements can be implemented. They describe how reasoning techniques can be applied to formal policy specifications, so as to infer information about the flow of data within the system and the confidentiality and integrity properties of the system.

Chapter 8, *"Generalization of Events and Rules to Support Advanced Applications"*, proposes extensions to ECA rule syntax and semantics in order to support the requirements of advanced applications, exemplifying these requirements with examples drawn from the specification and implementation of policies for access control to physical spaces. The ECA rule extensions include: specification of alternative actions to be undertaken when the event part of a rule is detected but the condition is false; generalisation of event specification and detection using constraints expressed on the attributes of events; and detection of partial and failed complex events. The chapter discusses event detection techniques for handling these extensions, possible alternative implementations for distributed event detection, previous work on reasoning with ECA rules, and open questions relating to reasoning using the extended ECA rules.

Chapter 9, *"Pattern Detection in Extremely Resource-Constrained Devices"*, focuses on the challenges of detecting event patterns in distributed resource-constrained devices comprising a wireless network of sensors and actuators. Such pervasive computing environments pose particular problems due to the resource constraints of the devices and the lack of reliable communication and synchronisation capabilities in the network. The chapter discusses online data mining approaches to detecting anomalous or novel event patterns in the sensor data. The authors also present their own approach which combines concepts from data mining, machine learning and statistics and offers a variety of algorithms for both (i) detecting known patterns of interest in the sensor data, as specified by users, and (ii) detecting previously unknown event patterns by a phase of training on normal sensor data and then online detection of deviations from the learned patterns.

Chapter 10, *"Smart Patient Care"*, describes a prototype system for monitoring patients in emergency care units with the aim of predicting if they will have a cardiac arrest in the next 24 hours. The system merges real-time patient data with historical data, medical knowledge in the form of rules, and data mining models, in order to generate appropriate alarms customised according to the patient and also doctor's preferences. The system combines several Oracle products to achieve this functionality: Total Recall for managing the history of data changes, Continuous Query Notification for detecting changes in the data, Oracle Data Mining (ODM) for detecting complex patterns in the data, and Business Rules Manager for complex event processing.

Chapter 11, *"The principle of immanence in event-based distributed systems"*, focuses on collaborative Grid-based environments, and on the emergence of the principle of 'immanence' from the activities of collaborating partners in virtual organisations. It discusses how the AGORA Grid-based platform has been extended with agent-based mechanisms to foster the emergence of collaborative behaviours and to reflect information about the emergence of such behaviours back into the system architecture, with the aim of improving the self-organisation capabilities of the system and the collaborating community that it supports.

Chapter 12, *"Context-based event processing systems"*, discusses the role of context in event processing systems, identifying four main uses of context in such systems: temporal context, spatial context, segmentation-oriented context and state-oriented context. A survey is given of these different types of context. Events may be processed differently depending on their context. Event processing applications may use combinations of such context, and the chapter also discusses composing contexts. A survey is given of how context is supported in five commercial event processing systems.

The final chapter, *"Event Processing over Uncertain Data"*, identifies alternative approaches for capturing uncertainty in the detection of simple event occurrences and their attribute values, and in the inference of derived events. A taxonomy for event uncertainty is presented as is an analysis of the various causes of event uncertainty. Two models are proposed for representing uncertainty in simple events, based on probability theory and fuzzy set theory. Handling uncertainty in the inference of derived events is discussed with respect to a simple event language employing a Bayesian network inference model. Several open research questions are highlighted: identifying the most suitable approach, or approaches, for specific application requirements; specifying or automatically deriving appropriate inference rules and the probabilities associated with them; and achieving scalable implementations of the inference algorithms.

4 Concluding Remarks

This book aims to give readers an insight into the rapidly evolving field of reasoning in event-based distributed systems. Due to its fast, and sometimes uninhibited, growth this area faces the challenge of having to consolidate existing approaches and paradigms while at the same time integrating newly emerging application requirements, such as uncertainty, security and resource constraints. Nevertheless, the developments in this field have shown great potential for building effective and efficient distributed systems.

Logic-based approaches are being used for the specification and detection of complex events, giving a sound basis for reasoning about event properties and for scalable implementations of event detection mechanisms, possibly in combination with event stream processing middleware. Defining denotational semantics for event query languages provides an implementation-independent reference point for proving the correctness of implementations and developing optimisations. Data mining, machine learning and statistical approaches are being used to discover event patterns in high-volume event streams. Rule-based specifications of security policies make possible context-aware reasoning about information access and flow in distributed applications. Probabilistic approaches provide promising foundations for handling uncertainty in event detection and inference. Agent-based processing is being used to foster the emergence of online collaboration in social organisations through the interaction of intelligent agents that can adapt and evolve their behaviour over time.

The general approach of event-based distributed processing allows developers to create loosely-coupled and reconfigurable systems that offer scalability and adaptability, two of the most critical properties of modern information systems. Event processing can be monitored with respect to specific quality factors and this information be fed back into the system for dynamic quality improvement. Declarative languages for specifying resource management policies offer the potential for formal policy analysis and policy evolution. Rule-based mechanisms can be used to capture different users' needs and preferences in order to support personalisation of users' interaction with the system. Another important goal is the provision of versatile and powerful middleware components that support the rapid development of reliable distributed applications. Finally, we must state that event-based distributed processing and reasoning may still need some more time to mature, but we believe that it will be an exciting area to work in for the years to come.

References

1. Akyildiz, I.F., Weilian, S., Sankarasubramaniam, Y.: A survey on sensor networks. IEEE Communications Magazine 40(8), 102–114 (2002)
2. Bodoff, S., Armstrong, E., Ball, J., Carson, D.B.: The J2EE Tutorial. Addison-Wesley Longman, Boston (2004)
3. Chakravarthy, S., Adaikkalavan, R.: The Ubiquitous Nature of Event-Driven Approaches: A Retrospective View. In: Event Processing, Dagstul Seminar Proceedings 07191 (2007)

4. Chan, A.: Transactional publish/subscribe: the proactive multicast of database changes. ACM SIGMOD Record 27(2), 521 (1998)
5. Chen, J., DeWitt, D., Tian, F., Wang, Y.: NiagaraCQ: A scalable continuous query system for internet databases. In: ACM SIGMOD Conf. 2000, Dallas, TX, pp. 379–390 (2000)
6. Harrison, T., Levine, D., Schmidt, D.: The design and performance of a real-time CORBA event service. In: 12th ACM SIGPLAN Conf. on Object-Oriented Programming, Systems, Languages and Applications (OOPSLA), Atlanta, GA, pp. 184–200 (1997)
7. Hinze, A., Sachs, K., Buchmann, A.: Event-Based Applications and Enabling Technologies. In: 3rd ACM Int. Conf. on Distributed Event-Based (DEBS), Nashville, TN, pp. 1–15 (2009)
8. Levene, M., Poulovassilis, A. (eds.): Web Dynamics — Adapting to Change in Content, Size, Topology and Use. Springer, Berlin (2004)
9. Paton, N.W. (ed.): Active Rules in Database Systems. Springer, New York (1999)
10. Roussos, G.: Networked RFID: Systems, Software and Services. Springer, London (2008)
11. Widom, J., Ceri, S. (eds.): Active Database Systems: Triggers and Rules for Advanced Database Processing. Morgan Kaufmann, San Francisco (1994)

Distributed Architectures for Event-Based Systems

Valentin Cristea, Florin Pop, Ciprian Dobre, and Alexandru Costan

Abstract. Event-driven distributed systems have two important characteristics, which differentiate them from other system types: the existence of several software or hardware components that run simultaneously on different inter-networked nodes, and the use of events as the main vehicle to organize component intercommunication. Clearly, both attributes influence event-driven distributed architectures, which are discussed in this chapter. We start with presenting the event-driven *software* architecture, which describes various logical components and their roles in events generation, transmission, processing, and consumption. This is used in early phases of distributed event-driven systems' development as a blueprint for the whole development process including concept, design, implementation, testing, and maintenance. It also grounds important architectural concepts and highlights the challenges faced by event-driven distributed system developers. The core part of the chapter presents several *system* architectures, which capture the physical realization of event-driven distributed systems, more specifically the ways logical components are instantiated and placed on real machines. Important characteristics such as performance, efficient use of resources, fault tolerance, security, and others are strongly determined by the adopted system architecture and the technologies behind it. The most important research results are organized along five themes: complex event processing, Event-Driven Service Oriented Architecture (ED-SOA), Grid architecture, Peer-to-Peer (P2P) architecture, and Agent architecture. For each topic, we present previous work, describe the most recent achievements, highlight their advantages and limitations, and indicate future research trends in event-driven distributed system architectures.

Valentin Cristea · Florin Pop · Ciprian Dobre · Alexandru Costan
University Politehnica of Bucharest, 313 Splaiul Independentei,
060042 Bucharest, Romania
e-mail: valentin.cristea@cs.pub.ro, florin.pop@cs.pub.ro,
 ciprian.dobre@cs.pub.ro, alexandru.costan@cs.pub.ro

S. Helmer et al.: Reasoning in Event-Based Distributed Systems, SCI 347, pp. 11–45.
springerlink.com   © Springer-Verlag Berlin Heidelberg 2011

1 Introduction and Motivation

Many distributed systems use the event-driven approach in support of monitoring and reactive applications. Examples include: supply chain management, transaction cost analysis, baggage management, traffic monitoring, environment monitoring, ambient intelligence and smart homes, threat / intrusion detection, and so forth. In e-commerce applications, the business process can be managed in real time by generated events that inform each business step about the status of previous steps, occurrence of exceptions, and others. For example, events could represent order placements, fall of the inventory below a specific optimal threshold, high value orders, goods leaving the warehouse, goods delivery, and so forth. An event-driven system detects different events generated by business processes and responds in real time by triggering specific actions.

Event-driven solutions satisfy also the requirements of large scale distributed platforms such as Web-based systems, collaborative working and learning systems, sensor networks, pervasive computing systems, Grids, per-to-peer systems, wireless networking systems, mobile information systems, and others, by providing support for fast reaction to system or environment state changes, and offering high quality services with autonomic behavior. For example, large scale wireless sensor networks can be used for environment monitoring in some areas. A large number of different events can be detected (such as heat, pressure, sound, light, and so forth) and are reported to the base stations that forward them to event processors for complex event detection and publication. Appropriate event subscriber processes are then activated to respond to event occurrences. They can set alarms, store the event data, start the computation of statistics, and others.

Event-driven distributed architectures help in solving interoperability, fault tolerance, and scalability problems in these systems. For example, Grid systems are known for their dynamic behavior: users can frequently join and leave their virtual organizations; resources can change their status and become unavailable due to failures or restrictions imposed by the owners. To cope with resource failure situations, a Grid monitoring service can log specific events and trigger appropriate controls that perform adaptive resource reallocation, task re-scheduling, and other similar activities. A large number of different event data are stored in high volume repositories for further processing to obtain information offered to Grid users or administrators. Alternatively, the data collected can be used in automatic processes for predictive resource management, or for optimization of scheduling .

Most important, event-driven distributed architectures simplify the design and development of systems that react faster to environment changes, learn from past experience and dynamically adapt their behavior, are pro-active and autonomous, and support the heterogeneity of large scale distributed systems.

The objective of this chapter is to give the reader an up-to-date overview of modern event-driven distributed architectures. We discuss the main challenges, and present the most recent research approaches and results adopted in distributed system architectures, with emphasis on incorporating intelligent and reasoning techniques that increase the quality of event processing and support higher efficiency of the applications running on top of distributed platforms. Future research directions in the area of event-driven distributed architectures are highlighted as well.

2 Background

Events are fundamental elements of event-driven systems. An event is an occurrence or happening, which originates inside or outside a system, and is significant for, and consumed by, a system's component. Events are classified by their types and are characterized by the occurrence time, occurrence number, source (or producer), and possible other elements that are included in the event specification. Events can be **primitive**, which are atomic and occur at one point in time, or **composite**, which include several primitive events that occur over a time interval and have a specific pattern. A composite event has an **initiator** (primitive event that starts a composite event) and a **terminator** (primitive event that completes the composite event). The occurrence time can be that of the terminator (point-based semantics) or can be represented as a pair of times, one for the initiator event, and the other for the terminator event [43, 21]. The interval temporal logic [1] is used for deriving the semantics of interval based events when combining them by specific operators in a composite event structure.

Event streams are time-ordered sequences of events, usually append-only (events cannot be removed from a sequence). An event stream may be bounded by a time interval or by another conceptual dimension (content, space, source, certainty) or can be open-ended and unbounded. Event stream processing handles multiple streams, aiming at identifying the meaningful events and deriving relevant information from them. This is achieved by means of detecting complex event patterns, event correlation and abstraction, event hierarchies, and relationships between events such as causality, membership, and timing. So, event stream processing is focused on high speed querying of data in streams of events and applying transformations to the event data. Processing a stream of events in order of their arrival has some advantages: algorithms increase the system throughput since they process the events "on the fly"; more specifically, they process the events in the stream when they occur and send the results immediately to the next computation step. The main applications benefiting from event streams are algorithmic trading in financial services, RFID event processing applications, fraud detection, process monitoring, and location-based services in telecommunications.

Temporal and causal dependencies between events must be captured by specification languages and treated by event processors. The expressivity of the specification should handle different application types with various complexities, being able to capture common use patterns. Moreover, the system should allow complete process specification without imposing any limiting assumptions about the concrete event process architecture, requiring a certain abstraction of the modeling process. The pattern of interesting events may change during execution; hence the event processing should allow and capture these changes through a dynamic behavior. The usability of the specification language should be coupled with an efficient implementation in terms of runtime performance: near real-time detection and non-intrusiveness [40]. Distributed implementations of event detectors and processors often achieve these goals. We observe that, by distributing the composite event detection, scalability is also achieved by decomposing complex event subscriptions into sub-expressions and detecting them at different nodes in the system [4]. We add to these requirements the fault tolerance constraints imposed on event composition, namely that correct execution in the presence of failures or exceptions should be guaranteed, based on formal semantics. One can notice that not all these requirements can be satisfied simultaneously: while a very expressive composite event service may not result in an efficient or usable system, a very efficient implementation of composite event detectors may lead to systems with low expressiveness. In this chapter, we describe the existing solutions that attempt to balance these trade-offs.

Composite events can be described as hierarchical combinations of events that are associated with the leaves of a tree and are combined by operators (specific to an event algebra) that reside in the other nodes. Another approach is continuous queries, which consists of applying queries to streams of incoming data [17]. A derived event is generated from other events and is frequently enriched with data from other sources. The representation of the event must completely describe the event in order to make this information usable to potential consumers without the need to go back to the source to find other information related to the event.

Event-driven systems include components that produce, transmit, process, and consume events. Events are generated by different sources — event producers, and are propagated to target applications — event consumers. Producers and consumers are loosely coupled by the asynchronous transmission and reception of events. They do not have to know and explicitly refer each other. In addition, the producers do not know if the transmitted events are ever consumed. What the system does is to offer the events to the interested consumers. To do this, other components are used such as the event channel and the event processing engine. To them must be added components for event development, event management, and for the integration of the event-driven system with the application (Figure 1). We next describe briefly the roles of these components.

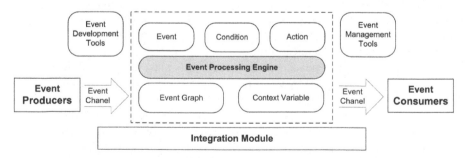

Fig. 1 Components of the event-driven systems

Event producers vary from one application to another. For example, mobile location sensors (GPS) or readers of mobile RFID tags are frequently used in location detection for context aware applications [18]. Web tracking services crawl the Internet to find new postings and generate events accordingly. RFID tag readings combined with information from a company's database can generate events for supply chain management. Mobile sensors can be used for health monitoring systems. Stationary sensors for detecting the presence of persons or sensing the ambient conditions are used in smart home applications. The event producer can also be an application, data store, service, business process, medical equipment, IC card, PC, phone, TV, PDA, notebook, smart phone, and so forth. It is not surprising that various event categories correspond to these sources: temporal, absolute or logical positioning events, change of status events, interval events, event strings, homogeneous or heterogeneous composite events, and others. In addition, events carry contextual information such as social context and group membership [27]. The producer can be a collaboration support service such as Instant Messaging or email. As we can see, there is a large variety of event producers, which generate events in different formats. Consequently, the events need to be converted, prior transmission, to the format accepted by the event channel. In distributed systems, the producer could also include a local event processor that has the role of selecting events for transmission over the event channel based on specific rules, or of detecting composite events. For example, if a temperature sensor generates an event each minute, a local event filter could select and send on the event channel only the events with temperatures greater than a threshold T. Since, in distributed systems, event producers might be spread over a large geographic area, filtering helps to reduce the traffic injected in the transport network that supports the event channels. Also, the producer can generate composite events out of workflows or message streams. For example, the application might want to know when the average temperature for each sequence of ten events "greater than T" events becomes larger than a limit T'. So, an event processor local to the producer calculates the average for each group of ten events and generates a new event when this average is greater than T'. This approach aims to reduce the network traffic and the load of the event processing engine by performing, at the place of the event source, some simple event processing operations.

Event consumers are system components such as software, services, humans, intelligent applications, business processes, performance dashboards, automated agents, actuators, or effectors. They receive events communicated by the event processing engine and react to these events. The reaction might include the generation of new events, so that consumers can be event producers as well.

The **event processing** engine receives events via defined channels, processes them according to specific rules, and initiates actions on behalf of event consumers. Processing can be event aggregation, classification, and abstraction, which are used for generating higher level complex events. For example, event combinations can be a disjunction or a conjunction of two events, a sequence of two events, occurrence of an aperiodic event in a time interval, periodic occurrence of an event, nonoccurrence, and temporal [12]. To these operations, event monitoring, tracking, and discovery can be added. Finally, event handling (for example event rating) can be used for situation detection and prediction purposes.

Engines can process the events individually, one at a time, or in the context of former events. The second approach is known as Complex Event Processing (CEP) and aims at identifying event patterns by processing multiple events. Detecting event patterns can determine the generation of new events or trigger actions of event consumers. Intelligent event processing engines base their decisions on AI techniques and knowledge processing, in which case their behavior adapts continuously to the changing environment they monitor. Engine actions might include event logging, which keeps information about the time at which the event occurred. For composite events, the log preserves the constituent events in the order they were produced. The event processing can be centralized in a single engine or can be distributed to a network of cooperative event processing agents. Both solutions are used in distributed systems and have advantages and disadvantages that will be discussed further in this chapter.

Several models have been proposed for complex event detection. An early approach was extending finite state automata with support for temporal relationships between events [40]. A complex event detection automaton has a finite set of states, including an initial state in which the detection process is started, and *generative* states, each one corresponding to a complex event. Transitions between states are fired by events from an event input alphabet, which are consumed sequentially by the automaton. When the automaton reaches a generative state a complex event is detected. This approach provides efficiency through its direct support for event processing distribution since detectors can subscribe to composite events detected by other automata. Finite automata have several advantages: they are a well-understood model with simple implementations; their restricted expressive power has the benefit of limited, predictable resource usage; regular expression languages have operators that are tailored for pattern detection, which avoids redundancy and incompleteness when defining a new composite event language; complex expressions in a regular language may easily be decomposed for distributed detection.

Many event processing engines are built around the Event, Condition, Action (ECA) paradigm [13], which was firstly used in Data Base Management Systems (DBMS) and was then extended to many other categories of system. These elements are described as a rule that has three parts: the **event** that triggers the rule invocation; the condition that restricts the performance of the action; and the **action** executed as a consequence of the event occurrence. To fit this model, the event processing engine includes components for complex **event detection, condition evaluation**, and **rule management**. In this model, event processing means detecting complex events from primitive events that have occurred, evaluating the relevant context in which the events occurred, and triggering some actions if the evaluation result satisfies the specified condition. Event detection uses an event graph, which is a merge of several event trees [15]. Each tree corresponds to the expression that describes a composite event. A leaf node corresponds to a primitive event while intermediate nodes represent composite events. The event detection graph is obtained by merging common sub-graphs. When a primitive event occurs, it is sent to its corresponding leaf node, which propagates it to its parents. When a composite event is detected, the associated condition is submitted for evaluation. The context, which can have different characteristics (e.g. temporal, spatial, state, and semantic) is preserved in variables and can be used not only for evaluation of the condition but also in the performance of the action.

In distributed systems, **channels** are used for event notification or communication. Events are communicated as messages by using negotiated protocols and languages such as the FIPA ACL (FIPA Agent Communication Language) used for inter-agent communication. Protocols can be push-based, in which producers distribute the events to consumers, or pull-based, in which consumers request the information. In the **publish-subscribe** paradigm [27], producers place notifications into a channel while consumers subscribe for notifications to specific channels. When an event is published, all entities that subscribed to that specific event category are automatically notified. As an alternative, content-based publish-subscribe [47] mechanisms use the body of the event description for routing the information from producers to consumers. Finally, concept-based publish-subscribe [19] uses ontologies and context information to decide on the routing. The publish-subscribe model has at least two advantages: first, it supports event delivery to multiple consumers; second, it decouples the event consumers from the rest of the system, with beneficial effects on scalability and on simplifying systems' design and implementation.

Event development tools are used for the specification of events and rules. **Event management** tools are used for the administration and monitoring of the processing infrastructure and of the event flows. The **integration** module includes interface modules for the invocation and publish-subscribe actions, access to application data, and adapters for event producers.

3   Event-Driven Distributed System Architectures

The concrete realization of a system requires the instantiation of the abstract architecture, placement of components on real machines, protocols to sustain the interactions, and so forth, using specific technologies and products [53]. Such an instance is called **system architecture**.

Event-driven distributed system architectures must respond to users' quality requirements, and to problems raised by the distributed nature of platforms and applications. One challenge is the scale in number of users and resources that are distributed over large geographic areas. They generate a huge number of events that must be efficiently processed. For example, in a transaction cost analysis application, hundreds of thousands of events per second of transaction records must be processed [27]. To support such and similar tasks, the distributed system is required to process composite events with the minimum delay and with minimum resource consumption, which might require a very careful placement of event services on available resources, based on load balancing and replication techniques [58]. Another challenge is presented by fault tolerance, which requires component replication combined with recovery techniques and the use of persistent memory for events.

Yet another challenge is the dynamic nature of the context in which the applications are running. Typical context aware applications are those based on wireless sensor networks or mobile ad-hoc networks for traffic monitoring, which require a greater adaptivity that can be supported by AI and knowledge-based techniques. Last but not least, systems must respond to the needs of large communities of users with different profiles and backgrounds, by offering expressive tools for specifying complex events, and using intelligent techniques for the manipulation of event patterns.

3.1 Complex Events Detection

Distributed event processing is based on decomposing complex event expressions into parts that can be detected separately in a distributed approach. An optimal strategy must be used for the distribution of event processing services between clients and servers or among peers of a distributed system. This is guided by optimization criteria, which can be a tradeoff between a low latency in event processing and low resource consumption. It takes into account the characteristics of systems' infrastructure and of the applications, which in most cases are dynamically changing and ask for flexible and adaptive mechanisms. For example, the policy could encourage the reuse of an existing complex event detector for several other complex events. In other cases, it could be more useful to replicate services to support fault tolerance and reliability or to gain in performance by placing the event processing closer to the consumers.

In many systems, the generation of new inferred events is based on other events and some mechanisms for predefined event pattern specifications. A widespread model that supports the dynamic modification of complex events is the **rule based** paradigm, which currently relies on expressive definition of the relevant events and the update of rules over time. The model uses a set of basic events along with their inter-relationships and event-associated parameters. A mechanism for automating the inference of new events [54] combines partial information provided by users with machine learning techniques, and is aimed at improving the accuracy of event specification and materialization. It consists of two main repetitive stages, namely rule parameter prediction and rule parameter correction. The former is performed by updating the parameters using available expert knowledge regarding the future changes of parameters. The latter stage utilizes expert feedback regarding the event occurrences and the events materialized by the complex events processing framework to tune rule parameters. There are some important directions which are worth exploring, for example casting the learning problem as an optimization one. This can be achieved by attaching a metric to the quality of the learned results and by using generic optimization methods to obtain the best values. For example, transforming rule trees into Bayesian Networks enables the application of learning algorithms based on the last model [56].

Event workflows are a natural choice for the composition of tasks and activities, and are used to orchestrate event interactions in distributed event driven systems. They can be based on associating components into groups or scopes, which induce an hierarchical organization. Two components are visible to each other if there is a scope that contains them. When a component publishes an event, this is delivered to all visible subscribers. Each component relies on an interface specifying the types of events the component publishes (out-events), and the events the component subscribes to (in-events). Scopes are considered components as well. Each scope has an interface specifying the in- and out-events of the whole group. It regulates the interchange of events with the rest of the system. Exchange of events can be further controlled with selective and imposed interfaces that allow a scope to orchestrate the interactions between components within the scope. In event-driven workflow execution, ECA rules are fundamental for defining and enforcing workflow logic. Fault tolerance is supported by exception event notification mechanisms which, combined with rules, are used in reacting to workflow execution errors. However, further work remains to be done on how to specify and implement semantic recovery mechanisms based on rules. Also, the specific workflow life-cycle needs to be addressed, particularly with respect to long running workflows and organizational changes.

The advent of **stream** processing based on sensor and other data generated on a continuous basis enhances the role of events in critical ways. Currently, event stream technologies converge with classic publish/subscribe approaches for event detection and processing. The key applications of the event stream processing technologies rely on the detection of certain event patterns (usually corresponding to the applications domain). However, efficient evaluation of the pattern queries over streams requires new algorithms and optimizations since the conventional techniques for stream query processing have proved inadequate [9]. We note that applications may cope differently with performance requirements on event streams: while some applications require a strict notion of correctness that is robust relative to the arrival order of events, others are more concerned with high throughput. An illustration of such systems that integrate both event streams and publish/subscribe technologies and support different precise notions of consistency is CEDR, Complex Events Detection and Response [9]. The proposed model introduces a temporal data dimension with clear separation of different notions of time in streaming applications. This goal is supported by a declarative query language able to capture a wide range of event patterns under temporal and value correlation constraints. A set of consistency levels are defined to deal with inherent stream imperfections, like latency or out-of-order of delivery and to meet applications' quality of service demands. While most stream processing solutions rely on the notion of stream tuples seen as points, in CEDR each tuple has a validity interval, which indicates the range of time when the tuple is valid from the event provider's perspective. Hence, a data stream is modeled as a time varying relation with each tuple in the relation being an event. Stream processing faces some particular issues in the context of large-scale events processing. The high volume of streams reaches rates of thousands of detected events per second in large deployments of receptors. Also, extracting events from large windows is a difficult task when relevant events for a query are widely dispersed across the window.

In various event-driven systems, information is combined from different sources to produce events with a higher level of abstraction. When the sources are heterogeneous the events must be meaningfully enriched, possibly by adding meta-data that is often automatically extracted from semi-structured data [10, 26]. **Event enrichment** involves understanding the semantics of the events and of the external sources of information. Depending on the degree of abstract knowledge needed, the event-driven system might generate recommendations automatically, which in response might call for human involvement. Other approaches to addressing heterogeneity of event sources are based on the use of an intermediary ontology-based translation layer. Such systems include an additional layer of mediation that intelligently resolves semantic conflicts based on dynamic context information (history of events or state-context or any other information) and an ontology service [19]. The concept-based publish/subscribe mechanism is part of the event processing layer. If notifications are to be routed in the same context then no additional mediation is needed. If publisher and subscriber use different contexts, an additional mediation step is used. Therefore, concept-based publish/subscribe can be implemented on top of any other publish/subscribe methods (channel-based, content-based, subject-based).

3.2   Classes of Event-Driven Distributed Architectures

There are two large classes of event-driven distributed system architectures: **client-server** and **peer-to-peer**. Systems in the first category are based on the asymmetric relation between clients and servers that run on different machines. A client can actively invoke a service and waits for the response. The server, passively waits for invocations, executes the requested service and sends the results back to the client. Since the server can be client for another server, the possible client-server topologies can be very diverse. The client-server model has several sub-categories. Web-based applications use the browser as client and the Web server as broker for message exchanges with the application server. This sub-model has several advantages: there is no need to build special clients for application servers; new application services can be easily deployed without changing the client; the client's functionality can be enriched by downloading code from the server, and executing it in the browser. Both entities (client and server) can have an event-driven orientation as is illustrated in the sequel. A second sub-model is the event-driven Service Oriented Architecture (event-driven SOA) in which an event, possibly produced by a client's action, can trigger the invocation of one or several services. In turn, service execution can produce new events that are propagated to other services that contribute to solving the client's request. This sub-model is more flexible than the previous one since the client does not have to know the server in advance. Instead, the server publishes its interface, and a lookup service allows clients to retrieve the interface description and formulate service requests according to this description. In addition, the use of the publish-subscribe paradigm allows clients to be notified when new services are made available. Clearly, the use of the same paradigm for events and services simplifies the development of event-driven SOA systems, mainly in enterprise environments. Some authors [39] consider that event-driven SOA is nothing else than another style of SOA, two primary styles being "composite application" and "flow".

The event-driven SOA has proved to be very useful in Grid computing, for enhancing the performance of Grid services in several respects. One is for improving the collaborative activity in Virtual Organizations, in which partner processes can become more reactive to cooperation events. Another one is in monitoring and control of data Grid components engaged in performing complex processing on high volumes of data.

For the second architectural category, we mention multi-agent and peer-to-peer systems. Entities (agents or peers) have equal capabilities and develop symmetric relations. In multi-agent systems, both the event processing and business processing are distributed to agents that can be specialized to different functionalities and interact to perform the specific tasks that correspond to their roles. By definition, **agents** react to events that correspond to environment changes. So, agents are reactive, but they are also proactive, autonomous, adaptive, communicative, and mobile. Consequently, multi-agent systems are attractive for many applications in which these characteristics are important. **Peer-to-peer systems** can also base their collaboration, processing and content distribution activities on the event paradigm. They are very large scale systems capable of self-organizing in the presence of failures and fluctuations in the population of nodes. They have the advantage that ad hoc administration and maintenance are distributed among the users, which reduces the cost of collaboration, communication, and processing.

In the sequel, the discussion is focused on the architectures of intelligent event-driven distributed systems including service-oriented, Web-based, Grid, multi-agent, and P2P systems. It aims at analyzing research issues related to the development of these systems, and to the integration of intelligent and reasoning techniques in distributed platforms. The impact of these techniques on the efficiency, scalability and reliability of stand-alone and business integrated platforms is also presented.

3.3 Event-Driven SOA

The event-driven SOA architecture is an extension of the SOA architecture with event processing capabilities. Services are entities that encapsulate business functionalities, offered via described interfaces that are published, discovered and used by clients [57]. Complex distributed systems are built as collections of loosely coupled, technologically neutral, and location independent services that belong to middleware and application levels. Traditionally, enterprise distributed system components interact by sending invocations to other components and receiving responses. Complex interactions are controlled by orchestration (centralized coordination of services) and choreography (distributed collaboration among services that are aware of the business process). Events introduce a different interaction model in which event channels allow consumers to subscribe for specific events, and receive them when such events are published by producers. This mechanism is adopted in open standards (e.g. CORBA), and in products or platforms (such as .NET, websphere Business Events, Oracle CEP application server, and others) with the aim of simplifying the design of complex interactions and supporting interoperability.

Services can be event producers and consumers, but can also act as re-usable components for event processing, such as Rule service, Decision service, Invocation service, and Notification service (see Figure 2).

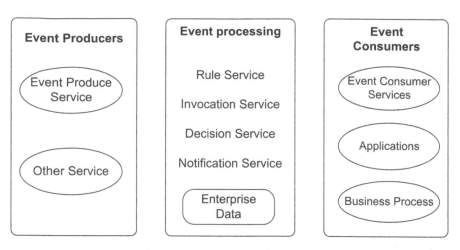

Fig. 2 Services of the event-driven SOA

The Enterprise Service Bus (ESB) architecture defines facilities for business events handling and complex event processing, along with message routing, transport protocol conversion, and message splitting / aggregation to support powerful, flexible, and real-time component interaction. ESB accommodates business rules, policy driven behavior (particularly at the service level), and advanced features such as pattern recognition. ESB has a hierarchical structure determined by horizontal causality of events that are produced and consumed by entities residing in the same architectural layer [23].

The ESB Core Engine is responsible for event processing and uses the Transformation, message Routing, and Exception Management modules depicted in Figure 3. Routing and addressing services provide location transparency by controlling service addressing and naming, and supporting several messaging paradigms (e.g. request / response or publish / subscribe).

The growing needs of the modern business environment have resulted in new standards, products and platforms. New emerging technologies are used in Event-Driven Business Process Management (EDBPM) as an enrichment of BPM with new concepts of Event Driven Architecture (EDA), Software as Service, Business Activity Monitoring, and Complex Event Processing (CEP). New standards for EDA have been defined or are under development by OASIS and OMG:

- Enterprise Collaboration Architecture is a model driven architecture approach for specifying Enterprise Distributed Object Computing systems; it supports event driven systems;

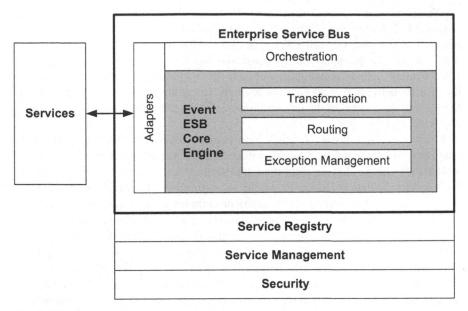

Fig. 3 ESB Components

- Common Alerting Protocol is a data interchange standard for alerting and event notification applications;
- WS-Notification is a family of three specifications (WS-BaseNotification, WS-BrokeredNotification, and WS-Topics) that define a Web services approach to notification using a topic-based publish/subscribe pattern;
- Notification / JMS Interworking refer to event message mapping, event and message filtering, automatic federation between a Notification Service channel concept and topic/queue concepts;
- Production Rules Representation relates to support for specifying Event - Condition - Action rule sets;
- Document Object Model Level 2 and 3 Events Specification refers to the registration of event handlers, describes event flows, and provides basic contextual information for events.

Specific products and platforms have been developed based on these standards. The Oracle Event-Driven Architecture Suite with Oracle Fusion Middleware products allow customers to sense, identify, analyze and respond to business events in real-time. Oracle®EDA is compliant with SOA 2.0, the next-generation of SOA that defines how events and services are linked together to deliver a truly flexible and responsive IT infrastructure [38]. Event-driven workload automation, added in IBM®Tivoli Workload Scheduler 8.4, performs on-demand workload automation and plan-based job scheduling [29]. This defines rules that can trigger on-demand workload automation.

Web services add to SOA their own set of event related standards, WS-Eventing [28] and WS-Addressing [7], targeting the implementation of event driven service-oriented ubiquitous computing systems. WS-Addressing offers endpoint descriptions of service partners for synchronous and asynchronous communication. WS-Eventing defines messaging protocols for supporting the publish/subscribe mechanism between web service applications. Event notifications are delivered via SOAP, and the content of the notifications can be described without restrictions for a specific application.

Much research is directed towards increasing the Web's (and Web applications') reactivity, which means disseminating information about data updates. This can be realized with events that are combined, transmitted, detected, and used by different Web servers. Events could be as simple as posting new discounts for flights that should be notified to interested customers or complex combinations of events that could happen in a more complex Web-based service. One solution to cope with the complexity and scale of the Web environment is the use of event-driven declarative approaches and languages. XChange [44] is a language and an associated runtime environment that supports the detection of complex events on the web and the separation between two data categories, namely *persistent data* (XML or HTML documents) and *volatile data* (event data communicated between XChange programs).

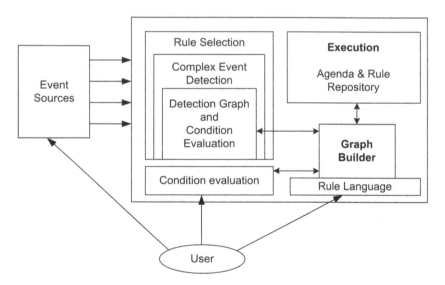

Fig. 4 RIA client architecture

Events can increase the reactivity of **Rich Internet Applications** (RIAs). These are the Web-based counterparts of many applications that are available on desktops. Clearly, to compete with local desktop environments, they have very high QoS requirements concerning the client - server interactivity. To respond to these requirements, RIAs adopt the fat client model, which implements in the browser the user interface, and uses an asynchronous client-server interaction with reduced waiting times on both client and server.

Adding the capabilities of event processing and declarative rule execution on the client side leads to intelligent RIA (IRIA) that benefits from increased reactivity, greater adaptability to complicated requirements, and higher scalability. The system's architecture (Figure 4) presented in [49] supports the event-condition-action (ECA) paradigm by including a complex event detector, condition evaluator, rule engine, together with adapters for the event sources and the rule language. A rule is implemented as an object in JSON, which contains the triple (event, condition, action). The event expression uses the operators defined in Snoop [14]. The condition part introduces restrictions on the set of composite events that are permitted by the event expression. It uses filters and joins similar to production systems. The action part includes one or more JavaScript code blocks. The system accepts events from local sources (resulting from user-browser interaction, Document Object Model events, and temporal events) and events coming from the network via servers (a stream of stock market events provided by a Comet server, and events resulting from polling RSS feeds). Incoming events are forwarded to the complex event detector, which uses an event detection graph. When a composite event is detected, the associated condition is evaluated. A Rete network [22] is used as the matching algorithm to find patterns in a set of objects contained in the Working memory. If a match for the condition is found, the action is triggered. The action may be the execution of JavaScript code, the triggering of a new event or the modification of the working memory used by the condition evaluator. The system has been tested and has shown that the use of declarative event patterns is able to process continuous event streams and makes RIAs more reactive and adaptive. Future work is needed to formalize the JSON ECA rule language. Also, the efficiency of using the active rules on the client side requires further experimentation in a larger application spectrum.

Event-driven SOA has moved into the sphere of ubiquitous computing. The first step in this direction was the integration of Web services in small devices and wireless network by the definition of a Universal Plug and Play (UPnP) architecture for direct device interconnections in home and office networks [42]. The second step was the addition of event-driven capabilities, which give support for context-based applications by using the sensing services offered by a multitude of ambient device types (sensors, mobile phones, PDAs, medical instruments, and so forth). Clearly, the highly heterogeneous devices handle various sets of data that are carried in the event parameters. The use of WS-Eventing for event notification in embedded services is shown in Figure 5 [30]. ECA rules are expressed in WS-ECA, an XML-based ECA rule description language for web service-enabled devices. Several event types are accepted: time, service, external, and internal. They can be combined in complex events with disjunction, conjunction, sequence, and negation operators. The condition is implemented by an XPath expression. The action part is a conjunction or disjunction of several primitive or composite actions. Primitive actions can be a service invocation, the creation and publishing of an external event or the creation of an internal event and triggering other rules on the same device. WS-ECA suffers of possible static or dynamic conflicts (several rules triggered by an event may execute conflicting service actions). Some solutions for conflict detection and resolution have been proposed [34].

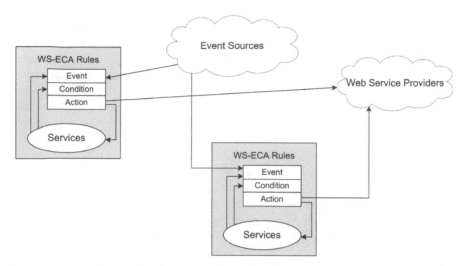

Fig. 5 The use of WS-ECA rules for embedded systems

When event processing is human-centered, the event description technique and the architecture supporting it must be carefully tailored to include context information in a readable form. In the architecture proposed in [31], a statement description, named Five W's and one H, includes context information which indicates: the device that created the statement (Who); the place in which the statement is valid (Where); the time or period in which the statement is valid (When); the name of the data, such as 'temperature' (What); the value domain of the data, for example the set of natural numbers (How); and the identifier of the previous statement on which the current statement causally depends. A primitive event is a sequence of one or more statements in which specific conditions are satisfied.

A composite event expression uses disjunction, conjunction, serialization, and negation operators. The framework architecture (Figure 6) includes the ubiquitous centralized server u-Server that receives reports from, and transmits commands to, several access points, APs, that connect to service nodes. Reports and commands are transmitted as statements. The event detector in the u-Server transmits detected events to a Context analyzer (the context is represented by statements received before the time of the event) which triggers the active rules. The Rule manager generates commands for the control part of the service nodes.

The statement descriptions can be mapped to the event-condition-action (ECA) model and can be integrated with WS-Eventing and WS-ECA event technologies for the implementation of event-driven SOA-based context-aware distributed platforms. While being focused on statement-driven event descriptions, this work opens

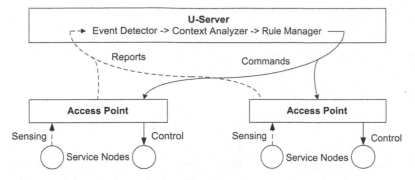

Fig. 6 Event processing framework architecture

new directions for further research towards an easier adaptation of the u-Server to dynamic changes of the context by adopting intelligent approaches in the u-Server functionality.

3.4 Event-Driven Collaboration

Events can be used also in Computer Supported Cooperative Work (CSCW) and Collaborative Working Environment (CWE), which have evolved from simple forms of groupware to more recent Virtual Organizations (VOs) used in both scientific and enterprise environments. A VO is composed of people from different organizations, working together on a common project, and sharing resources for this purpose. The collaboration between VO members can be supported by specific tools, which implement collaboration models adapted to the specific features of distributed systems, such as forums, chats, shared whiteboards, negotiation support, group building tools, and so forth. Such collaborations are routine and can take place according to well established **patterns**, which are recurring segments of collaboration that can be captured, encapsulated as distinct components, and used as solutions for further collaboration problems. Examples of such patterns could be "Team organization", "Project plan development", "Collaborative task execution", "Report elaboration", "Final result analysis" and others. Each pattern can be characterized by some triggering event (for example a specific time or the completion of a specific set of tasks), by the use of specific collaboration tools (forums, chats, videoconferences, shared repositories, and so forth), and by the nature and order of activities (one time, repetitive, scheduled, ad-hoc, and so forth). For example, when organizing a project team, the project leader might publish the number and skills of people needed, and then candidates make offers. The leader could interview the candidates, make a selection, notify selected people, and have a meeting with them. During the meeting, the leader could find out that some of the selected people are not available for the entire duration of the project. To replace them, the leader can restart the process from the selection activity.

Some collaborative activities are dynamic and cannot be captured in fixed pre-defined patterns. Instead, abstract high level patterns can be dynamically adapted to the continuous changes of the context they are used in by services that exploit collaboration knowledge bases and intercommunicate by events [55]. A Recommender service can provide the actions to be executed and the collaboration tools to be used. Awareness services process the events and give information about the collaboration work. Using the monitored events, Analytics services offer statistics about past and present collaborations. The collaboration patterns must be described in terms of the collaboration problems solved, the context they work in, the precondition and post condition of their use, the triggering event, and other relevant features.

The system architecture is presented in Figure 7. The Event and Service Bus links the event sensors to the Complex Event Processing Engine. Simple events are accepted by the Event Reasoner, which detects complex events using information from its Pattern Base. When a pattern is detected, the Event Reasoner notifies the Rule Engine. This engine uses facts about the collaboration, from the Collaboration Knowledge, and retrieves the collaboration pattern whose preconditions and triggers match these facts. Usually, collaboration patterns are combined in workflows and are mixed with user actions. Based on monitoring the collaboration, the system can make changes to the current collaboration pattern or recommend another pattern to replace it.

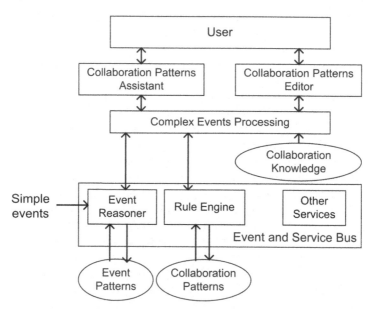

Fig. 7 Collaborative system architecture

The main contribution of this architecture [55] is the use of collaborative patterns in conjunction with knowledge-based event-driven architectures for coping with challenging problems of dynamic Virtual Organizations. Future experiments carried out on platforms that implement the above presented concepts will help to determine the viability of the approach. Also, the development of an ontology [45] that will provide different levels of abstraction of collaboration patterns could have an impact on pattern integration in collaborative platforms. Another issue is the exploitation of the knowledge accumulated in knowledge bases for synthesizing new high performance collaboration patterns.

3.5   Event-Driven Grids

Event Processing in Policy Oriented Data Grids. Event processing is useful in Data Grids, which allow users to access and process large amounts of diverse data (files, databases, video streams, sensor streams, and so forth) stored in distributed repositories. Data Grids include services and infrastructure made available to user applications for executing different operations such as data discovery, access, transfer, analysis, visualization, transformation, and others. Several specific features influence the Data Grid architecture: users and resources are grouped in Virtual Organizations, large collections of data must be shared by VO members, access to data can be restricted, a unified namespace is used to identify each piece of data, different meanings can be associated with the same data set due to the use of different metadata schemas, and others. In order to support data sharing, protection, and fault tolerance, several services are offered for concurrency, data replication, placement and backup, resource management, scheduling of processing tasks, user authentication and authorization and so on. In addition, data consumers and data providers can specify requirements and constraints on data access and use. The contextual information about data, users, resources, and services is stored in persistent databases and is used in management activities related to the data life cycle.

In the Integrated Rule-Oriented Data System, iRODS [46], a Data Grid complex operation is an event that triggers a sequence of actions and other events. An event has a name and is represented as an extended ECA-style rule:

$$A : -C|M_1,\ldots,M_n|R_1,\ldots,R_n,$$

in which A is the triggering event, C is the condition for activating the rule, M_i is an action (named a micro-service) or a "sub"-rule, and R_i is a recovery micro-service.

More than one rule can be defined for an event, in which case the rules are tried in a priority order. If the condition evaluation (based on the context) is successful, the sequence of actions is executed atomically. Subsequently, if the execution of an action fails, the recovery micro-services are executed to roll back the effect of the performed actions, and another rule is considered for activation.

These extended ECA rules give more flexibility to the system. Even if the set of rules that applies in a user's session is fixed, different users and groups can use different rules. In addition, users and administrators are permitted to define rules and publish them for use by other users or administrators. The conditions, which are part of the rules, adapt rule execution to the context. The following example, reproduced from [46], explains the role of contextual information for a rule that refers to an ingestion event for uploading a data set into the iRODS data grid:

```
(a) OnIngest :- userGroup == astro
    | findResource, storeFile, regInIcat, replFile
    | nop, removeFile, rollback, unReplicate.
(b) OnIngest :- userGroup == seismic && size > 1GB
    | findTapeResource, storeFile, regInIcat, seisEv1
    | nop, removeFile, rollback.
(c) OnIngest :- userGroup == seismic && size <= 1GB
    | findTinyResource, storeFile, regInIcat, seisEv2
    | nop, removeFile, rollback.
```

The format respects the rule structure previously described and includes the event and condition (on the first line), the action (the second line), and the recovery services (the third line). In the three rules, the context is represented by the "user group" and by the "size" of the data set being processed.

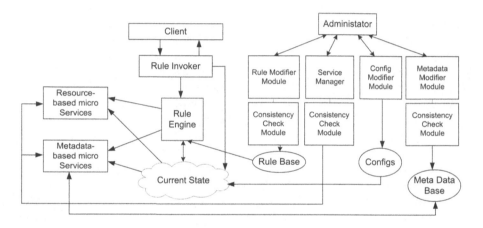

Fig. 8 iRODS architecture

The iRODS architecture is shown in Figure 8. When a user invokes a service, a rule is fired that activates micro-services. To do this, the rule engine uses information from the rule base and Current State. The micro-services can run in parallel or at different times and can intercommunicate by using a Messaging Server. They check the conditions and execute operations on data resources (for example copying a file) or on the MetaData Base. The modifications of the MetaData Base are persistent and can be viewed by other services and by other subsequently executed rules. Micro-services can also intercommunicate by means of a white-board that keeps the local context information. Micro-services can have side-effects outside the iRODS system. For example, the creation of a file can be such a side-effect. Sending an e-mail is another example. The two mentioned operations behave differently with respect to recovery: while a created file can be destroyed, a sent mail cannot be cancelled although a separate mail could be sent to ask the receiver to discard it.

While the intelligent event-driven paradigm can be found in other works related to processing high data volumes [51], iRODS is, to our knowledge, the first attempt to use this paradigm in a Data Grid. The rule-based event processing engine has been successfully integrated with the Data Grid at San Diego Supercomputer Center (SDSC), and it is expected that further experiments will help to improve its performance and adding new features to the platform.

Grid Event Driven Monitoring. The Global Grid Forum elaborated a Grid Monitoring Architecture (GMA) model [5] as a reference to encourage monitoring systems implementations in Grid environments. GMA has several components: a producer, which implements at least one Application Programming Interface (API) for providing events; a consumer that uses an implementation of at least one consumer API; a registry (or lookup service). After discovering each other through the registry, producers and consumers communicate directly. GMA defines several types of interactions between producers and consumers: publish/subscribe, notification, and query/response. It also defines a republisher and a schema repository. The republisher implements producer and consumer interfaces for filtering, aggregating, summarizing, broadcasting, and caching, which correspond to the reactive and processing component of composite event driven models. The schema repository holds the event schema, as a collection of defined event types. A system that supports an extensible event schema must have an interface for dynamic and controlled addition, modification and removal of event types.

A relevant implementation of this model is MonALISA [35], a system able to monitor and control large-scale distributed systems. MonALISA is designed as an ensemble of autonomous self-describing agent-based dynamic services. These services are able to collaborate and cooperate in performing a wide range of distributed event detection, filtering and processing. The system's architecture is based on four layers of services, closely coupled with the GMA model and the abstract model of composite event-driven systems. The first layer is the lookup services network, which provides dynamic registration and discovery for all other services and agents. The second layer represents the event producers. They provide the execution engine that accommodates many monitoring modules, event detectors and a variety of loosely coupled agents that analyze the collected information in real-time. Dynamically loadable agents and filters are able to process the events locally and communicate with other services or agents in order to perform global optimization tasks according to some sets of specified rules. The use of dynamic remote event subscription allows a service to register an interest in a selected set of event types, even in the absence of a notification provider at registration time. Proxy services make up the third layer of the MonALISA framework (Figure 9). They provide intelligent multiplexing of the events requested by the clients or other services and are used for reliable communication among agents. Higher-level services and clients (the event consumers) access the detected events using the proxy layer and thus can obtain real-time or historical data by using a predicate mechanism for requesting or subscribing to selected events and for imposing additional conditions or constraints for interesting events. Once subscribed, consumers receive a stream of relevant events that are stored and processed. The high level services allow filtering of these events and implement a custom aggregation mechanism to support complex composite events and present global views.

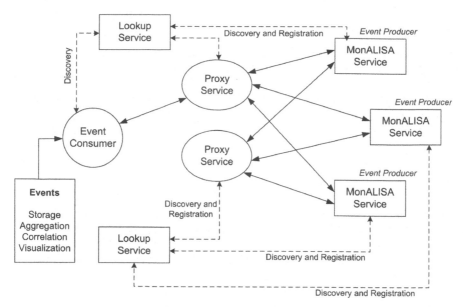

Fig. 9 MonALISA architecture

MonALISA further supports the reactive component of the abstract model, using the detected and processed events to improve the monitored system. The automated management framework implemented within MonALISA represents the first step toward the automation of decisions that can be made based on the monitored events. Actions can be performed at two key points: locally, close to the event producers (in the MonALISA service) where simple actions can be executed; and globally, in a central event consumer (client) where the logic for triggering the actions can be more sophisticated, as it can depend on several flows of events. Hence, the central consumer is equipped with several decision-making agents that help in operating complex systems: restarting remote services when they do not pass functional tests, sending alerts when automatic restart procedures do not fix problems, managing the DNS-based load balancing of the central machines, automatically executing standard applications when CPU resources are idle, and supporting scheduling decisions based on real-time events.

3.6 P2P Systems

Peer-to-peer (P2P) systems consist of interconnected nodes that have similar functions and execute similar tasks. Peers directly share resources such as content, CPU cycles, storage and bandwidth, without requiring the support of a global centralized server. Instead, they cooperate by means of events that take the form of messages exchanged between peers. P2P systems are capable of adapting to failures and dynamic populations of nodes while maintaining acceptable performance. P2P systems are used to support application services for communication and collaboration, distributed computation, content distribution, and so forth, and middleware services like routing and location, anonymity, and privacy.

While resource sharing is based on direct communication between peers, lookup and locating the peer to communicate with are supported by different mechanisms, which use an overlay network that connects all peers and supports the exchange of events between peers. The overlay is built on top of a physical computer network such as the Internet but has a different topology. Also, the mechanisms depend on the category of P2P network. In unstructured networks, the placement of resources is not related to the overlay topology. By contrast, structured networks map keys that reflect resource characteristics (or content) to node addresses where the resource is located [3].

In unstructured P2P networks, a flooding mechanism is used for event transmission. Each event is transmitted by a peer to all neighboring peers in the overlay. Each receiving peer processes the event (for example, discards the event if it is not interested in it or stores it for further tracking). If the event is addressed to the receiving peer, the event detector processes it, decides on the rule to be executed, and performs the corresponding actions. If the event should be made known to other peers then the current peer forwards it to its neighbors in the overlay network. This approach is used in Gnutella [60] and other similar P2P systems. Routes can be computed by a central event dispatcher (ED). Peers are autonomous computational units that interact with other peers by explicitly producing and consuming events. An event is generated by a peer and sent to the ED, which computes the route that includes all subscriber nodes. The event then traverses this predetermined path. Since the solution is based on a central ED node, is not scalable and does not tolerate faults [20].

Better approaches are offered by structured P2P networks, consisting of transmitting an event only to those neighbors situated on the path towards its subscribers. Traffic reduction is particularly important when events need to be transmitted to a small number of subscriber peers. One solution is to arrange the subscribers into logical clusters such that the event routing is performed by a small number of nodes.

When an event is produced by node A (see Figure 10), it is transmitted to one node in the cluster of subscriber peers (peer B in the figure), which in turn transmits it further to other subscribers (nodes C and D in the figure).

A publish-subscribe architecture based on node clustering (see Figure 11) could include two layers. One is concerned with the management of subscription groups; the other deals with routing events within the network. Subscription groups are formed based on the events' content (the content-based model) or the category they should belong to (the topic-based model). In each node, an Event Handler transmits events, publishes them and notifies subscribers. A Subscription Lookup &

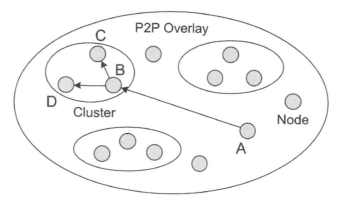

Fig. 10 The topology of the event-driven overlay network systems

Partition Merging component uses a catalogue to map event topics to node addresses for transmitting events. A Subscription Handler performs the node clustering mentioned earlier. The Multicast component connects the event clustering logic to the P2P network overlay underneath.

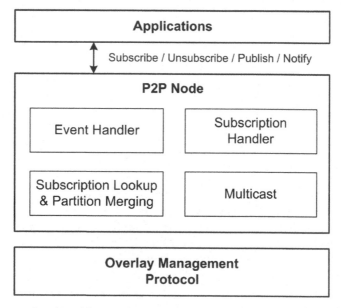

Fig. 11 A general node architecture

Scribe [11] implements this publish/subscribe architecture for managing subscription groups and the multicast communication necessary to send an event to its subscribers. It is constructed on top of Pastry, a P2P location and routing platform that achieves peer clustering based on the similarity of 128-bit keys used as node identifiers. The routing is performed by always sending an event message to the neighbor identified by a key being numerically closest to the key of the destination subscriber. In Scribe each subscription group has a unique *groupID*. The node with the ID numerically closest to the *groupID* acts as the rendezvous point for the associated group. Each node having an interest in receiving a particular flow of events joins a corresponding group. Each event is routed to the rendezvous node, which further sends it in the form of a multicast message to all members of that group. A similar approach is used in Bayeux [61]. These systems suffer from high cost of maintaining the publish/subscribe groups. Also, each group is accessed through one rendezvous node, which is a communication bottleneck and a single-point-of-failure. TERA [6] avoids this disadvantage by introducing several Access Point Lookup components per group, which are able to receive and route events to the appropriate subscribers.

In mobile environments, producers must be able to send events even to subscribers that are permanently on the move. A notification service has to store all events while subscribers are offline. Once a peer is reachable again, large amounts of data associated with the saved events have to be delivered to it. The peer would have to process them locally in order to extract relevant information [50]. However, in some mobile environments, such as Intelligent Transportation Systems, peers have only a few seconds of connectivity and very limited bandwidth [59]. This may lead to loss of high priority events, such as safety-critical driver warnings, with effects on the system's effectiveness and efficiency. Solutions have been proposed such as using a combination of Distributed Hash Tables (DHTs) with Aspect-oriented Space Containers [33]. The Space Container is a storage and retrieval component for structured, spatial-temporal distributed data. Aspects are components with customizable application logic executed either before or after the operation on the Space Container for event processing. Aspects are executed on the peer where the Space Container is located and can be triggered by operations on the Space Container. A peer (e.g. a vehicle) subscribes for events by deploying a Space Container, installing an Aspect, and publishing it in the DHT network. The Aspect registers itself as a subscriber and, independently of the connectivity mode of the original subscriber, receives events, processes them and stores the results in the Space Container for the use of the original peer.

Most of the available P2P systems are research prototypes, which concentrate on scalability and reliability rather than on durability in P2P environments. Durability refers to the capability to correctly send the events to *all* subscribers, even if nodes or links in the underlying communication layer *fail*. In these P2P systems, some nodes may be involved in subscription groups receiving many events, while others may rarely receive any event. So, load balancing of subscriber groups among all peers in the network is also an issue. The separation of the communication layer from the subscription management layer presents another problem: in these systems events are sent to nodes that are not necessarily interested in receiving them.

3.7 Agent Systems

Software agents react in response to other agents and to environment changes, and can act independently (are autonomic). In addition, agents initiate actions that affect the environment (are pro-active), are flexible (able to learn) and cooperate with other agents in multi-agent systems. Some agents are mobile (can migrate from one place to another). Agent architectures are distributed, robust, and fault tolerant.

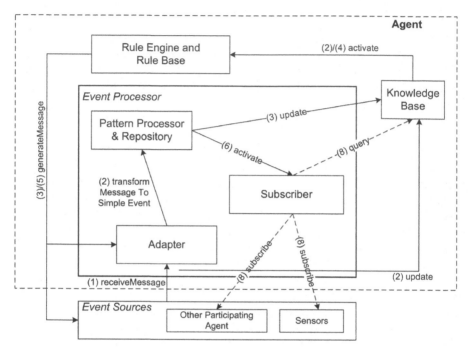

Fig. 12 Event-driven Agent architecture

These characteristics make agents suitable for distributed system middleware, more specifically for parallel execution of tasks, event monitoring, load balancing, trust evaluation, intrusion detection, routing, and other tasks that benefit from the combination of agent distribution and intelligent techniques to obtain optimal solutions.

This section presents relevant architectural features of agents and multi-agent platforms in relation to their role in event-driven distributed systems. The general event-driven agent architecture is presented in Figure 12. The main components are the event processor (EP), a rule base (RB), and a knowledge base (KB). The occurrence of an event is detected by the event processor and is stored in the knowledge base, which then activates a specific rule or set of rules from the rule base. The actions that correspond to the activated business rules are then executed. Complex control functions or computations result from cooperation among agents in the multi-agent system.

In **reactive agent** architectures, the event processor includes modules that are also typically found in other distributed architectures: The **adapter** transforms messages, coming from the environment or other agents, to the format required by the event processor. The **pattern matcher** receives events from the adapter, combines them and detects complex events with the help of a pattern repository. To ease this process, the subscriber inspects the pattern repository, determines the set of interesting possible future events, and subscribes for them. Other features might be present in specific contexts, for example those related to **exception patterns** [16]. An exception pattern E_{ex} for E is obtained by complementing E and eliminating events that cannot happen or are not relevant for the complemented pattern. Thus, E_{ex} represents all the conditions that make E not happen. The agent must include internal policies that are triggered when exceptions are detected. Pattern matching is performed by residuating all event patterns found in the pattern repository with the event that has occurred. For example, for the event pattern $E = a \cdot b \cdot c$ (meaning that event a must be followed by b and then by c), the residual with a is $E/a = b \cdot c$, which is the pattern left to be satisfied after the occurrence of the event a.

Intelligent **pro-active** agents have additional capabilities for interpreting perceptions, drawing inferences, taking decisions, planning actions, and scheduling their execution. They act upon the environment to fulfill the goal for which they were designed. This goal-oriented reasoning allows the agent to commit to the course of action that best accomplishes its task. Intelligent agents have additional capabilities such as supporting contradictory rules, learning, social abilities, natural language processing capabilities and others.

In multi-agent architectures, capabilities are distributed, giving rise to **different agent types**. The authors of [32] present a design in which three primary types of intelligent agents are used: reasoning, learning and evolving agents. Reasoning agents have the ability to make inferences by following a chain of predefined rules, and can be proactive in their behavior. Learning agents are capable of following a set of rules, and also improving their responses by learning from their experience (for example, by dynamically weighting their decisions). Evolving agents improve their behavior with each successive generation. Agents can play other roles too in event-driven architectures [25]. For example, event processing agents [36, 37] act as event detectors, while routing agents are used to construct different types of event channels [24].

Learning is at the base of agent adaptation as response to unexpected events or to dynamic environments. For example, multi-agent systems should adapt to agent failures. They should also support events that occur randomly, events with fluctuating priorities (importance), the inclusion of new information sources and agents, and so forth. The architecture of CyberARIES [48] has an agent part, and a distribution layer. The distribution layer manages the event flows between agents and determines which agents should assist other agents during perception. Agents receive continuous flows of images from one or more camera, filter images and select motion events. They use a motion detection algorithm to acquire a model of the environment. The first step is the construction of a background model from successive frames, using an Auto Regressive filter. In addition, the motion detection algorithm classifies moving objects either as people (single or in groups), vehicles or unknown objects. The agents act together as a neural network, permanently adapting the classification algorithm based on their previous experience. The final result can be configured dynamically, for example calling a service if a person is noticed or adjusting the movement trajectory based on obstacles noticed. Improvements have been added to this work in [41] who include a surveillance system composed of mobile path-planning robots. The system, called CyberScout, produces a timely interpretation of the environment using feedback from perception processes.

Cooperation abilities (mentioned previously) of agents are important in many other systems, such as those for tracking moving targets. In a sensor-based agent infrastructure for tracking [52], each sensor, which is fixed at a specific physical location, collaborates with neighboring sensors to triangulate their measurements and obtain an accurate estimate of the position and velocity of the mobile targets passing through their coverage area. As more targets appear in the area, the sensors need to decide which ones to track and when to track them, always being aware of the status and usage of sensor resources. Since accurate target tracking requires triangulation, an agent that finds a potential target must contact other agents to ask for their help. AI methods are used to optimize the selection of objects to track and also of the neighbors to help triangulating positions while minimizing resource consumption.

Scalability is an important requirement for interactive intelligent multi-agent systems. In EVA (Evolutionary Virtual Agent) conversational system, a virtual assistant, also called a conversational creature [2], undertakes a dialogue in natural language. EVA has a 3D face with real-time animations and is able to support natural language interactions. EVA's goals are very ambitious: to correctly answer the user's questions in minimal time, avoid some inappropriate questions, achieve tasks that the user seems interested in, and build a user profile. The cognitive part of EVA consists of a natural language processing module, a reasoning module, and a learning module. The reasoning module is a multi-agent system with a pro-active architecture based on a combination of an active layer and a reactive layer. The active layer includes a plan agent (which creates sequences of actions for achieving a goal) and a strategy agent (which adapts the layer's behavior to the environment). To ensure pro-activeness, the active layer reconfigures the priorities of the agents in the reactive layer when specific events occur. The decisions are based on measuring the values of some parameters that characterize the environment. For learning, a classifier system is introduced to express agent behaviors and activate specific procedures accordingly.

Integrating the event-driven component with the business part of the system can be a challenge especially when both parts have high performance requirements. This is the case, for example, in a virtual reality environment for training or computer games, in which AI and Virtual Reality techniques are used to simulate the real world inhabited by autonomous intelligent entities [8]. The system incorporates capabilities for simulating intelligent autonomous entities and, at the same time, for responding to high performance demands of visualizing the virtual world. In a high performance implementation, each agent can have its own computer to run on independently of the visualization module. This is made possible due to a framework that includes three modules: a FIPA compliant multi-agent platform that acts as middleware, a multi-agent system (MAS) that runs on top of the multi-agent platform, and a visualization module. MAS uses two classes of agents: inhabitant agents simulate beings in the virtual environment and execute actions that change the state of the virtual world; and a simulation controller maintains consistency and synchronization between the inhabitant agents and the virtual world.

The simulation controller has a 3-layer architecture. The simulation layer contains the *world's knowledge base* that maintains the data representing the virtual world state, and the simulator's logic manager module that controls the simulation. The reactive layer contains the sensory responder module that captures events from the environment, performs agents' actions, modifies the *world's knowledge base* and sends the changes to other agents involved; it also has a second component which sends information to several graphic viewers that are connected to the simulation. The social layer supports interaction with other agents.

Inhabitant agents have also a 3-layer architecture. The physical environment layer connects to the virtual world through sensors (that capture events in the virtual world) and effectors (that send actions to the simulation controller). The *cognitive* layer has a memory module (knowledge base), a decision module (with a reactive sub-module for immediate reactions and a deliberative sub-module for better solutions based on the use of the knowledge base), and a perform module with a list of tasks, a scheduler and a dispatcher. The social layer supports interaction with other agents.

Since many results presented previously refer to pilot implementations of event-driven distributed systems, further work is needed to reinforce the results obtained so far. Important research issues include: ensuring platform independence of event-based multi-agent systems, supporting high scalability, resolving issues related to uncertain environments by new facilities for trust estimation, increasing reliability, confidence support, resistance to security attacks, and others. More efforts are needed in understanding the role of agent mobility and self-replication for ensuring multi-agent systems resistance to external attacks. Also, improved collaborative methods leading to better perception of environment changes are needed.

4   Conclusions and Future Work

This chapter focuses on distributed intelligent event processing in Web, Grid, P2P, and agent-based systems. Previous work and research for solving scalability, interoperability, and fault tolerance problems are discussed, with emphasis on those solutions that ensure high reactivity and adaptability to environment changes, proactive and autonomous behavior, learning and social abilities. For each major topic the impact, strengths, weaknesses, and possible improvements are presented.

Adopting event processing in distributed systems is supported by specific models such as the ECA rules paradigm. The capacity to describe the composition and derivation of complex events supports reasoning over event relationships and distribution of the event detection functionality. In addition, the declarative nature of the rules facilitates adaptation to new and evolving situations.

Event-driven capabilities have been added to distributed systems by extensions to traditional architectures. The extension of SOA has benefited from the publish / subscribe mechanism included in the original SOA model. In addition, the SOA orientation towards open standards has stimulated the development of standards related to events and event services, as well.

The Web has been extended by adding event processing capabilities to servers and clients (browsers) with the aim of making them more responsive to dynamic changes of Web resources (data) and increasing interactivity in user dialogues. Web monitoring services allow users to express their interests and respond by sending alerts or executing other activities. For example, Google Alerts sends users email notifications of events related to their interests.

Event driven capabilities are used in collaborative VOs to help users cooperate for achieving common tasks, or for Grid performance optimization. P2P networks have adopted new models for distributed event transmission and routing, and for event detection that exploit the collaboration of nodes with similar capabilities. Last but not least, multi-agent systems have innate capabilities (pro-activeness, learning, social abilities, and so forth) that make them suitable for perceiving and processing environment events.

Event-based distributed systems are an active research field with many contributors. New ideas have been recently proposed and tested, which show the feasibility of solutions based on the new concepts. Nevertheless, these proposals need further evaluation studies to confirm their validity and performance in more significant, larger scale environments, and for a larger application spectrum. This will require the development of specific evaluation models and metrics for event-based distributed systems. While some results have been reported in the literature [27], more efforts and collaboration with neutral benchmark organizations, like TPC and SPECS, will be needed.

Clearly, testing and evaluation are just two steps of the complex software and system development process for event-driven distributed systems. An important trend will be moving the interest from the development of individual pilot systems to methodologies, software engineering methods, models, and frameworks for the whole software process, which includes requirements specification, design, implementation, deployment, maintenance, policy statement, and system administration. Techniques and methods to develop high performance distributed event detectors and rule-based systems are important subjects for future research. Also, since the design patterns approach has been successfully used in different domains, it is expected that more effort will be directed towards understanding and formalizing the architectural features of the event-driven distributed systems developed so far, and deriving design patterns for different application domains.

Further work on event formalization is also needed, including formal specification and verification of models used in complex event processing. Research and development in several other directions could also be of interest. One is related to information produced by heterogeneous sources. In order to combine them for deriving meaningful complex events, context information (metadata) might be added to better understand the semantics of events and also of the event sources. Context information is also needed in adaptive pervasive systems in which event and context semantics play an important role for the discovery and composition of services. More work is needed in the development of event, context, and service ontologies. Also, future research will be focused on classifying events and developing discriminant functions for event classes. Developing new methods for exploiting knowledge bases and learning processes could help in improved event detection and replace the human intervention that is used in some systems. Another issue is related to enhancing the event life cycle model with new approaches for event replication, logging, disregarding, consumption, and others to develop a common consistent framework for the operational semantics of event-driven systems [27].

Since wireless and mobile event-driven systems are expected to cover large-area applications, issues related to high variations of connectivity and unreliable data communication will be an important research subject. New policies, event semantics, state synchronization methods on reconnection, late event delivery, security, and others will have to be considered in the design of event-driven systems based on wireless and low capability devices.

Another topic of interest, which goes beyond the borders of event-based systems, will be security such as privacy and protection of producers and consumers. These issues are augmented by the use of profiling techniques for enhancing the performance and precision of event processing engines, and of portable devices used in tracking services. More research will be focused on techniques and methods for ensuring anonymity and for controlling access to sensitive information. Also, more work will be required in finding solutions for reducing the vulnerabilities due to the distribution of system components over large geographic areas, the broadcast communication, and the reduced capabilities of low-end equipment used frequently as event producers or consumers.

References

1. Allen, J., Gerguson, G.: Action and Events in Interval Temporal Logic. Journal of Logic and Computation 4(5), 31–79 (1994)
2. Ameur, R., Heudin, J.-C.: Interactive Intelligent Agent Architecture. In: Proceedings of the 2006 IEEE/WIC/ACM International Conference on Web Intelligence and Intelligent Agent Technology (WI-IATW 2006), pp. 331–334. IEEE Computer Society, Washington (2006)
3. Androutsellis-Theotokis, S., Spinellis, D.: A Survey of Peer-to-Peer Content Distribution Technologies. ACM Computing Surveys 36(4), 335–371 (2004)
4. Anicic, D., Fodor, P., Stojanovic, N., Stühmer, R.: Computing complex events in an event-driven and logic-based approach. In: Proceedings of the Third ACM international Conference on Distributed Event-Based Systems (DEBS 2009), Nashville, Tennessee, USA, pp. 1–2 (2009)

5. Aydt, R., Smith, W., Swany, M., Taylor, V., Tierney, B., Wolski, R.: A Grid Monitoring Architecture. GWDPerf-16-3, Global Grid Forum (2001), http://wwwdidc.lbl.gov/GGF-PERF/GMA-WG/papers/GWD-GP-16-3.pdf (retrieved on February 02, 2010)
6. Baldoni, R., Beraldi, R., Quema, V., Querzoni, L., Tucci Piergiovanni, S.: A Scalable p2p Architecture for Topic-Based Event Dissemination. Technical report, Universita di Roma "La Sapienza" (2007)
7. Bank, D.: Web Services Eventing, W3C Member Submission (2006), http://www.w3.org/Submission/WS-Eventing (retrieved February 26, 2010)
8. Barella, A., Carrascosa, C., Botti, V.: Agent Architectures for Intelligent Virtual Environments. In: 2007 IEEE/WIC/ACM International Conference on Intelligent Agent Technology (IAT 2007), pp. 532–535 (November 2007)
9. Barga, R.S., Goldstein, J., Ali, M., Hong, M.: Consistent Streaming Through Time: A Vision for Event Stream Processing. In: Proc. of the 3rd Biennial Conference on Innovative Data Systems Research (CIDR), Asilomar, California, USA, pp. 363–374 (2007)
10. Blanco, R., Wang, J., Alencar, P.: A metamodel for distributed event based systems. In: Proceedings of the Second international Conference on Distributed Event-Based Systems (DEBS 2008), vol. 332, pp. 221–232. ACM, New York (2008)
11. Castro, M., Druschel, P., Kermarrec, A., Rowstron, A.: SCRIBE: A large-scale and decentralized application-level multicast infrastructure. IEEE JSAC 20(8), 1489–1499 (2002)
12. Chakravarthy, S., Adaikkalavan, R.: Provenance and Impact of Complex Event Processing (CEP): A Retrospective View. In: Buchmann, A., Koldehofe, B. (eds.) Special Issue of IT - Complex Event Processing, vol. 51(5), pp. 243–249. Oldenbourg Publications (September 2009)
13. Chakravarthy, S., Adaikkalavan, R.: Ubiquitous Nature of Event-Driven Approaches: A Retrospective View (Position Paper). In: Proceedings of the Dagstuhl Seminar 07191 (2007), http://drops.dagstuhl.de/volltexte/2007/1150/pdf/07191.ChakravarthySharma.Paper.1150.pdf (retreived January 10, 2010)
14. Chakravarthy, S., Mishra, D.: Snoop: An expressive event specification language for active databases. Data Knowledge Engineering 14(1), 1–26 (1994)
15. Chakravarthy, S., Krishnaprasad, V., Anwar, E., Kim, S.-K.: Composite Events for Active Databases: Semantics, Contexts and Detection. In: Proceedings of the 20th International Conference on Very Large Data Bases, pp. 606–617. Morgan Kaufmann Publishers Inc., San Francisco (1994)
16. Chakravarty, P., Singh, M.P.: An event-driven approach for agent-based business process enactment. In: Proceedings of the 6th International Joint Conference on Autonomous Agents and Multiagent Systems (AAMAS), Article No.: 214, Honolulu, Hawaii, pp. 1261–1263 (May 2007)
17. Chandrasekaran, S., Franklin, M.: Streaming queries over streaming data. In: Proc. of the 28th Int. Conference on Very Large Data Bases (VLDB 2002), pp. 203–214 (2002)
18. Cheng, S., Jih, W., Hsu, J.Y.: Context-aware Policy Matching in Event-driven Architecture. In: AAAI 2005 Workshop: Contexts and Ontologies: Theory, Practice and Applications, Pittsburgh, Pennsylvania, USA, pp. 140–141 (2005)
19. Cilia, M., Antollini, M., Bornovd, C., Buchman, A.: Dealing with heterogeneous data in pub/sub systems: The Concept-Based approach. In: International Workshop on Distributed Event-Based Systems (DEBS 2004), Edinburgh, Scotland (2004), http://www.dvs.tu-darmstadt.de/publications/pdf/Concept-based04.pdf (retrieved 10 January, 2010)

20. Cugola, G., Di Nitto, E., Fuggetta, A.: The jedi event-based infrastructure and its application to the development of the OPSS WFMS. IEEE Trans. Softw. Eng. 27(9), 827–850 (2001)

21. Dasgupta, S., Bhat, S., Lee, Y.: Event Semantics for Service Composition in Pervasive Computing. In: Intelligent Event processing - AAAI Spring Symposium 2009, pp. 27–37. AAAI Press, Menlo Park (2009)

22. Doorenbos, R.B.: Production Matching for Large Learning Systems, PhD Thesis (1995), http://reports-archive.adm.cs.cmu.edu/anon/1995/CMU-CS-95-113.pdf (retrieved March 11, 2010)

23. Ermagan, V., Krüger, I.H., Menarini, M.: Aspect-oriented modeling approach to define routing in enterprise service bus architectures. In: Proceedings of the 2008 International Workshop on Models in Software Engineering (MiSE 2008), Leipzig, Germany, pp. 15–20 (2008)

24. Etzion, O.: Event Cloud. Encyclopedia of Database Systems, 1034–1035 (2009)

25. Fortino, G., Garro, A., Mascillaro, S., Russo, W.: Using event-driven lightweight DSC-based agents for MAS modelling. International Journal on Agent Oriented Software Engineering (IJAOSE) 4(2), 113–140 (2010)

26. Hinze, A., Michel, Y., Schlieder, T.: Approximative filtering of XML documents in a publish/subscribe system. In: 29th Australasian Computer Science Conference, ACSC 2006, pp. 177–185 (2006)

27. Hinze, A., Sachs, K., Buchmann, A.: Event-Based Applications and Enabling Technologies. In: Proc. of the 3rd ACM International Conference on Distributed Event-Based Systems (DEBS 2009), Nashville, TN, USA (2009), Session Keynote papers, Article No.: 1. http://delivery.acm.org/10.1145/1620000/1619260/a1-buchmann.pdf?key1=1619260&key2=9544530821&coll=GUIDE&dl=GUIDE&CFID=98611992&CFTOKEN=93216417 (retrieved January 15, 2010)

28. Huang, Y., Gannon, D.: A Comparative Study of Web Services-based Event Notification Specifications. In: Proceedings of the 2006 international Conference Workshops on Parallel Processing (ICPPW), pp. 7–14. IEEE Computer Society, Washington (2006)

29. IBM. IBM Tivoli Workload Scheduler Version 8.2: New Features and Best Practices. IBM Press (2004)

30. Jung, J., Park, J., Han, S., Lee, K.: An ECA-based framework for decentralized coordination of ubiquitous web services. Inf. Softw. Technol. 49(11-12), 1141–1161 (2007)

31. Jung, J.-Y., Hong, Y.-S., Kim, T.-W., Park, J.: Human-Centered Event Description for Ubiquitous Service Computing. In: Proc. of International Conference on Multimedia and Ubiquitous Engineering, International Conference on Multimedia and Ubiquitous Engineering (MUE 2007), Seoul, Korea, pp. 1153–1157 (2007)

32. Khalifa, Y.M.A., Okoene, E., Al-Mourad, M.B.: Autonomous Intelligent Agent-Based Tracking Systems, Recent Developments. ICGST-ACSE Journal 7(1), 21–31 (May 2007)

33. Kühn, E., Mordinyi, R., Keszthelyi, L., Schreiber, C., Bessler, S., Tomic, S.: Aspect-Oriented Space Containers for Efficient Publish/Subscribe Scenarios in Intelligent Transportation Systems. In: Meersman, R., Dillon, T., Herrero, P. (eds.) OTM 2009. LNCS, vol. 5870, pp. 432–448. Springer, Heidelberg (2009)

34. Lee, W.-s., Lee, S.-y., Lee, K.-c.: Conflict Detection and Resolution method in WS-ECA framework. In: Proc. of The 9th International Conference on Advanced Communication Technology, vol. 1, pp. 786–791 (2007)

35. Legrand, I.C., Cirstoiu, C., Grigoras, C., Betev, L., Costan, A.: Monitoring, accounting and automated decision support for the alice experiment based on the MonALISA framework. In: Proceedings of the 2007 Workshop on Grid Monitoring (GMW 2007), Monterey, California, USA, pp. 39–44 (2007)
36. Luckham, D.: The Power of Events: An Introduction to Complex Event Processing in Distributed Enterprise Systems, May 18. Addison-Wesley Professional, Reading (2002)
37. Luckham, D., Schulte, R. (eds.): Event Processing Glossary - Version 1.1, Event Processing Technical Society (July 2008), http://www.ep-ts.com/ (retrieved January 10, 2010)
38. Memon, A., Xie, Q.: Using Transient/Persistent Errors to Develop Automated Test Oracles for Event-Driven Software. In: Proceedings of the 19th IEEE international Conference on Automated Software Engineering. ASE, pp. 186–195. IEEE Computer Society, Washington (2004)
39. Michelson, B.M.: Event-Driven Architecture Overview. Patricia Seybold Group / Business-Driven ArchitectureSM, February 2, pp. 1–8 (2006), http://soa.omg.org/Uploaded%20Docs/EDA/bda2-2-06cc.pdf (Retrieved January 10, 2010)
40. Mühl, G., Fiege, L., Pietzuch, P.: Distributed Event-Based Systems. Springer, Heidelberg (2006)
41. Oliver, C.S.: Autonomous Mission Planning for a Distributed Surveillance System. Master Thesis: Department of Electrical and Computer Engineering. Carnegie Mellon University, USA (2000)
42. OMA. OMA Web Services Enabler (OWSER): Overview. OMA-AD-OWSER Overview-V1 1-20060328-A (2006), http://www.openmobilealliance.org/releaseprogram/owserv11.html (retrieved March 20, 2010)
43. Paschke, A.: Design Patterns for Complex Event Processing. In: Proceedings of the 2nd International Conference on Distributed Event-Based Systems (DEBS 2008), Rome, Italy (2008), http://arxiv.org/ftp/arxiv/papers/0806/0806.1100.pdf (retrieved Januaruy 15, 2010)
44. Pătrânjan, P.L.: The Language XChange: A Declarative Approach to Reactivity on the Web. PhD thesis. University of Munich, Germany (September 2005)
45. Pattberg, J., Fluegge, M.: Towards an ontology of collaboration patterns. Lecture Notes in Informatics, vol. 120 (2007), pp. 85–96 (2009), http://subs.emis.de/LNI/Proceedings/Proceedings120/gi-proc-120-007.pdf (retrieved February 1, 2010)
46. Rajasekar, A., Moore, R., Wan, M.: Event Processing in Policy Oriented Data Grids. In: Proc. of Intelligent Event Processing AAAI Spring Symposium, Stanford, California, USA, pp. 61–66 (2009)
47. Rosenblum, D., Wolf, A.: A design framework for internet-scale event observation and notification. ACM SIGSOFT Software Engineering Notes 22(6), 344–360 (1997)
48. Saptharishi, M., Bhat, K., Diehl, C., Oliver, C., Savvides, M., Soto, A., Dolan, J., Khosla, P.: Recent Advances in Distributed Collaborative Surveillance. In: Proceedings of SPIE's 14 Annual Conference on Aerospace-Defense Sensing, Simulation and Controls, AeroSense, Orlando, USA, pp. 129–208 (2000)
49. Schmidt, K.-U., Stühmer, R., Stojanovic, L.: Gaining Reactivity for Rich Internet Applications by Introducing Client-side Complex Event Processing and Declarative Rules. In: Proc. of the Intelligent Event Processing - AAAI Spring Symposium, pp. 67–72. Stanford University, USA (2009)

50. Schwiderski-Grosche, S., Moody, K.: The SpaTeC composite event language for spatio-temporal reasoning in mobile systems. In: Proceedings of the Third ACM international Conference on Distributed Event-Based Systems (DEBS 2009), Nashville, Tennessee, USA, pp. 1–12 (2009)

51. Seufert, A., Schiefer, J.: Enhanced Business Intelligence - Supporting Business Processes with Real-Time Business Analytics. In: Proceedings of the 16th International Workshop on Database and Expert Systems Applications (DEXA 2005), pp. 919–925 (2005)

52. Soh, L., Tsatsoulis, C.: Reflective Negotiating Agents for Real-Time Multisensor Target Tracking. International Journal Conference on Artificial Intelligence, 1121–1127 (2001)

53. Tanenbaum, A.S., van Steen, M.: Distributed Systems. Principles and paradigms, 2nd edn. Prentice-Hall, Englewood Cliffs (2007)

54. Turchin, Y., Gal, A., Wasserkrug, S.: Tuning complex event processing rules using the prediction-correction paradigm. In: Proceedings of the Third ACM international Conference on Distributed Event-Based Systems (DEBS 2009), Nashville, Tennessee, USA, pp. 1–12 (2009)

55. Verginadis, Y., Apostolou, D., Papageorgiou, N., Mentzas, G.: Collaboration Patterns in event-driven environments for Virtual Organizations. In: Intelligent Event Processing - AAAI Spring Symposium 2009, Atlanta, US, pp. 92–97 (2009)

56. Vijayakumar, N., Plale, B.: Missing Event Prediction in Sensor Data Streams Using Kalman Filters. In: Ganguly, A.R., Gama, J., Omitaomu, O.A., Gaber, M.M., Vatsavai, R.R. (eds.) Knowledge Discovery From Sensor Data, pp. 149–170. CRC Press, Boca Raton (2009)

57. von Ammon, R., Emmersberger, C., Ertlmaier, T., Etzion, O., Paulus, T., Springer, F.: Existing and future standards for event-driven business process management. In: Gokhale, A., Schmidt, D.C. (eds.) Proceedings of the Third ACM International Conference on Distributed Event-Based Systems 2009, pp. 1–5. ACM, New York (2009)

58. Xhafa, F., Paniagua, C., Barolli, L., Caballé, S.: A Parallel Grid-based Implementation for Real Time Processing of Event Log Data in Collaborative Applications. Int. J. Web and Grid Services, IJWGS 6(2) (2010) (in press)

59. Zaera, M.: Wave-based communication in vehicle to infrastructure real-time safety-related traffic telematics. Master's thesis, Telecommunication Engineering. University of Zaragoza (August 2008)

60. Zhao, S., Stutzbach, D., Rejaie, R.: Characterizing files in the modern Gnutella network: A measurement study. In: Proc. Multi-media Computing and Networking Conf., San Jose, CA, USA, pp. 267–280 (2006)

61. Zhuang, S.Q., Zhao, B.Y., Joseph, A.D., Katz, R.H., Kubiatowicz, J.D.: Bayeux: an architecture for scalable and fault-tolerant wide-area data dissemination. In: Proc. of the 11th International Workshop on. Network and Operating Systems Support for Digital Audio and Video (NOSSDAV 2001), Danforts on the Sound, Port Jefferson, New York, USA, pp. 11–20 (2001)

A CEP Babelfish: Languages for Complex Event Processing and Querying Surveyed

Michael Eckert, François Bry, Simon Brodt, Olga Poppe, and Steffen Hausmann

Abstract. Complex Event Processing (CEP) denotes algorithmic methods for making sense of events by deriving higher-level knowledge, or complex events, from lower-level events in a timely fashion and permanently. At the core of CEP are queries continuously monitoring the incoming stream of "simple" events and recognizing "complex" events from these simple events. Event queries monitoring incoming streams of simple events serve as specification of situations that manifest themselves as certain combinations of simple events occurring, or not occurring, over time and that cannot be detected solely from one or parts of the single events involved.

Special purpose Event Query Languages (EQLs) have been developed for the expression of the complex events in a convenient, concise, effective and maintainable manner. This chapter identifies five language styles for CEP, namely *composition operators, data stream query languages, production rules, timed state machines,* and *logic languages,* describes their main traits, illustrates them on a sensor network use case and discusses suitable application areas of each language style.

1 Introduction

Event-driven information systems demand a systematic and automatic processing of events. Complex Event Processing (CEP) encompasses methods, techniques, and tools for processing events *while they occur*, i.e., in a continuous and timely fashion. CEP derives valuable higher-level knowledge from lower-level events; this

Michael Eckert
TIBCO Software, Balanstr. 49, 81669 Munich, Germany
e-mail: meckert@tibco.com

François Bry · Simon Brodt · Olga Poppe · Steffen Hausmann
Institute for Informatics, University of Munich, Oettingenstr. 67, 80538 Munich, Germany
e-mail: {bry,brodt,poppe,hausmann}@pms.ifi.lmu.de

S. Helmer et al.: Reasoning in Event-Based Distributed Systems, SCI 347, pp. 47–70.
springerlink.com   © Springer-Verlag Berlin Heidelberg 2011

knowledge takes the form of so called complex events, that is, situations that can only be recognized as combinations of several events.

The term Complex Event Processing was popularized in [51]; however, CEP has many independent roots in different research fields, including discrete event simulation, active databases, network management, and temporal reasoning. Only in recent years, has CEP emerged as a discipline of its own and as an important trend in the industry. The founding of the Event Processing Technical Society [23] in early 2008 underlines this development.

Important application areas of CEP are the following:

Business activity monitoring aims at identifying problems and opportunities in early stages by monitoring business processes and other critical resources. To this end, it summarizes events into so-called key performance indicators such as, e.g., the average run time of a process.

Sensor networks transmit measured data from the physical world to, e.g., Supervisory Control and Data Acquisition systems that are used for monitoring of industrial facilities. To minimize measurement and other errors, data of multiple sensors frequently has to be combined. Further, higher-level situations (e.g., fire) usually have to be derived from raw numerical measurements (e.g., temperature, smoke).

Market data such as stock or commodity prices can also be considered as events. They have to be analyzed in a continuous and timely fashion in order to recognize trends early and to react to them automatically, for example, in algorithmic trading.

The situations (specified as complex events) that need to be detected in these applications and the information associated with these situations are distributed over several events. Thus CEP can only derive such situations from a number of correlated (simple) events. To this end many different languages and formalisms for querying events, the so called Event Query Languages (EQLs), have been developed in the past.

There are also some surveys in the realm of CEP. For example, in [61, 60], rule-based approaches for reactive event processing are classified according to their origins. In [12], EQLs are divided into groups depending on the kind of system architecture they are used in. The survey of EQLs described in [70] distinguishes between a non-logic and logic-based view on handling the event triggered reactivity. There are also comparisons of different single CEP products, e.g., [35]. Both the multitude of EQLs and the diversity of surveys on event processing and reactivity can be attributed in part to the fact that CEP has many different roots and is only now recognized as an independent field.

To the best of our knowledge, there are no comprehensive surveys so far that (1) classify different EQLs into groups according to the language "style" or "flavor" and (2) compare the groups by means of the same example queries with respect to their expressivity, ease of use and readability, formal semantics, success in the industry and some other features. This chapter surveys the state of the art in CEP regarding these two points. Since CEP is a field that is very broad and without clear-cut boundaries, this chapter focuses strongly on languages specified for querying events. Other, less developed aspects of CEP such as detecting unknown events using approaches like machine learning and data mining on event streams, are not discussed here.

The contributions of this chapter are:

1. Identification and abstract description of five language styles, namely *composition operators, data stream query languages, production rules, timed state machines,* and *logic languages*
2. Illustration of each language style on a sensor network use case
3. Discussion on suitable application areas of each language style
4. Abstract description of some of the combined approaches

2 Terminology

Since CEP has evolved from many different research areas, a standard terminology has not yet established and found broad adoption. For example, what is called a (complex) event query might also be called a complex event type, an event profile, or an event pattern, depending on the context. We will therefore devote this section to the basic notions and our informal definitions of them.

An **event** is a message indicating that something of interest happens, or is contemplated as happening. Events can be represented in different data formats such as relational tuples, XML documents or objects of an object-oriented programming language (e.g., Java).

In this chapter, we use the following presentation of events: *event type* (*attribute name$_1$* (*attribute value$_1$*), ..., *attribute name$_n$* (*attribute value$_n$*)). An **event type** specifies an event structure, similar to a relational database schema specifying the structure of tuples of a relation. For example, *high_temp(area)* is an event type of an event *high_temp(area(a))* indicating high temperature in area a. In this event, *area* is an attribute and a is its value. (In the following, capital letters denote variables and small letters denote literals.) An **event attribute** is a component of the structure of an event. It can be an entry of a tuple, an XML fragment, or a field of an object, depending on the event representation. The set of attribute values of an event is called **event data**.

The formalism introduced here is by no means compulsory. One could prefer to use an unnamed perspective identifying attribute values by their positions or tuse any alternative event representation instead. Since in all languages proposed so far, events are flat or structured records or tuples, the formalism adopted for this chapter is no restriction.

Since events happen at particular time which is essential for event processing, all events must have a possibly implicit attribute called event occurrence time. An **event occurrence time** is a time point or time interval indicating when this event happens. A time interval is described by two timestamps indicating its bounds. A time point is described by a single timestamp. We shall see below that using time points or time intervals has far reaching consequences for event processing.

Timing and event order are difficult issues of distributed systems. Each node (computer, device, etc.) in a distributed system has its own local clock and the clocks of different nodes are hard to be synchronized perfectly [19]. Furthermore the transmission time of messages varies depending on sender and receiver, routing, network load, and other factors. Therefore the reception order of some events may differ from their emission order [48]. These issues are ignored in this chapter for the sake of simplicity.

Another characteristic feature of events is event identification. For example, one can assign an identifier t to the event $high\_temp(area(a))$, written as $t : high\_temp$ $(area(a))$. We will see the advantages of this feature below.

Events are sent by event producers (e.g., sensors) to event consumers (e.g., Supervisory Control and Data Acquisition system) on so called **event streams**.

In order to react to an event e (e.g., turn on air conditioning in an area if an event indicating high temperature in the area arrives) or to derive a new event from another event e (e.g., derive an event indicating high temperature from an event containing a temperature measurement if the measurement is considered to be high), an event query which matches e is specified in an event query language. An **Event Query Language (EQL)** is a high level programming language (possibly of limited expressivity) for querying events. A **simple event query** is a specification of a certain kind of *single* events by means of an event query language. A **complex event query** is a specification of a certain *combination* of events using multiple simple event queries and conditions describing the correlation of the queried events.

A **simple event** is either an event arriving on the event stream or an event derived by a simple event query (i.e., from a *single* event). A **complex event** is an event derived by a complex event query (i.e., from a certain *combination* of at least two events occurring or not occurring over time). In the EPTS Glossary [23] many other kinds of events are defined, such as composite event, virtual event, derived event, raw event and some others.

Note that the occurrence time of a complex event e comprises the occurrence time of all events e it has been derived from. For example, a complex event $f:fire(area(a))$ indicating fire can be derived from two simple events $s:smoke(area(a))$ and $t:$ $high\_temp(area(a))$ indicating smoke and high temperature respectively. f begins as soon as s or t begins and it ends as soon as both simple events are over.

The derivation of complex events is called Complex Event Processing. **Complex Event Processing (CEP)** denotes algorithmic methods for making sense of base events (low-level knowledge) by deriving complex events (high-level knowledge) from them in a timely fashion and over periods of time.

These are the most important notions in the field of event processing. In this chapter we will also need some other notions which will be informally introduced before using them.

3 Identification of Language Styles

To bring some order into the multitude of EQLs, we try to group languages with a similar "style" or "flavor" together. We will focus on the general style of the languages and the variations within a style, rather than discussing each language and its constructs separately. It turns out most approaches for querying events fall into one of the following five categories:

1. languages based on composition operators (sometimes also called composite event algebras or event pattern languages),
2. data stream query languages (usually based on SQL),
3. production rules,
4. timed (finite) state machines, and
5. logic languages.

As we will see, the first, the second and the fifth approaches are languages explicitly developed for specifying event queries, while the third one is only a clever way to use the existing technologies of production rules to implement event queries. Similarly, the fourth approach is the use of an established technology to model event queries in a graphical way.

In Sections 4–8, we will describe each language style individually, mentioning the respective important languages from the research and industry. We will also discuss the strengths and weaknesses of each style and illustrate them on a sensor network use case which can be implemented using, e.g., TinyDB [52]. Section 9 summarizes the comparison by a discussion on suitable application areas of each language style. It is further worth mentioning that many industry products follow approaches where several languages of different flavors are supported or a single language combines aspects of several flavors. Section 10 will therefore be devoted to hybrid approaches. Section 11 concludes this chapter.

4 Composition Operators

4.1 General Idea

The first group of languages that we discuss builds complex event queries from simple event queries using composition operators. Historically, these languages have their roots primarily in Active Database Systems [64], though newer systems like Amit [4] run independently from a database. Some examples include: the COMPOSE language of the Ode active database [32, 33, 34], the composite event detection language of the SAMOS active database [30, 31], Snoop [17] and its successor SnoopIB [2, 3], GEM [53], SEL [76], CEDR [7], ruleCore [72, 57], the SASE Event Language [75], the Cayuga Event Language [20], the original event specification language of XChange [21, 13, 14], and the unnamed languages proposed in the following papers: [66], [55], [37], [15], [8], [68, 67].

Complex event queries are expressed by composing single events using different composition operators. Typical operators are conjunction of events (all events must happen, possibly at different times), sequence (all events happen in the specified order), and negation within a sequence (an event does not happen in the time between two other events). Consider the use of the operators in the sensor network use case below.

4.2 Sensor Network Use Case

Since different composition-operator-based EQLs have very different and rather unreadable syntax we formulate the example queries in pseudo code in Figure 1. The pseudo code illustrates the idea of this kind of EQLs but it does not mean that each of the queries in Figure 1 can be analogously formulated in every composition-operator-based EQL.

Composition	$fire(area(A)) = (\ smoke(area(A)) \wedge high\_temp(area(A))\ )_{1\ min}$
Sequence	$fire(area(A)) = (\ smoke(area(A));\ high\_temp(area(A))\ )_{1\ min}$ or $fire(area(A)) = s{:}smoke(area(A));\ high\_temp(area(A));\ s{+}1\ min$
Negation	$failure(sensor(S)) = t{:}temp(sensor(S));\ not\ temp(sensor(S));\ t{+}12\ sec$
Aggregation	–

Fig. 1 Example queries in pseudo code for composition operators

The first query in Figure 1 triggers fire alarm for an area when smoke and high temperature are both detected in the area within 1 minute, in other words the query derives a complex event *fire(area(A))* from the two events *smoke(area(A))* and *high_temp(area(A))*. The events *smoke(area(A))* and *high_temp(area(A))* are joined on variable A. Their order does not matter but it is important that both events appear within 1 minute, indicated by the time window specification $(\ldots)_{1\ min}$. This is a typical example of event composition realized by the conjunction operator $\wedge$ and a time window specification.

The second example is similar to the first one but the events in the event query are connected by the sequence operator *;* denoting that the order of events is important, i.e., the event *smoke(area(A))* must appear before the event *high_temp(area(A))*. Only if the events appear within 1 minute and in the right order is the complex event *fire(area(A))* derived.

Alternatively if a composition-operator-based EQL supports event identification and relative timer events, this query can be formulated by means of the event identifier *s* for the event *smoke(area(A))* and a relative timer event *s+1 min*. In this case an EQL must decide whether the complex event $fire(area(A))$ is derived after the event $high\_temp(area(A))$ or after the event $s + 1$ min.

The sequence operator is not as intuitive as it seems at first sight. Let A, B and C be simple event queries. Under time point semantics $(A;B);C$ is not equivalent to $A;(B;C)$, i.e., both queries do not yield the same answers. Let b,a,c be events arriving in this order and matching B, A, and C, respectively. They yield an answer for the query $A;(B;C)$ since b and c satisfy $(B;C)$ with the occurrence time (point) of c which is later than that of a. b happens before a which is not allowed by the query $(A;B);C$.

Under time interval semantics $A;(B;C)$ and $(A;B);C$ are equivalent. They both match events a,b,c arriving only in this order. $(B;C)$ matches b,c and has the occurence time interval starting as soon as b begins and ending as soon as c ends. Furthermore $A;(B;C)$ requires that a is over before b begins. This query matches a,b,c arriving exclusively in this order. Analogously $(A;B)$ matches a,b and has the time interval described by two time points, namely the begin of a and the end of b. Consequently $(A;B);C$ requires that c begins after b is over. This query can also match only the events a,b,c arriving in this order. Hence, using time points or time intervals has far reaching consequences [28].

The third example in Figure 1 shows how negation can be expressed by means of composition operators. The query uses event identification and relative timer events. It demonstrates the necessity for event identification if two events of the same type are used within one query and it has to be distinguished between them.

Assume all sensors of our network send temperature measurements every 12 seconds. The third query detects a failure of a sensor when its measurement is missing, i.e., the query derives a complex event $failure(sensor(S))$ when there is an event $temp(sensor(S))$ which is not followed by another event $temp(sensor(S))$ within 12 seconds.

Another feature which must be supported by an EQL is aggregation. Aggregation means collection of data satisfying certain conditions, analysis of the data and construction of new data containing the result of the analysis. An example of aggregation in our use case is the computation of the average temperature reported by a sensor during the last minute every time a temperature measurement from the sensor arrives. Such a query is unfortunately not expressible by means of composition operators (compare Figure 1).

Nesting of expressions makes it possible to specify more complicated queries but we restrict ourselves to simple examples which should illustrate the main ideas of the language styles without embracing their whole expressivity.

4.3   Summary

Many composition-operator-based EQLs support restrictions on which events should be considered for the composition of a complex event. Event instance selection, for example, allows selection of only the first or last event of a particular type [77, 4, 37].

Event instance consumption prevents the reuse of an event for further complex events if it has already been used in another, earlier complex event [31, 77].

Composition operators offer a compact and intuitive way to specify complex events. Particularly temporal relationships and negation are well-supported. Event instance selection and consumption are features that are not present in the other approaches. Yet, there are hidden problems with the intuitive understanding of operators sometimes, e.g., several variants of the interpretation of a sequence (amongst others, interleaved with other events or not). Further, event data (i.e., access to the attribute values of an event) is often neglected in languages of this style, in particular regarding composition and aggregation.

Currently only very few CEP products are based on composition operators, among them IBM Active Middleware Technology (Amit) [4] and ruleCore [72, 57].

5 Data Stream Query Languages

5.1 General Idea

The second style of languages has been developed in the context of relational data stream management systems. Data stream management systems are targeted at situations where loading data into a traditional database management system would consume too much time. They are particularly targeted at nearly real-time applications where a reaction to the incoming data would already be useless after the time it takes to store it in a database. A typical example of data stream query languages is the Continuous Query Language (CQL) that is used in the STREAM system [6]. The general ideas behind CQL apply to a number of open-source and commercial languages and systems including Esper [24], StreamSQL [73, 40, 1], PIPES [42, 43], TelegraphCQ [18, 65], the CEP and CQL component of the Oracle Fusion Middleware [58], and Coral8 [56]. See also [44, 50] for recent research in the field of data stream query languages.

Data stream query languages are based on the database query language SQL and the following general idea: Data streams carry events represented as tuples. Each data stream corresponds to exactly one event type. The streams are converted into relations which essentially contain (parts of) the tuples received so far. On these relations a (almost) regular SQL query is evaluated. The result (another relation) is then converted back into a data stream. Conceptually, this process is done at every point of time. Note that this implies a discrete time axis. (See however [39] for variations.)

For the conversion of streams into relations, stream-to-relation operators like time windows such as "all events of the last hour" or "the last 10 events" are used. For the conversion of the result relation back into a stream there are three options: "Istream" stands for "insert stream" and contains the tuples that have been added to the relation compared to the previous state of the relation, "Dstream" stands for "delete stream" and contains the tuples that have been removed from the relation compared to its previous state, and "Rstream" stands for "relation stream" and contains simply every tuple of the relation. In the following we only use "Istream".

5.2 Sensor Network Use Case

Figure 2 shows equivalent example queries as Figure 1 but in Continuous Query Language (CQL). A CQL query is very similar to an SQL query. The FROM part of a CQL query is a cross product of relations, the optional WHERE part defines selection conditions, and the SELECT part is a usual projection.

For example, the FROM part of the first query in Figure 2 joins two relations *smoke* and *high_temp* which were generated out of event streams of type *smoke* and *high_temp* respectively by means of time windows. Generally there are several types of time windows. For the sake of brevity only two of them are explained here.

The first one is a simple sliding window. The resulting relation contains all stream tuples of a particular type between *now–d* and *now* where *now* is the current time point and *d* is a duration such as "1 Minute" or "12 Seconds". The syntax for a sliding window of duration *d* is *T [Range d]* where *T* is an event type and the name of the resulting relation. For example, the notation *smoke [Range 1 Minute]* produces the relation *smoke* containing tuple representations of all events of type *smoke* which happened in the last minute.

The second time window that we explain here is a now window. The resulting relation contains only the stream tuples of a particular type with the occurrence time *now* where *now* denotes the current time point. The syntax for this window is *T [Now]* where *T* is the event type and the name of the resulting relation. For example, the result of the expression *high_temp [Now]* is the relation *high_temp* containing tuple representations of all events of type *high_temp* which happened at the current moment. Note that *T [Range 0 Minutes]* is equivalent to *T [Now]*.

Consider the first example in Figure 2. Remember that the first query triggers fire alarm for an area when smoke and high temperature were both detected in the area within one minute. This temporal condition can be intuitively formulated by means of 1 minute-long simple sliding windows restricting the *smoke* and the *high_temp* streams. The join condition is specified in the WHERE block of the query.

When the order of queried events is important the same query becomes less intuitive. Consider the second example in the figure. The query triggers fire alarm for an area when high temperature is being measured in the area now and smoke has been detected in the same area during the last minute.

The definition of correct time windows is essential as it has semantic consequences such as differentiation between an unordered composition and a sequence. Observe that a sequence of more than two events can only be expressed by means of rule chaining. For example, the sequence of three events e_1, e_2, e_3 can be expressed in the following way: The first query guarantees that e_1 happens before e_2 and generates a complex event e as an intermediate result. The second rule queries events e and e_3 in this order and derives the resulting event.

Negation is hard to express in CQL (as well as in SQL) because the negated tuples have to be queried by an auxiliary query which is nested in the WHERE block of the main query and must be empty to let the main query produce an answer. For example, the third rule in the figure reports a failure of a sensor when it does not send a temperature measurement every 12 seconds.

Composition

```
SELECT Istream s.area
FROM smoke [Range 1 Minute] s,
     high_temp [Range 1 Minute] t
WHERE s.area = t.area
```

Sequence

```
SELECT Istream s.area
FROM smoke [Range 1 Minute] s,
     high_temp [Now] t
WHERE s.area = t.area
```

Negation

```
SELECT Istream t1.sensor
FROM temp [Now] t1
WHERE NOT EXISTS ( SELECT *
                   FROM temp [Range 12 Seconds] t2
                   WHERE t1.sensor = t2.sensor )
```

Aggregation

```
SELECT Istream t1.sensor, avg(t1.value)
FROM temp [Range 1 Minute] t1,
     temp [Now] t2
WHERE t1.sensor = t2.sensor
```

Fig. 2 Example queries in Continuous Query Language

Aggregation is well supported by the language as shown by the last example in Figure 2. Every time a temperature measurement from a sensor arrives the query computes the average temperature reported by the sensor during the last minute.

5.3 Summary

Data stream query languages are very suitable for aggregation of event data, as particularly necessary for market data, and offer a good integration with databases. Expressing negation and temporal relationships, on the other hand, is often cumbersome. The conversion from streams to relations and back may be considered somewhat unnatural, as may the prerequisite of a discrete time axis.

SQL-based data stream query languages are currently the most successful approach commercially and are supported in several efficient and scalable industry products. The better known ones are Oracle CEP, Coral8, StreamBase, Aleri and the open-source project Esper. However, there are big differences between the various projects and there also exist important extensions that go beyond the general idea that has been discussed here.

6 Production Rules

6.1 General Idea

Production rules are not an event query language as such, however they offer a fairly convenient and very flexible way of implementing event queries. The first successful production rule engine has been OPS [26], in particular in the incarnation OPS5 [25]. Since then, many others have been developed in the research and industry, including systems like Drools (also called JBoss Rules) [41], ILOG JRules [38], and Jess [69]. While the general ideas of production rules will be explained here, we refer the reader to [9] for a deeper introduction.

Production rules, which nowadays are mainly used in business rule management systems like Drools or ILOG JRules, are not EQLs in the narrower sense. The rules are usually tightly coupled with a host programming language (e.g., Java) and specify actions to be executed when certain states are entered [9]. The states are expressed as conditions over objects in the so-called working memory. These objects are also called facts.

Besides their use in business rule management systems that are not focused on events, production rules are also an integral part of the CEP product TIBCO Business Events, which also offers more CEP-specific features such as support for temporal aspects or modelling of event types and data.

The incremental evaluation (e.g., with Rete [27]) of production rules makes them also suitable for CEP. Whenever an event occurs, a corresponding fact must be created. Event queries are then expressed as conditions over these facts. In doing so, the programmer has much freedom but little guideline.

6.2 Sensor Network Use Case

Figure 3 contains our four example queries in the open source production rule system Drools. In Drools all events are represented as Java objects. Every time an event arrives some Java method has to convert it into an object, insert the object into the working memory, and call the rule engine to perform the rule evaluation (more precisely, fire all rules until no rule can fire). Note that in CEP-tailored systems such as TIBCO Business Events this happens automatically. If a complex event is derived by a rule it is also saved as an object in the working memory. We assume that in this case the *insert*-method sets the occurrence time of a complex event.

The occurrence time is a usual attribute of an object. This is actually a problem because every method can change every occurrence time inadvertently. This in turn leads to incorrect answers.

For the sake of simplicity we use time point semantics, assume that timestamps are given in seconds since the epoch (i.e., since the midnight of January 1, 1970) and we do not perform any garbage collection (i.e., deletion of events). These assumptions are not suitable for real-life applications but they help to keep the examples simple. Under the above assumptions we can express the temporal relations between events as simple comparisons of numbers. In real-life applications temporal relations would have to be programmed as Java methods that are called in Drools rules.

A Drools rule consists of two parts. The WHEN part is an event query, it specifies both the types of queried events and conditions on the events. The THEN part derives an object representing the complex event, sets its occurrence time, and saves the object into the working memory. This newly asserted object can then also activate further rules.

```
Composition when  s: Smoke()
                  High_temp(area == s.area &&
                            timestamp >= (s.timestamp - 60) &&
                            timestamp <= (s.timestamp + 60))
             then insert(new Fire(s.area));
```

```
Sequence     when  s: Smoke()
                   High_temp(area == s.area &&
                             timestamp > s.timestamp &&
                             timestamp <= (s.timestamp + 60))
             then insert(new Fire(s.area));
```

```
Negation     when  t: Temp()
                   not(exists(Temp(sensor == t.sensor &&
                                   timestamp >= t.timestamp &&
                                   timestamp <= (t.timestamp + 12))))
             then insert(new Failure(t.sensor));
```

```
Aggregation when  t: Temp()
                  a: Avg() from accumulate(
                                 Temp(sensor == t.sensor &&
                                      timestamp >= (t.timestamp - 60) &&
                                      timestamp <= t.timestamp &&
                                      v: value),
                                 average(v))
            then insert(new Avg_temp(t.sensor, a));
```

Fig. 3 Example queries in Drools

Remember that the first rule detects fire in an area when smoke and high temperature are both detected in this area within one minute (consider the first rule in Figure 3). These conditions are coded into the specification of a *High_temp* object. Its attribute values are compared with the respective attribute values of a *Smoke* object *s*. In particular a *High_temp* event may happen at most one minute before or after a *Smoke* event.

In the second rule of the figure the order of the queried events is relevant. Smoke appears before high temperature is measured in the area. This is expressed by changing one of the conditions on the occurrence time of a *High_temp* object.

Negation is supported in Drools as shown by the third query. Recall that the query reports a failure of a sensor when the sensor does not send a temperature measurement every 12 seconds.

Aggregation of events is also supported. Consider the last rule in Figure 3. Every time a sensor sends a temperature measurement the query computes the average temperature reported by the sensor during the last minute. As this example illustrates aggregation is hard to express in Drools because the result of aggregation must be represented as an object in the WHEN part of a rule (an *Avg()* object in this case) to be used as a parameter of an object representing the complex event in the THEN part of a rule (an *Avg_temp()* object in this case).

As the examples show all relations between events must be programmed manually and even simple temporal conditions (already in our strongly simplified time model) require low-level code which is hard to read.

6.3 Summary

CEP with production rules is very flexible and well integrated with existing programming languages. However, it entails working on a low abstraction level that is — since it is primarily state and not event oriented — somewhat different from other EQLs. Especially aggregation and negation are therefore hard to express. Garbage collection, i.e., the removal of events from the working memory, has to be programmed manually. (See however [74] for work towards an automatic garbage collection.) Production rules are considered to be less efficient than data stream query languages; this is however tied to the flexibility they add in terms of combining queries (in rule conditions) and reactions (in rule actions).

7 Timed State Machines

7.1 General Idea

State machines are usually used to model the behavior of a stateful system that reacts to events. The system is modelled as a directed graph. The nodes of the graph represent the possible states of the system. Directed edges are labeled with events and temporal conditions on them. The edges specify the transitions between states that occur in reaction to in-coming events.

State machines are founded formally on deterministic or non-deterministic finite automata (DFAs or NFAs). Since states in a state machine are reached by particular sequences of multiple events occurring over time, they implicitly define complex events. Timed Büchi Automata (TBA) [5] were the first attempt to extend automata to temporal aspects for modelling real-time systems. In a TBA each transition between states depends not only on the type of arriving events but also on their occurrence time. For this, temporal conditions are added to transitions. Other examples of this kind of EQLs are UML state diagrams and regular real-time languages [36]. Many representatives of this language style were developed to achieve a particular task or solve a problem of real-time distributed systems, examples being Timed abstract state machine language for real-time system engineering [59], Timed automata approach to real time distributed system verification [47], and Timed-constrained automata for reasoning about time in concurrent systems [54].

7.2   Sensor Network Use Case

In this chapter we do not describe different kinds of real-time automata but explain their common principle. Figure 4 contains our example queries in a pseudo code for timed state machines. The pseudo code is an extension of Timed Büchi Automata [5]. The first extension is the consideration of event data. The second extension is the representation of complex events as automata in such a way that only if the end state of an automaton is reached the respective complex event is derived. A complex event can determinate a transition between states of another automaton so that arbitrary levels of abstraction can be achieved.

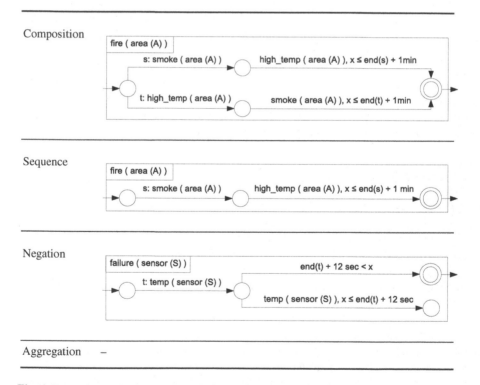

Fig. 4 Example queries in pseudo code for timed state machines

Remember that our first example derives a complex event *fire(area(A))* out of two events *smoke(area(A))* and *high_temp(area(A))* if these events happen within one minute. Their order does not matter. Since an automaton implicitly describes an ordered sequence we have to specify both acceptable orders of queried events. Consider the first query in Figure 4. The longer the composition of events the more acceptable orders (all possible permutations of events) must be considered by the machine, i.e., a simple composition query provokes a complicated automaton (exponential blow-up).

The events *smoke(area(A))* and *high_temp(area(A))* must happen within one minute. This condition is expressed using event identifiers, an auxiliary function *end(i)* which returns the end timestamp of event *i* and a global clock *x*. (As mentioned above, we do not consider such problems as clock synchronization in this chapter and refer the reader to [49].) Note that both events *smoke(area(A))* and *high_temp(area(A))* are joined upon the value of attribute *area*. If the end state of the state machine is reached the complex event *fire(area(A))* is derived.

The second query describes the sequence of events *smoke(area(A))* and *high_temp (area(A))*. The latter must happen at most one minute after the former to let the automaton reach its end state, i.e., to derive the complex event *fire(area(A))*. This is a very intuitive presentation.

Aggregation is not supported by timed state machines. Negation is also not supported but can be simulated by a failure state without outgoing edges and with an incoming edge which is labeled by a temporal condition and an event which should not arrive for the query to return an answer. For example, the third machine in Figure 4 detects a failure of a sensor when it does not send a temperature measurement every 12 seconds. If a temperature measurement comes within 12 seconds after the last measurement the state machine goes into the failure state, meaning that the end state is unreachable and the complex event *failure(sensor(S))* cannot be derived anymore. If 12 seconds since the last temperature measurement are over (consider the temporal condition of the incomimg edge of the end state) and no new measurement has arrived during this time, the state machine goes into the end state and derives the complex event *failure(sensor(S))*.

7.3 Summary

Though timed state machines provide intuitive visualization of complex events their expressivity is limited. They do not support aggregation. Negation and even composition of events are cumbersome. Conditions on the event data which are more complex than equi-joins (e.g., an attribute value must grow) cannot be expressed.

To overcome deficits of the theoretical automata, state machines are usually combined with languages of other styles. An example of this is the combination of state machines with production rules in TIBCO Business Events. There, a transition between two states is specified with a production rule. The condition of the production rule expresses when the transition is activated. Frequently reactions to the complex events that are implicit in a state machine are desirable. These can be specified for a transition (in the action part of the production rule) as well as for the entry or exit of states.

8 Logic Languages

8.1 General Idea

Logic languages express event queries in logic-style formulas. An early representative of this language style is the event calculus [45]. While event calculus is not an event query language per se, it has been used to model event querying and reasoning tasks in logic programming languages such as Prolog or Prova [46]. The latter combines the benefits of declarative and object-oriented programming by merging the syntaxes of Prolog and Java. Prova is used as a rule-based backbone for distributed Web applications in biomedical data integration. One of the key advantages of Prova is its separation of logic, data access, and computation.

XChange^{EQ} [10, 22] also adopts some ideas from event calculus-like approaches, but extends and tailors them to the needs of an expressive high-level event query language. XChange^{EQ} identifies and supports the following four complementary dimensions (or aspects) of event queries: data extraction, event composition, temporal (and other) relationships between events, and event accumulation. Its language design enforces a separation of the four querying dimensions.

A further example of this language style is Reaction RuleML [63, 62] combining derivation rules, reaction rules and other rule types such as integrity constraints into the general framework of logic programming.

8.2 Sensor Network Use Case

Figure 5 contains our four example queries in XChange^{EQ}. An XChange^{EQ} rule consists of two parts. The ON part, i.e., the rule body, is a complex event query which is a conjunction or disjunction of simple or complex event queries and an optional WHERE block containing temporal and other conditions on the queried events. The DETECT part, i.e., the rule head, is a construction of a complex event using the variable bindings returned by the respective event query.

Note that events are neither converted to relational tuples nor to objects of an object-oriented programming language. Furthermore, it is not possible to manipulate event timestamps neither consciously nor unwittingly. Finally, relative timer events are supported by XChange^{EQ}.

Event query specifications are very intuitive and flexible in XChange^{EQ}. There are four types of event queries charaterized by different kinds of brackets. Single brackets denote a complete event query, i.e., the query matches only those events which do not have attributes other than the ones specified in the query. In contrast double brackets denote an incomplete event query, i.e., events matched by the query may have additional attributes. Curly brackets denote an unordered query, i.e., the order of attributes does not matter. Square brackets denote an ordered event query.

Hence, there are four possible combinations of brackets, i.e., four types of event queries (ordered complete, unordered complete and so on).

Consider the first rule in Figure 5. Its complex event query is a conjunction of two simple incomplete and unordered event queries *event s: smoke*{{ *area*{{ *var A* }} }} and *event t: high_temp*{{ *area*{{ *var A* }} }} where variable *A* is bound to the value of attribute *area*. Since the same variable is used in both queries the queried events are joined on the value of this variable.

The WHERE block of the first rule in Figure 5 contains the additional temporal condition that both events, i.e., smoke and high temperature, appear within one minute. Note the use of event identifiers *s* and *t*. Note also that the temporal conditions (like *before* and *within*) are built-in into the language and must not be manually programmed.

The second query contains the additional temporal condition that the smoke event must appear before the high temperature event. The effect that the *additional* temporal condition is mapped to an *additional* statement in the query is an outstanding feature of XChange^{EQ}.

Composition | `DETECT` `fire { area { var A } }`
`ON and { event s: smoke {{ area {{ var A }} }},`
`          event t: high_temp {{ area {{ var A }} }}`
`    } where { {s,t} within 1 min }`
`END`

Sequence | `DETECT` `fire { area{ var A } }`
`ON and { event s: smoke {{ area {{ var A }} }},`
`          event t: high_temp {{ area {{ var A }} }}`
`    } where { s before t, {s,t} within 1 min }`
`END`

Negation | `DETECT` `failure { sensor { var S } }`
`ON and { event t: temp {{ sensor {{ var S }} }},`
`          event i: timer:from-end [ event t, 12 sec ],`
`          while i: not temp {{ sensor {{ var S }} }} }`
`END`

Aggregation | `DETECT` `avg_temp { sensor{ var S }, value { avg(all var T) } }`
`ON and { event t: temp {{ sensor {{ var S }} }},`
`          event i: timer:from-start-backward [ event t, 1 min ],`
`          while i: collect temp {{ sensor {{ var S }},`
`                                    value {{ var T }} }} }`
`END`

Fig. 5 Example queries in XChange^{EQ}

Negation and aggregation of events are supported as shown by the last two examples in Figure 5. Both negation and aggregation are restricted to finite time intervals. In the examples, the time intervals are given by relative timer events which are defined as follows:

- *timer:from-end[event e, d]* the relative timer *t* extends over the length of duration *d* starting at the end of *e*, i.e., $begin(t):=end(e), end(t):=end(e)+d$
- *timer:from-start-backward[event e, d]* the relative timer *t* extends over the length of duration *d* ending at the start of *e*, i.e., $begin(t):=begin(e)-d, end(t):=begin(e)$

In the above we write *begin(t)* and *end(t)* to denote the beginning and the end of event *t* respectively. There are of course many other relative timer events which are not discussed here, see [22].

Recall that the third example detects a failure of a sensor when it does not send a temperature measurement every 12 seconds, i.e., the query derives a complex event *failure{ sensor{ var S } }* when there is an event *temp{{ sensor{{ var S }} }}* which is not followed by another *temp{{ sensor{{ var S }} }}* event within 12 seconds.

The last query of the figure computes average temperature reported by a sensor during the last minute every time the sensor sends a temperature measurement. More precisely, every time an *event t: temp{{ sensor{{ var S }} }}* arrives, a relative timer event *i* denoting the time interval of one minute before *t*, is defined, all events happening during *i* and matched by the query *temp{{ sensor{{ var S }},* *value{{ var T }} }}* are collected and a complex event *avg_temp{ sensor{ var S },* *value{ avg(all var T) } }* containing the average temperature from the sensor *S*, is derived.

8.3 Summary

As the simple examples above demonstrate, logic languages offer a natural and convenient way to specify event queries. The main advantage of logic languages is their strong formal foundation, an issue which is neglected by many languages of other styles. (The chapter "Two Semantics for CEP, no Double Talk" in this volume describes a general, easily transferable approach for defining both the declarative and operational semantics of an EQL). Thanks to the separation of different dimensions of event processing, logic languages are highly expressive, extensible and easy to learn and use. Some languages of this style, e.g. XChange^{EQ} support an automatic garbage collection of events [11].

9 Application Areas of the Language Styles

Having described the strengths and weaknesses of the five language styles, we summarize the comparison by a discussion on suitable application areas of each language style.

Composition operators allow an intuitive specification of event patterns. This makes them attractive in scenarios where business users should be allowed to define event patters, such as real-time promotions and upselling (e.g., send three text messages within one hour to receive a free ringtone).

Data stream query languages are very suitable for aggregation of event data, as particularly necessary for applications involving market data (e.g., average price over 21 day sliding window), such as algorithmic trading. They also usually offer a good integration with databases, sharing in particular the common basis of SQL.

Production rules are very flexible and well integrated with existing programming languages. Since they allow the specification of actions to be executed when certain states are reached, they are particularly useful for applications involving tracking of stateful objects such as track and trace in logistics (maintain and react upon changes of the state of packages, containers, etc.) or monitoring of business processes and objects (also called Business Activity Monitoring). Due to their wide-spread use in business rules management systems, production rules often offer some support for exposing part of the logic to business users such as decision tables or trees.

Timed state machines also offer an easy and convenient way to maintain the current state. However they are limited to a finite set of states (e.g., "shipped", "delivered"). This makes them suitable, e.g., for monitoring of processes (which typically have a well-defined, finite number of states), but not suitable for applications involving infinite state spaces (e.g., a temperature control system where the temperature is a numeric value).

Logic languages have strong formal foundations, allow an intuitive specification of complex temporal conditions and account for event data. They could be successfully used in medical applications or emergency management in critical infrastructures.

Combination of different language styles in one approach allows to benefit from their strengths. This is the main reason why hybrid approaches are most successful in the industry. The next section is devoted to the combined approaches.

10   Combination of Different Language Styles

A comparison of the different language styles shows that so far there is no one-fits-all approach to querying events. Hence particularly industry products tend towards hybrid approaches, where several languages of different styles are supported or aspects of different styles are combined within one language. Hybrid approaches include the introduction of pattern matching into data stream query languages as in Oracle CEP [58], Esper [24], and some CQL dialects like the one used in [71], the use of composition operators on top of data stream queries [29, 16], the addition of composition operators to production rules [74], the combination of production rules and state machines, e.g., in TIBCO Business Events (see Section 7), and the decoupled use of different languages (and possibly evaluation engines) that communicate only by means of exchanging events (derived as answers to queries).

11 Conclusion

CEP is an industrial growth market as well as an important research area that is emerging from coalescing branches of other research fields.

Even though the prevalent event query languages can be categorized roughly into five families as we have done in this article, there are significant differences between the individual languages of a family. Whether a convergence to a single, dominant query language for CEP is possible and advisable is currently in no way agreed upon.

Efforts towards a standard for a SQL-based data stream query language are on the way [39], but not yet within an official standardization body. A standardized XML syntax for production rules is being developed by the W3C as part of the Rule Interchange Format (RIF); however, the special requirements of CEP are not considered there yet. The same applies to the Production Rule Representation (PRR) by the OMG.

Activities of the Event Processing Technical Society (EPTS) [23] aim at a coordination and harmonization, with the work on a glossary of CEP notions, the interoperability analysis of Event Processing systems from different vendors, a common reference architecture or framework of architectures that handles current and envisioned Event Processing architectures, the analysis of the application areas of CEP, and the creation of business value for users in order to increase the adoption of Event Processing in the business and industry. The EPTS has also a working group for the analysis of EQLs.

Acknowledgements

This research has been founded in part by the European Commission within the the the project "EMILI — Emergency Management in Large Infrastructures" under grant agreement number 242438 and by the German Research Foundation (Deutsche Forschungsgemeinschaft) within the project "QONCEPT — Query Optimization in Complex Event Processing Technologies" under reference number BR 2355/1-1.

References

1. Abadi, D.J., Carney, D., Çetintemel, U., Cherniack, M., Convey, C., Lee, S., Stonebraker, M., Tatbul, N., Zdonik, S.B.: A new model and architecture for data stream management. The VLDB Journal 12(2), 120–139 (2003)
2. Adaikkalavan, R., Chakravarthy, S.: Formalization and detection of events using interval-based semantics. In: Proc. Int. Conf. on Management of Data (COMAD), pp. 58–69. Computer Society of India (2005)
3. Adaikkalavan, R., Chakravarthy, S.: SnoopIB: Interval-based event specification and detection for active databases. Data and Knowledge Engineering 1(59), 139–165 (2006)
4. Adi, A., Etzion, O.: Amit — the situation manager. The VLDB Journal 13(2), 177–203 (2004)

5. Alur, R., Dill, D.: Automata for modeling real-time systems. In: Paterson, M. (ed.) ICALP 1990. LNCS, vol. 443, pp. 322–335. Springer, Heidelberg (1990)
6. Arasu, A., Babu, S., Widom, J.: The CQL continuous query language: Semantic foundations and query execution. The VLDB Journal 15(2), 121–142 (2006)
7. Barga, R.S., Caituiro-Monge, H.: Event correlation and pattern detection in CEDR. In: Grust, T., Höpfner, H., Illarramendi, A., Jablonski, S., Fischer, F., Müller, S., Patranjan, P.-L., Sattler, K.-U., Spiliopoulou, M., Wijsen, J. (eds.) EDBT 2006. LNCS, vol. 4254, pp. 919–930. Springer, Heidelberg (2006)
8. Bernauer, M., Kappel, G., Kramler, G.: Composite events for XML. In: Proc. Int. Conf. on World Wide Web, pp. 175–183. ACM, New York (2004)
9. Berstel, B., Bonnard, P., Bry, F., Eckert, M., Pătrânjan, P.-L.: Reactive rules on the web. In: Antoniou, G., Aßmann, U., Baroglio, C., Decker, S., Henze, N., Patranjan, P.-L., Tolksdorf, R. (eds.) Reasoning Web. LNCS, vol. 4636, pp. 183–239. Springer, Heidelberg (2007)
10. Bry, F., Eckert, M.: Rule-Based Composite Event Queries: The Language XChange^{EQ} and Its Semantics. In: Marchiori, M., Pan, J.Z., de Sainte Marie, C. (eds.) RR 2007. LNCS, vol. 4524, pp. 16–30. Springer, Heidelberg (2007)
11. Bry, F., Eckert, M.: On static determination of temporal relevance for incremental evaluation of complex event queries. In: Proc. Int. Conf. on Distributed Event-Based Systems, pp. 289–300. ACM, New York (2008)
12. Bry, F., Eckert, M., Etzion, O., Paschke, A., Riecke, J.: Event processing language tutorial. In: 3rd ACM Int. Conf. on Distributed Event-Based Systems, ACM, New York (2009)
13. Bry, F., Eckert, M., Pătrânjan, P.-L.: Querying composite events for reactivity on the web. In: Shen, H.T., Li, J., Li, M., Ni, J., Wang, W. (eds.) APWeb Workshops 2006. LNCS, vol. 3842, pp. 38–47. Springer, Heidelberg (2006)
14. Bry, F., Eckert, M., Pătrânjan, P.-L.: Reactivity on the Web: Paradigms and applications of the language XChange. J. of Web Engineering 5(1), 3–24 (2006)
15. Carlson, J., Lisper, B.: An event detection algebra for reactive systems. In: Proc. ACM Int. Conf. On Embedded Software, pp. 147–154. ACM, New York (2004)
16. Chakravarthy, S., Adaikkalavan, R.: Events and streams: Harnessing and unleashing their synergy! In: Proc. Int. Conf. on Distributed Event-Based Systems, pp. 1–12. ACM, New York (2008)
17. Chakravarthy, S., Krishnaprasad, V., Anwar, E., Kim, S.-K.: Composite events for active databases: Semantics, contexts and detection. In: Proc. Int. Conf. on Very Large Data Bases, pp. 606–617. Morgan Kaufmann, San Francisco (1994)
18. Chandrasekaran, S., Cooper, O., Deshpande, A., Franklin, M.J., Hellerstein, J.M., Hong, W., Krishnamurthy, S., Madden, S., Raman, V., Reiss, F., Shah, M.A.: Telegraphcq: Continuous dataflow processing for an uncertain world. In: CIDR (2003)
19. Coulouris, G., Dollimore, J., Kindberg, T.: Distributed Systems: Concepts and Design, 3rd edn. Addison-Wesley, Reading (2001)
20. Demers, A.J., Gehrke, J., Panda, B., Riedewald, M., Sharma, V., White, W.M.: Cayuga: A general purpose event monitoring system. In: CIDR, pp. 412–422 (2007)
21. Eckert, M.: Reactivity on the Web: Event Queries and Composite Event Detection in XChange. Master's thesis (Diplomarbeit), Institute for Informatics. University of Munich (2005)
22. Eckert, M.: Complex Event Processing with XChange^{EQ}: Language Design, Formal Semantics and Incremental Evaluation for Querying Events. PhD thesis, Institute for Informatics. University of Munich (2008)
23. Event Processing Technical Society (EPTS), http://www.ep-ts.com

24. EsperTech Inc. Event stream intelligence: Esper & NEsper,
    `http://esper.codehaus.org`
25. Forgy, C.: OPS5 user's manual. Technical Report CMU-CS-81-135, Carnegie Mellon
    University (1981)
26. Forgy, C., McDermott, J.P.: OPS, a domain-independent production system language.
    In: Proc. Int. Joint Conf. on Artificial Intelligence, pp. 933–939. William Kaufmann, San
    Francisco (1977)
27. Forgy, C.L.: Rete: A fast algorithm for the many pattern/many object pattern match prob-
    lem. Artificial Intelligence 19(1), 17–37 (1982)
28. Galton, A., Augusto, J.C.: Two approaches to event definition. In: Hameurlain, A., Cic-
    chetti, R., Traunmüller, R. (eds.) DEXA 2002. LNCS, vol. 2453, pp. 547–556. Springer,
    Heidelberg (2002)
29. Garg, V., Adaikkalavan, R., Chakravarthy, S.: Extensions to stream processing architec-
    ture for supporting event processing. In: Bressan, S., Küng, J., Wagner, R. (eds.) DEXA
    2006. LNCS, vol. 4080, pp. 945–955. Springer, Heidelberg (2006)
30. Gatziu, S., Dittrich, K.R.: Events in an active object-oriented database system. In: Proc.
    Int. Workshop on Rules in Database Systems, pp. 23–39. Springer, Heidelberg (1993)
31. Gatziu, S., Dittrich, K.R.: Detecting composite events in active database systems us-
    ing petri nets. In: Proc. Int. Workshop on Research Issues in Data Engineering: Active
    Database Systems, pp. 2–9. IEEE, Los Alamitos (1994)
32. Gehani, N.H., Jagadish, H., Shmueli, O.: Event specification in an active object-oriented
    database. In: Proc. Int. ACM Conf. on Management of Data (SIGMOD), pp. 81–90.
    ACM, New York (1992)
33. Gehani, N.H., Jagadish, H.V., Shmueli, O.: Composite event specification in active
    databases: Model & implementation. In: Proc. Int. Conf. on Very Large Data Bases,
    pp. 327–338. Morgan Kaufmann, San Francisco (1992)
34. Gehani, N.H., Jagadish, H.V., Shmueli, O.: Compose: A system for composite specifica-
    tion and detection. In: Adam, N.R., Bhargava, B.K. (eds.) Advanced Database Systems.
    LNCS, vol. 759, pp. 3–15. Springer, Heidelberg (1993)
35. Gualtieri, M., Rymer, J.R.: The Forrester Wave^{TM}: Complex Event Procecessing (CEP)
    Platforms,
    `http://www.forrester.com/rb/Research/wave%26trade%B_`
    `complex_event_processing_cep_platforms%2C_q3/q/id/48084/t/2`
36. Henzinger, T.A., Raskin, J.-F., Schobbens, P.-Y.: The regular real-time languages. In:
    Larsen, K.G., Skyum, S., Winskel, G. (eds.) ICALP 1998. LNCS, vol. 1443, pp. 580–
    591. Springer, Heidelberg (1998)
37. Hinze, A., Voisard, A.: A parameterized algebra for event notification services. In: Proc.
    Int. Symp. on Temporal Representation and Reasoning, pp. 61–65. IEEE, Los Alamitos
    (2002)
38. ILOG. ILOG JRules, `http://www.ilog.com/products/jrules`
39. Jain, N., Mishra, S., Srinivasan, A., Gehrke, J., Widom, J., Balakrishnan, H., Çetintemel,
    U., Cherniack, M., Tibbetts, R., Zdonik, S.: Towards a streaming SQL standard. In: Proc.
    Int. Conf. on Very Large Data Bases. VLDB Endowment, vol. 1, pp. 1379–1390 (2008)
40. Jain, N., Mishra, S., Srinivasan, A., Gehrke, J., Widom, J., Balakrishnan, H., Çetintemel,
    U., Cherniack, M., Tibbetts, R., Zdonik, S.B.: Towards a streaming sql standard.
    PVLDB 1(2), 1379–1390 (2008)
41. JBoss.org. Drools, `http://www.jboss.org/drools`
42. Krämer, J., Seeger, B.: Pipes: a public infrastructure for processing and exploring
    streams. In: Proceedings of the 2004 ACM SIGMOD International Conference on Man-
    agement of Data, SIGMOD 2004, pp. 925–926. ACM, New York (2004)

43. Krämer, J., Seeger, B.: Semantics and implementation of continuous sliding window queries over data streams. ACM Trans. Database Syst. 34, 4:1–4:49 (2009)
44. Kersten, M., Liarou, E., Goncalves, R.: A query language for a data refinery cell. In: Proc. Int. Workshop on Event-Driven Architecture, Processing and Systems (2007)
45. Kowalski, R.A., Sergot, M.J.: A logic-based calculus of events. New Generation Compututing 4(1), 67–95 (1986)
46. Kozlenkov, A., Penaloza, R., Nigam, V., Royer, L., Dawelbait, G., Schröder, M.: Prova: Rule-based java scripting for distributed web applications: A case study in bioinformatics. In: Grust, T., Höpfner, H., Illarramendi, A., Jablonski, S., Fischer, F., Müller, S., Patranjan, P.-L., Sattler, K.-U., Spiliopoulou, M., Wijsen, J. (eds.) EDBT 2006. LNCS, vol. 4254, pp. 899–908. Springer, Heidelberg (2006)
47. Krákora, J., Waszniowski, L., Hanzálek, Z.: Timed automata approach to real time distributed system verification. In: Proc. of EEE Int. Workshop on Factory Communication Systems (WFCS), pp. 407–410 (2004)
48. Lamport, L.: Time, clocks, and the ordering of events in a distributed system. Communications of the ACM 21(7), 558–565 (1978)
49. Li, Q., Rus, D.: Global clock synchronization in sensor networks. IEEE Transactions on Computers 55(2), 214–226 (2006)
50. Liarou, E., Goncalves, R., Idreos, S.: Exploiting the power of relational databases for efficient stream processing. In: Int. Conf. on Extending Database Technology (EDBT), vol. 360, pp. 323–334. ACM, New York (2009)
51. Luckham, D.C.: The Power of Events: An Introduction to Complex Event Processing in Distributed Enterprise Systems. Addison-Wesley, Reading (2002)
52. Madden, S.R., Franklin, M.J., Hellerstein, J.M., Hong, W.: TinyDB: An acquisitional query processing system for sensor networks. ACM Transactions on Database Systems 30(1), 122–173 (2005)
53. Mansouri-Samani, M., Sloman, M.: GEM: A generalized event monitoring language for distributed systems. Distributed Systems Engineering 4(2), 96–108 (1997)
54. Merritt, M., Modugno, F., Tuttle, M.R.: Time-constrained automata. In: Groote, J.F., Baeten, J.C.M. (eds.) CONCUR 1991. LNCS, vol. 527, pp. 408–423. Springer, Heidelberg (1991)
55. Moreto, D., Endler, M.: Evaluating composite events using shared trees. IEE Proceedings — Software 148(1), 1–10 (2001)
56. Morrell, J., Vidich, S.D.: Complex Event Processing with Coral8. White Paper (2007), http://www.coral8.com/system/files/assets/pdf/ Complex_Event_Processing_with_Coral8.pdf
57. MS Analog Software. ruleCore(R) Complex Event Processing (CEP) Server, http://www.rulecore.com
58. Oracle Inc. Complex Event Processing in the real world. White Paper, http://www.oracle.com/technologies/soa/docs/ oracle-complex-event-processing.pdf
59. Ouimet, M., Lundqvist, K.: The timed abstract state machine language: Abstract state machines for real-time system engineering. Journal of Universal Computer Science 14(12), 2007–2033 (2008)
60. Paschke, A., Boley, H.: Rules capturing events and reactivity. In: Handbook of Research on Emerging Rule-Based Languages and Technologies: Open Solutions and Approaches, pp. 215–252. IGI Global (2009)
61. Paschke, A., Kozlenkov, A.: Rule-based event processing and reaction rules. In: Governatori, G., Hall, J., Paschke, A. (eds.) RuleML 2009. LNCS, vol. 5858, pp. 53–66. Springer, Heidelberg (2009)

62. Paschke, A., Kozlenkov, A., Boley, H.: A homogenous reaction rule language for Complex Event Processing. In: In Proc. 2nd Int. Workshop on Event Drive Architecture and Event Processing Systems (2007)
63. Paschke, A., Kozlenkov, A., Boley, H., Tabet, S., Kifer, M., Dean, M.: Reaction RuleML (2007), http://ibis.in.tum.de/research/ReactionRuleML/
64. Paton, N.W. (ed.): Active Rules in Database Systems. Springer, Heidelberg (1998)
65. Reiss, F., Stockinger, K., Wu, K., Shoshani, A., Hellerstein, J.M.: Enabling real-time querying of live and historical stream data. In: SSDBM, p. 28 (2007)
66. Roncancio, C.: Toward duration-based, constrained and dynamic event types. In: Andler, S.F., Hansson, J. (eds.) ARTDB 1997. LNCS, vol. 1553, pp. 176–193. Springer, Heidelberg (1999)
67. Sánchez, C., Sankaranarayanan, S., Sipma, H., Zhang, T., Dill, D.L., Manna, Z.: Event correlation: Language and semantics. In: Alur, R., Lee, I. (eds.) EMSOFT 2003. LNCS, vol. 2855, pp. 323–339. Springer, Heidelberg (2003)
68. Sánchez, C., Słanina, M., Sipma, H.B., Manna, Z.: Expressive completeness of an event-pattern reactive programming language. In: Wang, F. (ed.) FORTE 2005. LNCS, vol. 3731, pp. 529–532. Springer, Heidelberg (2005)
69. Sandia National Laboratories. Jess, the rule engine for the Java(TM) platform, http://herzberg.ca.sandia.gov/
70. Schmidt, K.-U., Anicic, D., Stühmer, R.: Event-driven reactivity: A survey and requirements analysis. In: SBPM2008: 3rd Int. Workshop on Semantic Business Process Management in Conjunction with the 5th European Semantic Web Conf. (ESWC 2008). CEUR Workshop Proceedings (2008)
71. Seeger, B.: Kontinuierliche kontrolle. IX: Magazin für Professionelle Informationstechnik 2 (2010)
72. Seiriö, M., Berndtsson, M.: Design and implementation of an ECA rule markup language. In: Adi, A., Stoutenburg, S., Tabet, S. (eds.) RuleML 2005. LNCS, vol. 3791, pp. 98–112. Springer, Heidelberg (2005)
73. StreamBase Systems. StreamSQL Guide (2011), http://streambase.com/developers/docs/latest/streamsql/index.html
74. Walzer, K., Breddin, T., Groch, M.: Relative temporal constraints in the Rete algorithm for complex event detection. In: Proc. Int. Conf. on Distributed Event-Based Systems, pp. 147–155. ACM, New York (2008)
75. Wu, E., Diao, Y., Rizvi, S.: High-performance Complex Event Processing over streams. In: Proc. Int. ACM Conf. on Management of Data (SIGMOD), pp. 407–418. ACM, New York (2006)
76. Zhu, D., Sethi, A.S.: SEL, a new event pattern specification language for event correlation. In: Proc. Int. Conf. on Computer Communications and Networks, pp. 586–589. IEEE, Los Alamitos (2001)
77. Zimmer, D., Unland, R.: On the semantics of complex events in active database management systems. In: Proc. Int. Conf. on Data Engineering, pp. 392–399. IEEE, Los Alamitos (1999)

Two Semantics for CEP, no Double Talk: Complex Event Relational Algebra (CERA) and Its Application to XChange^{EQ}

Michael Eckert, François Bry, Simon Brodt, Olga Poppe, and Steffen Hausmann

Abstract. Complex Event Processing (CEP) denotes algorithmic methods for deriving higher-level knowledge, or complex events, from a stream of lower-level events in a continuous and timely fashion. High-level Event Query Languages (EQLs) are designed for expressing complex events in a convenient, concise, effective and maintainable manner. CEP differs fundamentally from traditional database or Web querying, as CEP continuously evaluates standing queries against a stream of incoming event data whereas traditional querying evaluates incoming ad hoc queries against (more or less) standing data.

However EQLs and traditional query languages share a need for clear formal semantics which typically consist of two parts: A declarative semantics specifying *what* the answer of a query should be and an operational semantics telling *how* this answer is actually computed. The declarative semantics serves as reference for the operational semantics which is the basis for query evaluation and optimization.

While formal semantics is well-understood for traditional query languages it has been rather neglected for EQLs so far. In this chapter we use the EQL XChange^{EQ} to demonstrate a general, easily transferable approach for defining both, the declarative and operational semantics of an EQL. The operational semantics on the one hand, bases on CERA, a tailored variant of relational algebra, and incremantal evaluation of query plans. Although the basic idea might sound familiar from previous approaches like [3, 12, 16], the way it is realized here is significantly different. The declarative semantics on the other hand, is defined using a Tarski-style model theory with accompanying fixpoint theory.

Michael Eckert
TIBCO Software, Balanstr. 49, 81669 Munich, Germany
e-mail: meckert@tibco.com

François Bry · Simon Brodt · Olga Poppe · Steffen Hausmann
Institute for Informatics, University of Munich, Oettingenstr. 67, 80538 Munich, Germany,
e-mail: {bry,brodt,poppe,hausmann}@pms.ifi.lmu.de

S. Helmer et al.: Reasoning in Event-Based Distributed Systems, SCI 347, pp. 71–97.
springerlink.com   © Springer-Verlag Berlin Heidelberg 2011

1 Introduction

In databases, relational algebra describes the order in which operators are applied to relations to compute answers for queries. It serves as a theoretical fundament for the operational semantics of database query languages (e.g., SQL) and query optimization.

A recent trend in information systems are continuous queries against event (or data) streams. This continuous querying of events is fundamentally different from the traditional ad hoc querying of databases or Web data, since event queries are standing queries evaluated continuously over time against changing event data received as an incoming stream.

Querying events often involves the notion of Complex Event Processing (CEP) which denotes algorithmic methods for making sense of events by deriving higher-level knowledge, or complex events, from lower-level events in a timely fashion and permanently. We refer the reader to the chapter "A CEP Babelfish: Languages for Complex Event Processing and Querying Surveyed", in this volume, for a discussion of Event Query Languages (EQLs). Specific evaluation methods have been conceived for the efficient, stepwise evaluation of complex event queries against event data streams. We demonstrate the usage of one of these languages, XChange^{EQ} [5, 11], in a sensor network use case in Section 2.

We present Complex Event Relational Algebra (CERA), an extended and tailored variant of relational algebra, to represent execution plans for complex event queries in Section 3. Starting out with relational algebra and thus building on the foundation of database queries is not just helpful for understandability, it also lets event queries benefit from many results in database research (e.g., join algorithms, adaptive query evaluation). Further, the uniformity in the foundations of event queries and traditional queries is beneficial in systems and languages where event and non-event data is processed together — and queries should be optimized and evaluated together. This is quite common, especially for Event Condition Action (ECA) rules, where the E part is an event query, the C part a traditional query, and the parts share information through variable bindings.

The basic idea of transferring relational algebra to CEP is not new [3, 12, 16]. However we propose a significantly different way of doing so. Previous approaches like CQL [3, 12, 16] use stream-to-relation operators like time windows to conceptually convert the stream into a *finite* relation for *each point of time*. After that, quite ordinary relational algebra expressions are applied to this finite relation. In contrast to that, CERA views the whole stream as *one* potentially *infinite* relation. Tailored variants of relational operators are then applied to this infinite relation. The trick is that these operators are restricted in such a way that for each point of time it is sufficient to know the finite available part of the stream to compute the result up to that time point (see Section 3.4.1). Therefore CERA is suitable for an incremental, step-wise evaluation as required for complex event queries.

We illustrate CERA by its application to XChange^{EQ} [5, 11] which is one of the recently developed, expressive and easy-to-use high-level EQLs. We also provide details on how XChange^{EQ} rules are translated into CERA expressions and how query plans consisting of CERA expressions can be optimized and incrementally evaluated.

An important aspect of any query evaluation method is its correctness. Yet, this aspect has often been neglected in CEP so far. Proving the correctness of an operational semantics, which focuses on *how* an answer for a query is computed, entails the existence of a declarative semantics, which focuses on *what* the answers should be. To this end, in Section 4 we introduce a declarative semantics for XChange^{EQ} that is quite natural for event streams and in Section 5 we sketch a proof of the correctness of the operational semantics based on CERA with respect to the declarative semantics. Section 6 concludes this chapter.

The contributions of this chapter are:

1. Formal definition of CERA as the first corner stone of the operational semantics for EQLs
2. Description of incremental evaluation of query plans with materialization points as the second corner stone of the operational semantics for EQLs
3. Formal definition of the declarative semantics for EQLs by a Tarski-style model theory with accompanying fixpoint theory
4. Illustration of both semantics by an XChange^{EQ} program in the realm of sensor network use case
5. Proof of the correctness, i.e., soundness and completeness of the operational semantics for EQLs with respect to their declarative semantics

2   CEP Examples

Figure 1 contains an XChange^{EQ} program P which will be used as an example throughout this chapter. In this section we briefly describe the program and refer the reader to the chapter on "A CEP Babelfish: Languages for Complex Event Processing and Querying Surveyed", in this volume, for the informal definitions of the basic notions in the realm of CEP (Section 2), the explanation of the syntax of XChange^{EQ} (Section 8), and the description of the sensor network use case (Section 1).

The first rule of the program in Figure 1 triggers fire alarm for an area if smoke and high temperature were measured in the area within one minute. For the second rule, assume all sensors of our network send their temperature measurements every 12 seconds. The second rule of the program infers that a sensor has been burnt down if it measured high temperature and did not send its measurements afterwards. And the last query computes the average temperature reported by a sensor during the last minute every time the sensor sends a temperature measurement.

```
DETECT    fire { area { var A } }
ON and { event s: smoke {{ area {{ var A }} }},
          event t: temp {{ area {{ var A }}, value {{ var T }} }}
   } where { s before t, {s,t} within 1 min, var T > 40 }
END

DETECT    burnt_down { sensor { var S } }
ON and { event n: temp {{ sensor {{ var S }}, value {{ var T }} }},
          event i: timer:from-end [ event n, 12 sec ],
          while i: not temp {{ sensor {{ var S }} }}
   } where { var T > 40 }
END

DETECT    avg_temp { sensor { var S }, value { avg(all var T) } }
ON and { event m: temp {{ sensor {{ var S }} }},
          event i: timer:from-start-backward [ event m, 1 min ],
          while i: collect temp {{ sensor {{ var S }}, value {{ var T }} }} }
END
```

Fig. 1 XChange^{EQ} program P

3 CERA: An Operational Semantics for Event Query Languages

This section is devoted to the formal definition of the operational semantics for EQLs. In Section 3.1 we clarify the purpose of an operational semantics for query languages in general and describe its desiderata for an EQL in particular. Complex Event Relational Algebra (CERA) builds the first corner stone of the operational semantics. CERA is based on relational algebra and extends it with operators which are able to treat notions specific for events such as event occurrence time. In Section 3.2 we define the operators formally and illustrate their usage by translating the XChange^{EQ} rules in Figure 1 into CERA expressions. Section 3.3 is devoted to the formal specification of the translation of a single XChange^{EQ} rule into a CERA expression. But keep in mind that CERA is independent from a particular EQL. Finally, Section 3.4 explains how query plans consisting of CERA expressions can be optimized and incrementally evaluated (the second corner stone of the operational semantics).

3.1 Purpose and Desiderata

In general, the purpose of an operational semantics is to provide an abstract description of an implementation of the evaluation engine of a language. For EQLs in particular, an operational semantics must fulfill the following core desiderata:

1. It should be an incremental, data-driven evaluation method storing and updating intermediate results instead of computing them anew in each step. The notion

"incremental" derives from the idea that in each step we compute only the changes relative to the previous steps.

2. Since the incoming event stream is unbounded, a naive query evaluation engine storing all intermediate results forever, runs out of memory sooner or later. Hence an operational semantics must enable garbage collection of irrelevant events, i.e., events which cannot contribute to a new answer any more. For the sake of brevity, we do not formalize garbage collection of events in this chapter and refer the reader to [7].

3. An operational semantics should provide a framework for query optimization since it must be able to capture the whole space of different query plans. In this chapter we only provide a general idea of event query optimization and refer the reader to [6, 4, 10, 18] for more details.

4. An operational semantics for EQLs must be correct with respect to the declarative semantics of EQLs and it must terminate for each incremental step.

3.2 CERA: Complex Event Relational Algebra

For the operational semantics we will restrict ourselves to hierarchical rule programs, i.e., programs that are free of any recursion of rules (see [11] for the formal definition). Note that this is a common restriction not just for event queries but also database views and queries, and causes no problems in most practical applications.

We want to base CERA on traditional relational algebra. There are three problems arising when doing so: the treatment of XML data carried by events, the incorporation of the time axis and the infinity of streams. We approach these problems in the following way: Streams are regarded from an omniscient perspective, i.e. as if all events ever arriving on the stream were already known. From this point of view the (complete infinite) stream is just a relation with tuples representing the events arriving on the stream. Each tuple representing an event consists of two timestamp attributes representing the begin and end of the occurrence interval of the event and a data-term attribute representing the XML data carried by the event. Of course this relation is potentially infinite and never known completely at any point of time. However this can be ignored for now. We will see below that due to a special property of CERA called "temporal preservation" (Section 3.4.1) and the "finite differencing" technique (Section 3.4.3), the evaluation is nevertheless able to incrementally compute the desired result working only on the finite part of the relation known up to some point of time.

All three points, the integration of XML-data, the explicit representation of the time axis by means of timestamp attributes and the way to cope with the infinity of streams, are fundamentally different to previous approaches. The chapter "A CEP Babelfish: Languages for Complex Event Processing and Querying Surveyed", in this volume, shows that composition operator based languages, data stream languages and production rule languages have none or only weak support for XML. These language groups also lack an explicit representation of time (though time plays a role of course) limiting the temporal relations expressible in these approaches. Data stream languages like CQL also tend to use a kind of relational algebra for their semantics, however they approach the infinity of streams by (conceptually) *converting* parts of the stream into a finite relation for every point of time by, for example, time windows. After that, relational algebra is applied and the resulting relation is then converted back into a stream using another operator. (We refer the reader to the chapter "A CEP Babelfish: Languages for Complex Event Processing and Querying Surveyed", in this volume, for more details). This significantly differs from our approach as we do not do any conversion at all but just view the complete potentially infinite stream as a relation and directly apply relational algebra to that relation.

With regards to the XML integration we are not finished yet at this point as we need the possibility to access the data contained in the data-term attribute of an event (or its tuple respectively), for example, for selections or joins. Therefore we introduce the matching operator Q^X which extracts the desired data into attributes of the resulting relation. The complementary operator to Q^X is the construction operator C^X which is used to construct the new data-terms for the derived complex events from a number of attributes of a relation. It will be explained more closely on Page 80. In this way applying a complete CERA expression, i.e., an expression with Q^X at the bottom and C^X at the top, to an event stream, or more precisely to the relation representing the stream, results in a relation with the same schema, or an event stream, again. Thus CERA is answer-closed. In the following we will mainly describe the effect of Q^X and C^X on the schema of a relation, as XML matching and construction are not in the focus here. For details on the exact semantics of the XML matching in Q^X and XML construction in C^X see [11].

The **event matching** operator Q^X takes two arguments, an event stream (i.e. the corresponding relation) and a simple event query *event i* : *q*. The result of applying Q^X to the event stream E using the simple event query *event i* : *q* is the relation $R_i = Q^X_{[i:\,q]}(E)$. Each event (or tuple) in the stream (or relation) matched by *q* corresponds to one or many tuples in R_i (see the definition of Q^X in [11]).

The schema $sch(R_i)$ of R_i corresponds directly to the free variables of *q*. Each free variable of *q* is a data attribute of R_i. Furthermore R_i contains an event reference attribute *i.ref* identifying the event a tuple was derived from, and two timestamp attributes *i.begin*, *i.end* representing the occurrence time interval of this event. Consequently, R_i has the schema $sch(R_i) = \{i.begin, i.end, i.ref, X_1, \ldots, X_n\}$, where $X_1, \ldots, X_n$ are the free variables of *q*. We denote the set of data attributes of R_i with $sch_{data}(R_i)$, the set of timestamp attributes with $sch_{time}(R_i)$ and the set of event reference attributes with $sch_{ref}(R_i)$.

Note that we use the named perspective on relations here, i.e., tuples are viewed as functions that map attribute names to values. This is more intuitive than the unnamed perspective identifying attribute values by their positions in ordered tuples.

The XChange^{EQ} program P in Figure 1 gives rise to the relations in Figure 2. These relations will be the input for the CERA expressions into which we will translate the rules of P.

$$\text{Smoke}_s = Q^X_{[\,s:\,smoke\{\{area\{\{var\,A\}\}\}\}\,]}(E), \qquad sch(\text{Smoke}_s) = \{s.begin, s.end,$$
$$s.ref, A\}$$

$$\text{Temp}_t = Q^X_{[\,t:\,temp\{\{area\{\{var\,A\}\},value\{\{var\,T\}\}\}\}\,]}(E), \quad sch(\text{Temp}_t) = \{t.begin, t.end,$$
$$t.ref, A, T\}$$

$$\text{Temp}_n = Q^X_{[\,n:\,temp\{\{sensor\{\{var\,S\}\},value\{\{var\,T\}\}\}\}\,]}(E), \quad sch(\text{Temp}_n) = \{n.begin, n.end,$$
$$n.ref, S, T\}$$

$$\text{Temp}_v = Q^X_{[\,v:\,temp\{\{sensor\{\{var\,S\}\}\}\}\,]}(E), \qquad sch(\text{Temp}_v) = \{v.begin, v.end,$$
$$v.ref, S\}$$

$$\text{Temp}_m = Q^X_{[\,m:\,temp\{\{sensor\{\{var\,S\}\}\}\}\,]}(E), \qquad sch(\text{Temp}_m) = \{m.begin, m.end,$$
$$m.ref, S\}$$

$$\text{Temp}_w = Q^X_{[\,w:\,temp\{\{sensor\{\{var\,S\}\},value\{\{var\,T\}\}\}\}\,]}(E), \; sch(\text{Temp}_w) = \{w.begin, w.end,$$
$$w.ref, S, T\}$$

Fig. 2 Input relations for the CERA expressions for the program P in Figure 1

Besides the relations shown in Figure 2, the program P makes use of **relative timer events**. Figure 3 contains generic definitions of the relative timer events used in P. Relative timer events are expressed by means of auxiliary event streams S and auxiliary relations X. The definitions take an event stream E and a relative timer specification as parameters. Each auxiliary event stream S contains one relative timer event s for each event r of the event stream E. The timestamps of s are defined relatively to the timestamps of r. The matching operator Q^X sets the value of attribute $j.ref$ of X to a reference to event r (denoted $ref(r)$). We need the attribute $j.ref$ to join X with another relation to drop superfluous tuples of X.

$$X_{[\,i:\,from\text{-}end[j,d]\,]}(E) \quad := Q^X_{[\,i:\,rel\text{-}timer\text{-}event[var\,j.ref]\,]}(S_{[\,from\text{-}end[d]\,]}(E)), \text{ where}$$

$$S_{[\,from\text{-}end[d]\,]}(E) \quad := \{s \mid \exists\, r \in E \text{ with } s := \begin{cases} s(begin) = r(end), \\ s(end) = r(end) + d, \\ s(term) = rel\text{-}timer\text{-}event(ref(r)) \end{cases} \}$$

$$X_{[\,i:\,from\text{-}start\text{-}backward[j,d]\,]}(E) := Q^X_{[\,i:\,rel\text{-}timer\text{-}event[var\,j.ref]\,]}(S_{[\,from\text{-}start\text{-}backward[d]\,]}(E)), \text{ where}$$

$$S_{[\,from\text{-}start\text{-}backward[d]\,]}(E) \quad := \{s \mid \exists\, r \in E \text{ with } s := \begin{cases} s(begin) = r(begin) - d, \\ s(end) = r(begin), \\ s(term) = rel\text{-}timer\text{-}event(ref(r)) \end{cases} \}$$

Fig. 3 Generic definitions of the relative timer events used in the program P in Figure 1

Besides event matching, Q^X, CERA allows the following operators: natural join $\bowtie$, selection σ, projection π, temporal join $\bowtie_{i \sqsupseteq j}$, temporal anti-semi join $\overline{\bowtie}_{i \sqsupseteq j}$, merging of time intervals μ, renaming ρ, and event construction C^X. The definitions of natural join, selection and projection are just the same as in traditional relational algebra. There is only one important limitation of projection. It is allowed to discard data attributes (e.g., X), but it is not allowed to discard time attributes (e.g., $i.begin$) and event references (e.g., $i.ref$).

The translation of while/collect in XChange^{EQ} is expressed by temporal join. Temporal join is a new operator of CERA, it does not exist in traditional relational algebra and cannot be expressed as a combination of operators of traditional relational algebra. Let R and S be relations. In a temporal join $R \bowtie_\theta S$, the condition θ has the form $i.begin \leq j.begin \wedge j.end \leq i.end$, where $\{i.begin, i.end\} \subseteq sch_{time}(R)$ and $\{j.begin, j.end\} \subseteq sch_{time}(S)$. We abbreviate these conditions with $i \sqsupseteq j$. To achieve the right implicit groupng by event references (see the description of the event construction operator C^X below), the reference $j.ref$ must be dropped. This is possible because the temporal restriction $i \sqsupseteq j$ guarantees that the groups stay finite after $j.ref$ is dropped.

Definition 1 (Temporal join). *Let R and S be relations such that $\{i.begin, i.end\} \subseteq sch_{time}(R)$, $i.ref \in sch_{ref}(R)$, $\{j.begin, j.end\} \subseteq sch_{time}(S)$, and $j.ref \in sch_{ref}(S)$.*

$$R \bowtie_{i \sqsupseteq j} S = \{o \mid \exists r \in R \; \exists s \in S \text{ such that}$$
$$r(X) = s(X) \text{ if } X \in sch(R) \cap Sch,$$
$$r(i.begin) \leq s(j.begin) \text{ and } s(j.end) \leq r(i.end),$$
$$o(X) = r(X) \text{ if } X \in sch(R) \text{ and } o(X) = s(X) \text{ if } X \in Sch,$$
$$o(X) = \bot \text{ otherwise } \},$$

where $Sch := sch(S) \setminus \{j.begin, j.end, j.ref\}$,

$$sch(R \bowtie_{i \sqsupseteq j} S) = sch(R) \cup Sch.$$

In order to express negation of an event in CERA we introduce a θ-anti-semi-join that uses the θ condition to define the event accumulation window (analogously to the above definition of temporal join). The temporal anti-semi-join $R \overline{\ltimes}_{i \sqsupseteq j} S$ takes two relations R and S as input, where $\{i.begin, i.end\} \subseteq sch_{time}(R)$ and $\{j.begin, j.end\} = sch_{time}(S)$. (Note that it is "$\subseteq$" for the timestamps of the left side of the anti-semi-join and "$=$" for timestamps on the right side!) Its output is R with those tuples r removed that have a "partner" in S, i.e., a tuple $s \in S$ that agrees on all shared attributes with r and whose timestamps $s(j.begin), s(j.end)$ are within the time bounds $r(i.begin), r(i.end)$.

Definition 2 (Temporal anti-semi-join). *Let R and S be relations with $\{i.begin, i.end\} \subseteq sch_{time}(R)$ and $\{j.begin, j.end\} = sch_{time}(S)$.*

$$R \overline{\ltimes}_{i \sqsupseteq j} S = \{r \in R \mid \nexists s \in S \text{ such that } \forall X \in sch(R) \cap sch(S). \; r(X) = s(X)$$
$$\text{and } r(i.begin) \leq s(j.begin), \; s(j.end) \leq r(i.end)\},$$

$$sch(R \overline{\ltimes}_{i \sqsupseteq j} S) = sch(R).$$

In contrast to temporal join and temporal anti-semi-join, natural join maintains the time attributes of both input relations in order to ensure temporal preservation, an important property of CERA operators allowing them to work on finite available parts of streams.[1] But as the reader will see in Section 3.4.1, to guarantee temporal presevation, CERA operators must not maintain all timestamps of an input event but only the *greatest* timestamp of each input event. To reduce the number of timestamps of an event, a merging operator is used. It computes a single time interval of an event out of many time intervals it carries. Let R be a relation. A merging operator $\mu[[begin, end] \leftarrow j_1 \sqcup \cdots \sqcup j_n](R)$ computes a new time interval $[begin, end]$ from existing ones $j_1, \ldots, j_n$ so that the time interval $[begin, end]$ covers all the intervals $j_1, \ldots, j_n$, i.e., $begin = min\{j_1.begin, \ldots, j_n.begin\}$ and $end = max\{j_1.end, \ldots, j_n.end\}$. Further, the merging operator extracts the beginning *begin* and the end *end* of the new time interval as well as all attributes of R except $j_1.begin, j_1.end, \ldots, j_n.begin, j_n.end$ in the manner of a projection. Merging of time intervals is not really a new operation in CERA. It is equivalent to an extended projection [14], a common practical extension of relational algebra used to compute new attributes from existing ones.

Definition 3 (Merging of time intervals). *Let R be a relation and $j_1, \ldots, j_n$ time intervals such that $\{j_1.begin, j_1.end, \ldots, j_n.begin, j_n.end\} \subseteq sch_{time}(R)$. Merging of $j_1, \ldots, j_n$ is a relation defined as follows:*

$$\mu[[begin, end] \leftarrow j_1 \sqcup \cdots \sqcup j_n](R) = \{t \mid \exists r \in R \text{ with}$$
$$t(begin) = min\{r(j_1.begin), \ldots, r(j_n.begin)\},$$
$$t(end) = max\{r(j_1.end), \ldots, r(j_n.end)\},$$
$$t(X) = r(X) \text{ if } X \in sch(R) \setminus \{begin, end, j_1.begin, j_1.end, \ldots, j_n.begin, j_n.end\},$$
$$t(X) = \bot \text{ otherwise}\},$$

$$sch(\mu[[begin, end] \leftarrow j_1 \sqcup \cdots \sqcup j_n](R)) =$$
$$\{begin, end\} \cup (sch(R) \setminus \{j_1.begin, j_1.end, \ldots, j_n.begin, j_n.end\}).$$

For technical reasons (i.e., usage of natural join), the named perspective of relational algebra sometimes requires a renaming operator, which changes the names of attributes without affecting their values. Renaming is denoted by $\rho[a_1' \leftarrow a_1, \ldots, a_n' \leftarrow a_n](R)$. It renames attributes $a_1, \ldots, a_n$ of the relation R to $a_1', \ldots, a_n'$ respectively. Note that if an attribute a_i is a data attribute of R it must be a data attribute in the resulting relation, if it is a timestamp, then a_i' must also be a time attribute, and if it is an event reference so also a_i'. Note also that timestamps must always occur pairwise; accordingly, they can only be renamed pairwise.

Definition 4 (Renaming). *Let R be a relation such that $\{a_1, \ldots, a_n\} \subseteq sch(R)$ and $\{a_1', \ldots, a_n'\} \cap sch(R) = \emptyset$.*

$$\rho[a_1' \leftarrow a_1, \ldots, a_n' \leftarrow a_n](R) = \{t \mid \exists r \in R \text{ such that } t(a_i') = r(a_i) \text{ and}$$
$$\forall X \notin \{a_1, \ldots, a_n\}. t(X) = r(X)\},$$

---

[1] Temporal restriction of temporal join and temporal anti-semi-join ansures temporal preservation.

$$\forall i \in \{1, \ldots, n\}. \quad a_i \subseteq sch_{time}(R) \quad iff \, a_i' \subseteq sch_{time} \, (\rho[a_1' \leftarrow a_1, \ldots, a_n' \leftarrow a_n](R)),$$
$$a_i \subseteq sch_{ref}(R) \quad iff \, a_i' \subseteq sch_{ref} \, (\rho[a_1' \leftarrow a_1, \ldots, a_n' \leftarrow a_n](R)),$$
$$a_i \subseteq sch_{data}(R) \quad iff \, a_i' \subseteq sch_{data} \, (\rho[a_1' \leftarrow a_1, \ldots, a_n' \leftarrow a_n](R)),$$

$$j.s \leftarrow i.s \in \{a_1' \leftarrow a_1, \ldots, a_n' \leftarrow a_n\} \, iff \, j.e \leftarrow i.e \in \{a_1' \leftarrow a_1, \ldots, a_n' \leftarrow a_n\},$$
$$where \, \{i.s, i.e\} \subseteq sch_{time}(R) \, and \, \{j.s, j.e\} \subseteq sch_{time}(\rho[a_1' \leftarrow a_1, \ldots, a_n' \leftarrow a_n](R)),$$

$$sch(\rho[a_1' \leftarrow a_1, \ldots, a_n' \leftarrow a_n](R)) = (sch(R) \setminus \{a_1, \ldots, a_n\}) \cup \{a_1', \ldots, a_n'\}$$

We will see the necessity of the operator in Section 3.4.

Finally, the **event construction** operator C^X undertakes the construction of the data-terms carried by the derived complex events. The output schema of C^X is the same as the input schema of Q^X, i.e. the output relation of C^X can be regarded as a stream of derived events. The operator takes two arguments, a relation R with $sch_{time}(R) = \{begin, end\}$ and a rule head h. The result of applying C^X to the relation R using the rule head h is the stream (or relation) $E' = C^X_{[h]}(R)$. One event (or tuple) in E' corresponds to one or many tuples in R depending on whether h contains grouping and aggregation constructs or not.

If h does not contain grouping and aggregation constructs like $\texttt{all}$, then C^X constructs for each tuple $r \in R$ one event represented as the data term annotated with the time interval $[r(begin), r(end)]$. The data term results from substituting each free variable X of h by $r(X)$, i.e., by the value of attribute X of the tuple r.

$$C^X_{[\, fire\{area\{var\, A\}\}\,]}(\\
\mu[[begin, end] \leftarrow s \sqcup t](\\
\sigma[\max\{s.end, t.end\} - \min\{s.begin, t.begin\} \leq 1 \, min](\\
\sigma[s.end < t.begin](\\
\sigma[T > 40](\\
\textsf{Smoke}_s \bowtie \textsf{Temp}_t)))))$$

Fig. 4 CERA expression for the first rule of the program P in Figure 1

For example, the first rule of the program P in Figure 1 corresponds to the CERA expression in Figure 4. Remember that the query triggers fire alarm for an area when smoke and high temperature are both detected in the area within one minute. Note that all temporal conditions (such as s *before* t and $\{s,t\}$ *within 1 min*) of the query have been turned into selections. Because temporal information is simply data in tuples (as *s.begin, s.end,* etc.), no special temporal operators are needed as part of the algebra; e.g., there is no need for a sequence operator as found in many event algebras.

The second rule of the program P corresponds to the CERA expression in Figure 5. Recall that the rule infers that a sensor had burnt down if it reported high temperature and did not send its measurements afterwards any more. For this rule we assume that all temperature sensors send their measurements every 12 seconds.

$$C^X_{[\,burnt\_down\{sensor\{var\ S\}\}\,]}\,($$
$$\mu[[begin,end] \leftarrow n \sqcup i \sqcup v]\,($$
$$\sigma[T > 40]\,($$
$$(\mathsf{Temp_n} \bowtie \mathsf{X}_{[\,i:\ from\text{-}end[n,12\ sec]\,]}(\mathsf{E})) \; \overline{\bowtie}_{i \sqsupseteq v} \mathsf{Temp_v})))$$

Fig. 5 CERA expression for the second rule of the program P in Figure 1

If the rule head h contains grouping and aggregation constructs, the operator C^X does the required grouping of tuples and computes the values of aggregation functions. Note that explicit grouping happens after an implicit grouping of the tuples of R by event references and time attributes. (Since time attributes are functionally dependent on event references, they do not have any effect on the result of grouping but they must be part of each resulting tuple to guarantee temporal preservation.) As temporal joins are restricted to finite time intervals and projections may not discard event references, after the implicit grouping by event references and time attributes, each group is finite (but there may be of course infinitely many groups because the stream is potentially infinite). Therefore grouping initiated by h does not need any treatment specific for CEP.

$$C^X_{[\,avg\_temp\{sensor\{var\ S\},value\{avg(all\ var\ T)\}\}\,]}\,($$
$$\mu[[begin,end] \leftarrow m \sqcup i \sqcup w]\,($$
$$(\mathsf{Temp_m} \bowtie \mathsf{X}_{[\,i:\ from\text{-}start\text{-}backward[m,1\ min]\,]}(\mathsf{E})) \bowtie_{i \sqsupseteq w} \mathsf{Temp_w}))$$

Fig. 6 CERA expression for the third rule of the program P in Figure 1

Consider the last rule of P in Figure 1 corresponding to the CERA expression in Figure 6. Recall that the rule computes the average temperature reported by a sensor during the last minute every time the sensor sends a temperature measurement. C^X takes the tuples of the joined input relations, groups them, first implicitly by the event references and time attributes and then according to the value of attribute S (denoting a sensor). For each group the average value of attribute T (denoting a temperature measurement) is computed and saved as the value of data attribute *value* of the resulting event.

3.3   Translation into CERA

We now turn to the formal specification of the translation of a single XChange^{EQ} rule into a CERA expression. The rules are first normalized, which means that or is eliminated and the literals in the rule body are ordered in a specific way. Figure 7 shows the general structure of a normalized XChange^{EQ} rule.[2] Note that all rules of the program in Figure 1 are normalized.

---

[2] Note that the normalization of a single rule usually yields a set of rules not a single rule due to the elmination of or.

DETECT
    h
ON
    $\text{and}\{\underbrace{b_1,\ldots,b_{i_s}}_{\substack{\text{Simple}\\\text{Event}\\\text{Queries}}},\underbrace{b_{i_s+1},\ldots,b_{i_t}}_{\substack{\text{Relative}\\\text{Timer}\\\text{Spec}}},\underbrace{b_{i_t+1},\ldots,b_{i_a}}_{\substack{\text{Accumulation}\\\text{for}\\\text{Collection}}},\underbrace{b_{i_a+1},\ldots,b_n}_{\substack{\text{Accumulation}\\\text{for}\\\text{Negation}}}\}\ \text{where}\{c_1,\ldots,c_k\}$

END

Fig. 7 Normalized XChange^{EQ} rule

Figure 8 shows the translation of a single normalized XChange^{EQ} rule. The translation of rule sets requires additionally the notion of query plans which will be introduced in Section 3.4.2.

$$
\begin{aligned}
b_1 &\mapsto B_1 \\
b_1,\ldots,b_{i+1} &\mapsto B_{i+1} \\
\text{and}\{b_1,\ldots,b_n\} &\mapsto B_n \\
\text{and}\{b_1,\ldots,b_n\}\ \text{where}\{c_1,\ldots,c_k\} &\mapsto C \\
\text{DETECT }h\text{ ON and}\{b_1,\ldots,b_n\}\ \text{where}\{c_1,\ldots,c_k\}\text{ END} &\mapsto Q
\end{aligned}
$$

$$B_1 := Q^{\mathsf{X}}_{[j:q]} \qquad\qquad \text{for } b_1 = \text{event } j:q$$

$$B_{i+1} := \begin{cases} B_i \bowtie Q^{\mathsf{X}}_{[j:q]} & \text{if } b_{i+1} = \text{event } j:q \\ B_i \bowtie X_{[j:\textit{REL-TIMER-SPEC}[j',d]]} & \text{if } b_{i+1} = \text{event } j:\textit{REL-TIMER-SPEC}[j',d] \\ B_i \bowtie_{j\sqsupseteq i'} Q^{\mathsf{X}}_{[i':q]} & \text{if } b_{i+1} = \text{while } j:q \\ B_i \,\overline{\bowtie}_{j\sqsupseteq i'} Q^{\mathsf{X}}_{[i':q]} & \text{if } b_{i+1} = \text{while } j:\text{not } q \end{cases}$$

   where $1 \leq i < n$ and i' is a fresh event identifier

$C := \sigma[c'_1 \wedge \cdots \wedge c'_q](B_n)$ where c'_i is the translation of c_i (see Figure 13.4 in [11] for details)

$Q := C^{\mathsf{X}}_h(\mu[[begin,end] \leftarrow j_1 \sqcup \cdots \sqcup j_l](C))$

Fig. 8 Translation of a single normalized XChange^{EQ} rule into a CERA expression

3.4 Incremental Evaluation

Till now we pretended to have a kind of "omniscience": The relations contain conceptually all events that ever happen and are probably infinite for that reason. In the actual evaluation of event queries it is not possible to foresee future events. Event queries are evaluated *incrementally* on finite event histories. This section explains the details on incremental evaluation of programs in an EQL. We start with temporal preservation of CERA, a property allowing incremental evaluation of CERA expressions (Section 3.4.1). Then, we introduce the notion of query plans with materialization points (Section 3.4.2) and explain their incremental evaluation (Section 3.4.3).

3.4.1 Temporal Preservation

The restrictions that CERA imposes on expressions (compared to an unrestricted relational algebra) make this approach reasonable since we do not need any knowledge about future events when we want to obtain all results of an expression with an occurrence time until *now*. More precisely, to compute all results of a CERA expression Q with an occurrence time before or at time point *now*, we need to know (the finite part of) its input relations up to this time point *now*. In order to formally define and prove this property of CERA, called temporal preservation, we need the following auxiliary definition.

Definition 5 (Occurrence time of a tuple). *The occurrence time of a tuple r in the result of a CERA expression Q is a time interval given by*

$$occtime(r) = [\ \min\{r(i.begin) \mid i.ref \in sch_{ref}(Q)\},$$
$$\max\{r(i.end) \mid i.ref \in sch_{ref}(Q)\}\ ]$$

To refer to the end of the occurrence time of a tuple, i.e., $end(occtime(r))$ in selections we introduce the shorthand $\mathsf{END_Q} := max\{\ i_1.end,\ \ldots,\ i_n.end\ \}$ where $\{\ i_1.end,\ldots,i_n.end\ \} \cup \{\ i_1.begin,\ldots,i_n.begin\ \} = sch_{time}(Q)$.[3]

Theorem 1 (Temporal preservation). *Let Q be a CERA-expression with input relations $R_1,\ldots,R_n$. Then for all time points now : $\sigma[\mathsf{END_Q} \leq now](Q) = Q'$, where Q' is obtained from Q by replacing each R_k with $R'_k := \{r \in R_k \mid end(occtime(r)) \leq now\}$.*

Proof (Sketch). By induction. For event matching, event construction, selection, and projection, the claim is obvious since timestamps are not changed at all. By definition, natural join maintains the timestamps of both input relations without change. By definition, merging does not change the maximum value over all timestamps. Temporal join and temporal anti-semi-join are only allowed with temporal restrictions that also ensure that the maximum value is maintained. We refer to [11] for details.

3.4.2 Query Plans with Materialization Points

In traditional relational databases, a query plan describes the order in which operators are applied to base relations to compute answers for queries and serves as a basis for query optimization techniques such as "push selection" and storage and reuse of shared subqueries. In CEP, queries are evaluated continuously over time against changing event data received as an incoming stream and therefore a query plan should additionally account for storage of intermediate results of CERA expressions to avoid their re-computation in later evaluation steps. To this end, query plans with so-called materialization points are introduced in this section. A

---

[3] Note that $\mathsf{END_Q}$ is a syntactical expression which can be used in selections whereas $end(occtime(r))$ denotes a mathematical function on the semantic level.

materialization point is a relation which saves and updates the results of a CERA expression instead of computing them anew in each evaluation step.

Definition 6 (Query plan with materialization points). *A query plan is a sequence* $QP = \langle M_1 := Q_1, \ldots, M_n := Q_n \rangle$ *of materialization point definitions* $M_i := Q_i$. M_i *is called a materialization point.* Q_i *is either a basic stream,[4] a CERA (sub-) expression or a union* $R_1 \cup \ldots \cup R_n$ *of materialization points. Each materialization point* M_i *is defined only once in QP, i.e.,* $M_i \neq M_j$ *for all* $1 \leq i < j \leq n$. *The materialization point definitions must be acyclic, i.e., if* M_j *occurs in* Q_i *then* $j < i$ *for all* $1 \leq i \leq n$ *and all* $1 \leq j \leq n$.

Since a query plan is acyclic, its semantics is straightforward: compute the results of its expressions from left to right, replacing references to materialization points with their (already computed) result.

$$\mathsf{Fire_f} := \mathsf{C}^{\mathsf{X}}_{[\mathit{fire}\{\mathit{area}\{\mathit{var}\ A\}\}\ ]}(\\ \qquad \mu[[\mathit{begin},\mathit{end}] \leftarrow s \sqcup t](\\ \qquad\qquad \sigma[\max\{s.\mathit{end},t.\mathit{end}\} - \min\{s.\mathit{begin},t.\mathit{begin}\} \leq 1\ \mathit{min}](\\ \qquad\qquad\qquad \sigma[s.\mathit{end} < t.\mathit{begin}](\\ \qquad\qquad\qquad\qquad \sigma[T > 40](\\ \qquad\qquad\qquad\qquad\qquad \mathsf{Smoke_s} \bowtie \mathsf{Temp_t})))))$$

$$\mathsf{Burnt\_down_b} := \mathsf{C}^{\mathsf{X}}_{[\ \mathit{burnt\_down}\{\mathit{sensor}\{\mathit{var}\ S\}\}\ ]}(\\ \qquad \mu[[\mathit{begin},\mathit{end}] \leftarrow n \sqcup i \sqcup v](\\ \qquad\qquad \sigma[T > 40](\\ \qquad\qquad\qquad (\mathsf{Temp_n} \bowtie \mathsf{X}_{[\ i:\ \text{from-end}[n,12\ \text{sec}]\ ]}(\mathsf{E})) \, \overline{\bowtie}_{i \sqsupseteq v} \mathsf{Temp_v})))$$

$$\mathsf{Avg\_temp_a} := \mathsf{C}^{\mathsf{X}}_{[\ \mathit{avg\_temp}\{\mathit{sendor}\{\mathit{var}\ S\},\mathit{value}\{\mathit{avg}(\mathit{all}\ \mathit{var}\ T)\}\}\ ]}(\\ \qquad \mu[[\mathit{begin},\mathit{end}] \leftarrow m \sqcup i \sqcup w](\\ \qquad\qquad (\mathsf{Temp_m} \bowtie \mathsf{X}_{[\ i:\ \text{from-start-backward}[m,1\ \text{min}]\ ]}(\mathsf{E})) \bowtie_{i \sqsupseteq w} \mathsf{Temp_w}))$$

Fig. 9 Query plan for the program P in Figure 1

Figure 9 shows a query plan for the program P. The plan can be significantly improved. First, it does not account for the materialization of shared subqueries. Relations $\mathsf{Temp_n}$ in the second expression and $\mathsf{Temp_w}$ in the third expression are equal except for the names of time attributes and event references, i.e., $\mathsf{Temp_w} = \rho[w.\mathit{begin} \leftarrow n.\mathit{begin}, w.\mathit{end} \leftarrow n.\mathit{end}, w.\mathit{ref} \leftarrow n.\mathit{ref}](\mathsf{Temp_n})$. (Compare their simple event queries in Figure 2.) But the same tuples of the relations are computed and saved twice. To avoid this, we introduce a new relation $\mathsf{Temp_{n'}}$ (where n' is a fresh event identifier), save all the respective tuples only once in it, and use this relation in both expressions.

---

[4] When translating an XChange[EQ] program, the basic streams are the incoming event stream E and the auxiliary streams for relative timer events.

The same holds for the relations Temp_v in the second expression and Temp_m in the third expression, i.e., $\mathsf{Temp}_v = \rho\,[v.begin \leftarrow m.begin, v.end \leftarrow m.end, v.ref \leftarrow m.ref](\mathsf{Temp}_m)$. We use the relation $\mathsf{Temp}_{m'}$ (where m' is a fresh event identifier) in both expressions.

Second, selections should be as near to their respective relations as possible to reduce the number of tuples which must be further considered (e.g., joined with tuples of other relations). This optimization technique, usually called "push selection", is adopted from the traditional relational algebra. To this end we apply the selection $\sigma[T > 40]$ to the relation Temp_t before Temp_t is joined with the relation Smoke_s in the first expression of the query plan. We analogously modify the second expression of the query plan.

Third, intermediate results of a query should be materialized to avoid their recomputation in later evaluation steps. For example, in the first query, if a new event arrives and is saved in the relation Smoke_s we have to compute $\sigma[T > 40](\mathsf{Temp}_t)$ anew in order to join it with the changed relation Smoke_s. To avoid this recomputation we define a new materialization point $\mathsf{A}_t := \sigma[T > 40](\mathsf{Temp}_t)$ and join it with the relation Smoke_s. To avoid re-computations in the other expressions we introduce the materialization points $\mathsf{B}_{n',i}, \mathsf{C}_{n'}$, and $\mathsf{D}_{m',i}$. Consider Figure 10 for the improved query plan.

$$\mathsf{Fire}_f := C^X_{[\,fire\{area\{var\,A\}\}\,]}(\\
\mu[[begin, end] \leftarrow s \sqcup t](\\
\sigma[\max\{s.end, t.end\} - \min\{s.begin, t.begin\} \le 1\ min](\\
\sigma[s.end < t.begin](\\
\mathsf{Smoke}_s \bowtie \mathsf{A}_t)))),\ \text{where}$$

$$\mathsf{A}_t := \sigma[T > 40](\mathsf{Temp}_t)$$

$$\mathsf{Burnt\_down}_b := C^X_{[\,burnt\_down\{sensor\{var\,S\}\}\,]}(\\
\mu[[begin, end] \leftarrow n' \sqcup i \sqcup m'](\\
\mathsf{B}_{n',i}\ \overline{\bowtie}_{i \sqsupseteq m'}\ \mathsf{Temp}_{m'})),\ \text{where}$$

$$\mathsf{B}_{n',i} := \mathsf{C}_{n'} \bowtie X_{[\,i:\ \text{from-end}[n', 12\ \text{sec}]\,]}(\mathsf{E}),\ \text{where}$$

$$\mathsf{C}_{n'} := \sigma[T > 40](\mathsf{Temp}_{n'})$$

$$\mathsf{Avg\_temp}_a := C^X_{[\,avg\_temp\{sendor\{var\,S\}, value\{avg(all\,var\,T)\}\}\,]}(\\
\mu[[begin, end] \leftarrow m' \sqcup i \sqcup n'](\\
\mathsf{D}_{m',i} \bowtie_{i \sqsupseteq n'} \mathsf{Temp}_{n'})),\ \text{where}$$

$$\mathsf{D}_{m',i} := \mathsf{Temp}_{m'} \bowtie X_{[\,i:\ \text{from-start-backward}[m', 1\ \text{min}]\,]}(\mathsf{E}),$$

Fig. 10 Improved query plan for the program P in Figure 1

So far it might seem that this is just an insignificant change in notation. However, it will become clear in the next section that only those intermediate results are "materialized", i.e., remembered across individual evaluation steps, that have a materialization point. Therefore the query plans in Figures 9 and 10 are different in terms of incremental evaluation, although of course both yield the same results for the program P. Note that the efficiency of a query plan depends on characteristics of its event streams and there is no general principle to tell which one is more efficient.

3.4.3 Finite Differencing

Evaluation of an event query program, or rather its query plan QP, over time is a step-wise procedure. A step is initiated by some base event (an event which is not derived by a rule) happening at the current time, which we denote *now*. Then for each materialization point M in QP, the required output for this step is the set of all computed answers (tuples r representing materialized intermediate results and derived events) that "happen" at this current time *now*, i.e., where $end(occtime(r)) = now$. In other words in each step, we are not interested in the full result of M, but only in $\triangle M := \sigma[\mathsf{END}_Q = now](M)$.[5]

A naive, non-incremental way of query evaluation would be: Maintain a stored version of each base event relation across steps. In each step simply insert the new event into its base relation and evaluate the query plan from scratch according to its non-incremental semantics (previous section). Then apply the selection $\sigma[\mathsf{END}_Q = now]$ to each materialization point to output the result of the step. This is, however, inefficient since we compute not only the required result $\triangle M = \sigma[\mathsf{END}_Q = now](M)$, but also all results from previous steps, i.e., also $\sigma[\mathsf{END}_Q < now](M)$.

It is more efficient to use an incremental approach, where we (1) store not only base relations but also some intermediate results, namely those of each materialization point M across steps and then (2) in each step only compute the changes of M that result from the step. It turns out that due to the temporal preservation of CERA (see Theorem 1), the change to each M involves only inserting new tuples into M and that these tuples are exactly the ones from $\triangle M$.

We can compute $\triangle M$ efficiently using the changes $\triangle R_i$ of the input relations R_i of $M := Q$, together with $\circ R_i = \sigma[\mathsf{END}_Q < now](R_i)$, their materialized states from the previous evaluation steps.[6] Using finite differencing, we can derive a

---

[5] We assume for simplicity here that the base events are processed in the temporal order in which they happen, i.e., with ascending ending timestamps. Extensions where the order of events is "scrambled" (within a known bound) are possible, however. Note that while the time domain can be continuous (e.g., isomorphic to the real numbers), the number of evaluation steps is discrete since we assume a discrete number of incoming events.

[6] Note that some previous results can become irrelevant in later evaluation steps, i.e., they cannot contribute to new answers any more. Therefore they should be deleted to speed up the later evaluation steps. See [7] for the formal definition of garbage collection enabled by CERA.

CERA-expression $\triangle Q$ so that $\triangle Q$ involves only $\triangle R_i$ and $\circ R_i$ and $\triangle Q = \triangle M$ (for each step). Finite differencing pushes the differencing operator $\triangle$ inwards according to the equations in Figure 11.

$$
\begin{aligned}
\triangle Q^X(Q) &= Q^X(\triangle Q) & \circ Q^X(Q) &= Q^X(\circ Q) \\
\triangle C^X(Q) &= C^X(\triangle Q) & \circ C^X(Q) &= C^X(\circ Q) \\
\triangle \sigma_C(Q) &= \sigma_C(\triangle Q) & \circ \sigma_C(Q) &= \sigma_C(\circ Q) \\
\triangle \rho_A(Q) &= \rho_A(\triangle Q) & \circ \rho_A(Q) &= \rho_A(\circ Q) \\
\triangle \pi_P(Q) &= \pi_P(\triangle Q) & \circ \pi_P(Q) &= \pi_P(\circ Q) \\
\triangle \mu_M(Q) &= \mu_M(\triangle Q) & \circ \mu_M(Q) &= \mu_M(\circ Q) \\
\triangle(Q_1 \bowtie Q_2) &= \triangle Q_1 \bowtie \circ Q_2 \cup \triangle Q_1 \bowtie \triangle Q_2 \cup & \circ(Q_1 \bowtie Q_2) &= \circ Q_1 \bowtie \circ Q_2 \\
& \quad \cup \circ Q_1 \bowtie \triangle Q_2 \\
\triangle(Q_1 \bowtie_{i \sqsupseteq j} Q_2) &= \triangle Q_1 \bowtie_{i \sqsupseteq j} \circ Q_2 \cup \triangle Q_1 \bowtie_{i \sqsupseteq j} \triangle Q_2 & \circ(Q_1 \bowtie_{i \sqsupseteq j} Q_2) &= \circ Q_1 \bowtie_{i \sqsupseteq j} \circ Q_2 \\
\triangle(Q_1 \overline{\bowtie}_{i \sqsupseteq j} Q_2) &= \triangle Q_1 \overline{\bowtie}_{i \sqsupseteq j} \circ Q_2 \cup \triangle Q_1 \overline{\bowtie}_{i \sqsupseteq j} \triangle Q_2 & \circ(Q_1 \bowtie_{i \sqsupseteq j} Q_2) &= \circ Q_1 \bowtie_{i \sqsupseteq j} \circ Q_2
\end{aligned}
$$

Fig. 11 Equations for finite differencing

Finite differencing is a method originating in the incremental maintenance of materialized views in databases, which is a problem very similar to incremental event query evaluation. We refer the reader to [11, 7] for more information on incremental evaluation and garbage collection enabled by CERA.

4 A Declarative Semantics for Event Query Languages

Section 4.1 explains the purpose and necessity of a declarative semantics for a programming language in general and its desiderata for an EQL in particular. Section 4.2 is devoted to the formal definition of the declarative semantics of XChange^{EQ} with a model-theoretic approach in order to prove the correctness of its operational semantics (Section 5).

4.1 Purpose, Necessity and Desiderata

In general, a declarative semantics relates the syntax of a language to mathematical objects and expressions that capture the intended meaning. In other words, a declarative semantics focuses on expressing *what* a sentence in the language means, rather than *how* that sentence might be evaluated (which is the purpose of an operational semantics).

A declarative semantics thus provides a convenient basis to prove the correctness of various operational semantics. In particular in the area of query languages there are usually a myriad of equivalent ways to evaluate a given query, that is, of possible operational semantics. If, on the other hand, the formal semantics of a language were specified only in an operational way, proving the correctness of other operational semantics would be significantly harder: since an operational semantics focuses on how the result is computed not on what is the result, we have to reason about the equivalence of two computations. When we prove correctness of an operational semantics with respect to a declarative semantics, we instead just reason about properties of the output of one computation. This use of a declarative semantics to prove correctness of evaluation methods is particularly useful in research on optimization.

Declarative semantics have often been neglected in EQLs so far. Our goal is a declarative semantics that is natural on event streams, i.e., does not require a conversion from streams to relations and back, like SQL-based EQLs do [3, 12, 16], and is as declarative as possible and thus avoids any notion of state.

Because of these reasons we specify the declarative semantics by a Tarski-style model theory with accompanying fixpoint theory in Section 4.2. This approach has another important advantage, namely it accounts well for data in events and rule chaining, two aspects that have often been neglected in the semantics of EQLs till now.

4.2   Model Theory and Fixpoint Theory

While the model-theoretic approach is well-established for traditional, non-event query and rule languages, its application to EQLs is novel and we highlight the extensions that are necessary in this section. We also show that our declarative semantics is suitable for querying events that arrive over time in unbounded event streams and illustrate this statement by the declarative semantics of the XChange[EQ] program P in Figure 1.

The idea of a model theory, as it is used in traditional, non-event query languages [15, 1, 17], is to relate expressions to an *interpretation* by defining an *entailment relation*. Expressions are syntactic fragments of the query language such as rules, queries, or facts viewed as logic sentences. The interpretation contains all facts that are considered to be true. The entailment relation indicates whether a given interpretation entails a given sentence in the language, that is, if the sentence is logically true under this interpretation. For the semantics of a given query program and a set of base facts are those interpretations of interest that (1) satisfy all rules of the program and (2) contain all base facts. Because it satisfies all rules, such an interpretation particularly contains all facts that are derived by rules. We call these interpretations *models*.

When we replace facts that are true with events that happen, this approach can also be applied to EQLs. The problem, of course, is that events are associated with *occurrence times* and event queries are evaluated *over time* against a potentially *infinite event stream*. At each time point during the evaluation we know only which events have happened (i.e., been received in the event stream) so far, not any events that might happen in the future. We start off with some basic definitions that explain how we represent time and events in the semantics of XChange[EQ].

Time is represented by a linearly ordered **set of time points** $(\mathbb{T}, <)$. The **set of time intervals** is $\mathbb{TI} = \{t = [begin, end] \mid begin \in \mathbb{T}, end \in \mathbb{T}, begin \leq end\}$. For an interval t, $begin(t)$ denotes its beginning and $end(t)$ its end, i.e., $t =$

$[begin(t), end(t)]$. We omit t in the notation if the interval is clear from the context, i.e., we write $begin$ and end instead of $begin(t)$ and $end(t)$ respectively.

The **set of data terms** is denoted *DataTerms* (see [17, 13] for the full grammar of Xcerpt data terms). Recall that data terms are used to represent data and type information for events. An event is a tuple of a time interval t and a data term e, written e^t. The **set of events** is denoted *Events*; $Events = DataTerms \times \mathbb{TI}$. Let $E \subseteq Events$ denote an **event stream** and *EventIdentifiers* the **set of event identifiers**.

To explain how simple event queries are matched against incoming events and how events derived by rules are constructed, we have to explain some concepts of the Web query language Xcerpt, whose query and construct terms are used in XChange^{EQ}. We try to keep these explanations brief and refer the reader to [9, 17] for details.

An **Xcerpt query term** is a pattern that accesses data to extract relevant portions of it. An **Xcerpt construct term** is a pattern that constructs new data. Consider the first rule in Figure 1. *fire{area{var A}}* is a construct term, *smoke{{area{{var A}}}}* and *temp{{ area{{var A}},value{{var T}}}}* are query terms. Since query and construct terms of a rule contain variables (A and T in this case) that are bound to values during the application of the rule, we need the concept of substitution.

Let *Vars* denote the **set of variable names**. A **substitution** σ is a partial mapping from variable names to data terms, i.e., $\sigma : Vars \rightarrow DataTerms$. We write substitutions as $\sigma = \{X_1 \mapsto v_1, \ldots, X_n \mapsto v_n\}$, meaning that $\sigma(X_i) = v_i$ for $i \in \{1, \ldots, n\}$ and $\sigma(Y) = \bot$ for $Y \notin \{X_1, \ldots, X_n\}$.

The **application of a substitution** σ to a query term q replaces the occurrences of variables V in q with their values $\sigma(V)$. The result is denoted $\sigma(q)$. If $\sigma(q)$ is a **ground term**, i.e., a term without variables, we call σ a **grounding substitution** of q.

Simple event queries in XChange^{EQ} are Xcerpt query terms that are matched against data terms e of incoming events e^t. This matching of simple event queries is based on **simulation** between ground terms as defined for Xcerpt [9, 17]. Intuitively, a ground query term q simulates into a data term d, denoted $q \preceq d$, if the nodes and the structure of the graph that q represents, can be found in the graph of d. This simulation relationship of Xcerpt is especially designed for the variations and incompleteness in semi-structured data.

A non-ground query term q' simulates into a data term d, $q' \preceq d$, if there is a grounding substitution σ such that $\sigma(q') \preceq d$. Note that for a given non-ground query term q' and a given data term d, there are often several substitutions that allow a simulation between the two. We denote the **substitution set** of q' and d by $\Sigma := \{\sigma \mid \sigma(q') \preceq d\}$.

An XChange^{EQ} rule head contains an Xcerpt construct term h for constructing new, derived events. This construction uses the substitution set Σ obtained from the evaluation of the query in the respective rule body to replace variables with values. Application of Σ to h, defined in [17], returns a set of data terms representing derived events.

Now we can define interpretation and entailment, which are the core of the model theory of XChange^{EQ}.

$$I, \sigma, \tau \models q^t \qquad\qquad\qquad \text{iff } \exists e^{t'} \in I \text{ such that } t' = t \text{ and } \sigma(q) \preceq e$$

$$I, \sigma, \tau \models (\texttt{event } i : q)^t \qquad \text{iff } \exists e^{t'} \in I \text{ such that } \tau(i) = e^{t'}, t' = t \text{ and } \sigma(q) \preceq e$$

$$M \models (q_1 \wedge q_2)^t \qquad\qquad \text{iff } \exists t_1, t_2 \text{ such that } t = t_1 \sqcup t_2, M \models q_1^{t_1} \text{ and } M \models q_2^{t_2}$$

$$M \models (q_1 \vee q_2)^t \qquad\qquad \text{iff } M \models q_1^t \text{ or } M \models q_2^t$$

$$I, \sigma, \tau \models (Q \texttt{ where } C)^t \qquad \text{iff } I, \sigma, \tau \models Q^t \text{ and } W_{\sigma, \tau}(C) = true$$

$$I, \sigma, \tau \models (\texttt{while } j : \texttt{ not } q)^t \qquad \text{iff } \exists e^{t'} \in I \text{ such that } \tau(j) = e^{t'} \text{ and } t' = t,$$
$$\text{and } \not\exists t'' \sqsubseteq t \text{ such that } I, \sigma, \tau \models q^{t''}$$

$$I, \sigma, \tau \models (\texttt{while } j : \texttt{ collect } q)^t \text{ iff } \exists e^{t'} \in I \text{ such that } \tau(j) = e^{t'} \text{ and } t' = t,$$
$$\text{and } \exists t'' \sqsubseteq t \text{ such that } I, \sigma, \tau \models q^{t''}$$

$$I, \sigma, \tau \models (h \leftarrow B)^t \qquad\qquad \text{iff } \forall \tau' : \Sigma_{\tau'} = \emptyset \text{ or } \Sigma_{\tau'}(h) \subseteq I$$
$$\text{with } \Sigma_{\tau'} := \{\sigma' \mid I, \sigma', \tau' \models B^t\}$$

Fig. 12 Entailment relation defining the model theory for XChange^{EQ}

Definition 7 (Interpretation). *An interpretation for a given XChange^{EQ} query, rule, or program is a 3-tuple $M = (I, \sigma, \tau)$ where:*

1. *$I \subseteq$ Events is the set of events e^t that "happen," i.e., are either in the stream of incoming events or derived by some rule.*
2. *σ is a grounding substitution for (data) variables.*
3. *$\tau :$ EventIdentifiers $\rightarrow$ Events is a substitution for event identifiers.*

The substitution τ for event identifiers is, compared to model theories of traditional, non-EQLs, unusual. It is needed for evaluating temporal conditions and relative timer events. Since τ signifies the events that contribute to the answer of some query, we also call it an "event trace."

The **entailment** (or satisfaction) $M \models F^t$ of an XChange^{EQ} expression F over a time interval t in an interpretation M is defined in Figures 12 and 13. (We require the programs to be range restricted [8].)

Figure 12 defines the more salient cases of the model theory. For the sake of brevity, the expressions in this figure use binary "and" with symbol $\wedge$ and binary "or" with symbol $\vee$ instead of the multi-ary *and*$\{\dots\}$ and *or*$\{\dots\}$. Also, rules are written as $h \leftarrow B$ instead of *DETECT h ON B END*.

Figure 13 defines entailment of the relative timer events used in the program in Figure 1. (See [11] for the complete version of the figure.) For the sake of brevity, the prefix "*timer:*" and the keyword "*event*" within the relative timer specification have been skipped.

Our entailment relation uses a fixed interpretation W for all conditions that can occur in the *where*-clause of a query. This includes the temporal relations like *before*

$I, \sigma, \tau \models (\text{event } i : \textit{from-end}[j,d])^t$ iff exists $e^{t'}$ with $\tau(j) = e^{t'}$,
$$\tau(i) = \text{rel-timer-event}(e, t')^t,$$
$$begin(t) = end(t'), end(t) = end(t') + d$$

$I, \sigma, \tau \models (\text{event } i : \textit{from-start-backward}[j,d])^t$ iff exists $e^{t'}$ with $\tau(j) = e^{t'}$,
$$\tau(i) = \text{rel-timer-event}(e, t')^t,$$
$$begin(t) = begin(t') - d, end(t) = begin(t')$$

Fig. 13 Entailment of the relative timer events in XChange^{EQ} used in Figure 1

$W_{\sigma, \tau}(i \text{ before } j) = \textit{true}$ iff $end(\tau(i)) < begin(\tau(j))$

$W_{\sigma, \tau}(\{i_1, \ldots, i_n\} \text{ within } d) = \textit{true}$ iff $E - B \leq d$ with $E := \max\{end(\tau(i_1)), \ldots, end(\tau(i_n))\}$
and $B := \min\{begin(\tau(i_1)), \ldots, begin(\tau(i_n))\}$.

Fig. 14 Fixed interpretation for conditions in the *where* clause used in Figure 1

as well as conditions on data such as arithmetic comparisons. This fixed interpretation of the temporal conditions is another feature of our model theory that is not common in model theories for traditional, non-EQLs.

W is a function that maps a substitution σ, an event trace τ, and an atomic condition C to a Boolean value (true or false). We usually write σ and τ in the subscript of W. $W_{\sigma, \tau}$ extends straightforwardly to Boolean formulas of conditions. Figure 14 gives the definitions of W for the temporal conditions of XChange^{EQ} that have been used in Figure 1. (See [11] for the complete version of the figure.) The definition of W is deliberately left outside the "core model theory" to make it more modular and demonstrate that it is easy to integrate further conditions or even a separate, external temporal reasoner.

Recall our primary goal in specifying declarative semantics for XChange^{EQ}: given an XChange^{EQ} program P and an event stream E, we want to find out all events that are derived by the rules of P. This means that we must find an interpretation that contains the event stream E and satisfies all rules of P. Such an interpretation is called a model.

Definition 8 (Model). *Given an XChange^{EQ} program P and a stream of incoming events E, we call an interpretation $M = (I, \sigma, \tau)$ a model of P under E if*

- *M satisfies all rules $r = (h \leftarrow B) \in P$ for all time intervals t, i.e., $M \models r^t$ for all $t \in \mathbb{TI}$ and all $r = (h \leftarrow B) \in P$, and*
- *M contains the stream of incoming events, i.e., $E \subseteq I$.*

On close inspection of the entailment relation, we can see that σ and τ are actually irrelevant to whether a given interpretation M is a model or not; it depends only on I. We therefore can identify the notion of a model with just the I part of an interpretation, $M = I$.

Consider the XChange^{EQ} program P in Figure 1 and the event stream

$$E = \{ \ temp \ \{ \ area\{a\}, sensor\{s\}, value\{40\} \ \}^{[60,63]},$$
$$smoke \ \{ \ area \ \{a\} \ \}^{[65,68]},$$
$$temp \ \{ \ area\{a\}, sensor\{s\}, value\{41\} \ \}^{[70,80]} \ \}.$$

We assume timestamps to be time intervals. The bounds of the intervals denote minutes since the beginning of the current 10-minutes-long-window. In this case an event with timestamp [0,0] happens 10 minutes before an event with timestamp [599,599]. The lifetime of all events is restricted to the window they happened within. These assumptions are not suitable for real-life applications but they help to keep the example simple.

The interpretation

$M_1 = \{$
$temp\{area\{a\}, sensor\{s\}, value\{40\}\}^{[60,63]}, \ fire\{area\{a\}\}^{[65,80]},$
$smoke\{area\{a\}\}^{[65,68]}, \qquad\qquad\qquad\qquad burnt\_down\{sensor\{s\}\}^{[70,92]},$
$temp\{area\{a\}, sensor\{s\}, value\{41\}\}^{[70,80]}, \ avg\_temp\{sensor\{s\}, value\{40\}\}^{[10,80]}$
$\}$

is a model for P under E: by applying the recursive definition of $\models$ we can check that $M_1 \models r^t$ for all $t \in \mathbb{TI}$, $r \in P$, and we also have $E \subseteq M_1$. Note that each rule of P derives exactly one complex event of M_1 and the timestamp of a complex event comprises the timestamps of all events this complex event was derived from. Note also that the second temperature measurement does not fall into the time window for aggregation of the last rule of P. That is why the avarage temperature is 40, not 40,5.

The interpretation

$M_2 = \{$
$temp\{area\{a\}, sensor\{s\}, value\{40\}\}^{[60,63]}, \ fire\{area\{a\}\}^{[65,80]},$
$smoke\{area\{a\}\}^{[65,68]}, \qquad\qquad\qquad\qquad burnt\_down\{sensor\{s\}\}^{[70,92]},$
$temp\{area\{a\}, sensor\{s\}, value\{41\}\}^{[70,80]}, \ avg\_temp\{sensor\{s\}, value\{40\}\}^{[10,80]},$
$temp\{area\{b\}, sensor\{t\}, value\{20\}\}^{[1,2]}$
$\}$

where we "added" the event $temp\{area\{b\}, sensor\{t\}, value\{20\}\}^{[1,2]}$ in comparison to M_1, is also a model of P under E. Clearly, however, M_2 is not the model we intend for our program to have, because the "additional" event is "unjustified." More precisely, this event is neither in the event stream E nor derived by a rule of P. M_1 is the intended model, because all events in it are justified.

To unambiguously settle on a single, intended model, we will use the fixpoint theory which builds upon the model theory. Note that the problem of specifying the intended model out of the (infinitely) many possible models is also a common part of the traditional model-theoretic approach. It is not specific for EQLs.

The intended model is the (least) fixpoint of the immediate consequence operator, which derives new events from known events (based on the model theory). Non-monotonic features such as negation and aggregation introduce well-known issues when they are combined with recursion of rules. In particular, there might be no fixpoint or several. To ensure that a single fixpoint exists, we restrict XChange^{EQ} programs to be stratifiable. Stratification restricts the use of recursion in rules by ordering the rules of a program P into so-called strata (sets P_i of rules with $P = P_1 \uplus \cdots \uplus P_n$) such that a rule in a given stratum can only depend on (i.e., access results from) rules in lower strata (or the same stratum, in some cases).

Restriction to stratifiable programs is a common approach from logic programming introduced first in [2]. But in contrast to logic programming, in CEP three types of **stratification** are required:

1. *Negation stratification*: Events that are matched by a negative simple event query of a rule r (e.g., *not temp{{sensor{{var S}}}}* of the second rule in Figure 1) may only be constructed by rules in lower strata than the stratum of r. Events which are matched by a positive simple event query of a rule r may be constructed by rules in lower strata or the same stratum as that of r.
2. *Grouping stratification*: Rules r with grouping constructs like *all* in the construction may only query events constructed by rules in lower strata than the stratum of r. Therefore the last rule in Figure 1 must be in a higher stratum then all rules the head of which is matched by *temp{{sensor{{var S}}, value {{var T}} }}*.
3. *Temporal stratification*: If a rule r defines a relative timer event, e.g., *timer: from-end [event n, 12 sec]* in the second rule in Figure 1, then the anchoring event (here: n) may only be constructed by rules in lower strata than the stratum of r.

While negation and grouping stratification are fairly standard, temporal stratification is a requirement specific to complex event query programs like those expressible in XChange^{EQ}. We are not aware of former consideration of the notion of temporal stratification. See [11] for the formal definitions of the notions.

The basic idea for obtaining the fixpoint interpretation of a stratifiable XChange^{EQ} program is to apply the rules stratum by stratum: first apply the rules in the lowest stratum to the incoming event stream, then apply the rules in the next higher stratum to the result, and so on until the highest stratum. This requires the definition of the immediate consequence operator T_P for an XChange^{EQ} program.

Definition 9 (Immediate consequence operator). *The immediate consequence operator T_P for an XChange^{EQ} program is defined as:*

$$T_P(I) = I \cup \{e^t \mid \text{there exist a rule } h \leftarrow B \in P, \text{ and } \tau \text{ such that } e \in \Sigma_\tau(h)$$
$$\text{where } \Sigma_\tau := \{\sigma \mid I, \sigma, \tau \models B^t\} \qquad \}$$

The operator is obviously monotonic [8]. Hence according to the Knaster-Tarski theorem, it has a fixpoint [8]. The repeated application of T_P until a fixpoint is reached is denoted T_P^ω. A fixpoint means here an interpretation I such that $T_P(I) = I$.

Definition 10 (Fixpoint interpretation). *Let* $\overline{P}_i = \bigcup_{j \leq j} P_j$ *denote the set of all rules in strata* P_i *and lower. The fixpoint interpretation* $M_{P,E}$ *of an XChange^{EQ} program P with stratification* $P = P_1 \uplus \cdots \uplus P_n$ *under event stream E is defined by computing fixpoints stratum by stratum:*

$$M_0 = E = T_{\emptyset}^{\omega}(E),$$
$$M_1 = T_{\overline{P_1}}^{\omega}(M_0),$$
$$\ldots,$$
$$M_{P,E} = M_n = T_{\overline{P_n}}^{\omega}(M_{n-1}).$$

The fixpoint interpretation $M_{P,E}$ *is also called the intended model of P under E and specifies the declarative semantics.*

Consider, for example, the XChange^{EQ} program P in Figure 1 and the event stream E above.

$$
\begin{aligned}
M_0 = E = \quad & \{ \ temp \ \{ \ area\{a\}, \ sensor\{s\}, \ value\{40\} \ ^{[60,63]}, \\
& smoke \ \{ \ area \ \{a\} \ \} \ ^{[65,68]}, \\
& temp \ \{ \ area\{a\}, \ sensor\{s\}, \ value\{41\} \ \} \ ^{[70,80]} \ \} = T_{\emptyset}^{\omega}(E) \\
M_{P,E} = M_1 = M_0 \cup \{ \ & fire \ \{ \ area \ \{a\} \ \} \ ^{[65,80]}, \\
& burnt\_down \ \{ \ sensor \ \{s\} \ \} \ \} \ ^{[70,92]}, \\
& avg\_temp \ \{ \ sensor \ \{s\}, \ value\{40\} \ \} \ ^{[10,80]} \quad \} = T_{\overline{P_1}}^{\omega}(M_0)
\end{aligned}
$$

In addition to giving unambiguous semantics to stratifiable XChange^{EQ} programs, the fixpoint theory also describes an abstract, simple, forward-chaining evaluation method, which can easily be extended to work incrementally as it is required for event queries.

5   Two Semantics, no Double Talk

We now want to show that the semantics of XChange^{EQ}, the declarative one from Section 4.2 and the operational semantics given by the translation to CERA in Section 3.3, are equivalent. In other words we now want to show the correctness of the translation of a normalized rule $h \leftarrow B$.[7] We only give the main ideas here, for details see [11].

We consider only hierarchical programs with a single rule.[7] Recall that the declarative semantics for a program $\{h \leftarrow B\}$ with a single rule $h \leftarrow B$ is given by $M_{\{h \leftarrow B\}, E}$. Therefore we have

$$M_{\{h \leftarrow B\}, E} = T_{\{h \leftarrow B\}}^{\omega}(E) = T_{\{h \leftarrow B\}}(E)$$

---

[7] Note that the proof for a hierarchical program with a single rule immediately applies to arbitrary hierarchical programs. The reason is that there exits a topological ordering of the rules for hierarchical programs. Applying the proof for a single rule to the rules of the hierarchical program in topological order yields a proof for arbitrary hierarchical programs.

meaning that the semantics is given by applying the fixpoint operator to the program $\{h \leftarrow B\}$ and the event stream E once. That is, the declarative semantics of a single rule $h \leftarrow B$ is given by:

$$T_{\{h \leftarrow B\}}(E) = E \cup \{e^t \mid \exists \tau \text{ such that } e \in \Sigma_\tau(h) \text{ where } \Sigma_\tau := \{\sigma \mid I, \sigma, \tau \models B^t\}\}$$

Let Q be the CERA expression that translates the rule $h \leftarrow B$. We identify events e^t with tuples r of $Q(E)$ and E by $r(begin) = begin(t)$, $r(end) = end(t)$, $r(term) = e$. In the other direction we identify tuples r with events $r(term)^{[r(begin),r(end)]}$. With this identification correctness of the translation means that

$$Q(E) \cup E = T_{\{h \leftarrow B\}}(E).$$

For this it suffices to show that

$$Q(E) = \{e^t \mid \exists \tau \text{ such that } e \in \Sigma_\tau(h) \text{ where } \Sigma_\tau := \{\sigma \mid I, \sigma, \tau \models B^t\}\} \qquad (*)$$

Next, we have to find a correspondence between the elements r of the relations $S(E)$ generated by subexpressions S of Q, i.e., $B_1, \dots, B_n, C$ defined in Section 3.3, and the combined σ, τ used by the declarative semantics in Section 4.2. In other words we want to have a corresponding tuple $r_{\sigma,\tau}$ for each pair σ, τ and a corresponding pair σ_r, τ_r for each tuple r. Figure 15 shows the correspondence.

$$r_{\sigma,\tau} := \begin{cases} r(X) & = \sigma(X) & \text{for all } X \in sch_{data}(S) \\ r(j.ref) & = ref(\tau(j)) & \text{for all } i.ref \in sch_{ref}(S) \\ r(j.begin) & = begin(\tau(j)) & \text{for all } i.begin \in sch_{time}(S) \\ r(j.end) & = end(\tau(j)) & \text{for all } i.end \in sch_{time}(S) \\ r(X) & = \bot & \text{otherwise} \end{cases}$$

$$\sigma_r := \begin{cases} \sigma_r(X) = r(X) & \text{for all } X \in sch_{data}(S) \\ \sigma_r(X) = X & \text{otherwise} \end{cases}$$

$$\tau_r := \begin{cases} \tau_r(i) = s_i(term)^{[s_i(begin),s_i(begin)]} & \text{for all } i.ref \in sch_{ref}(S), \text{ where } s_i := ref^{-1}(i.ref) \\ \tau_r(i) = \bot & \text{otherwise} \end{cases}$$

Fig. 15 Correspondence between tuples and substitutions

We use this correspondence to show a lemma about each subexpression S of Q that translates a subexpression F of the rule body B (see Figure 8).

Lemma 1 (Equivalence of subexpressions). *Let F be the subexpression translated by S and be $t = occtime(r)$. Then following equivalence holds:*

$$r \in S(E) \quad \Leftrightarrow \quad E, \sigma_r, \tau_r \models F^t$$

The left to right part of Lemma 1 is soundness ("results produced by the operational semantics are results according to the declarative semantics"). The right to left part is completeness ("results according to the declarative semantics are actually produced by the operational semantics").

Consider the definitions of B_n and C in Figure 8. The proof of Lemma 1 is by induction on n over B_n, and then by additionally showing that Lemma 1 also holds for C.

The proof of (*) is by applying Lemma 1 to rule body B as used in (*) and the translation C of B as defined in Figure 8. The details of the proof are given in [11].

6 Conclusion and Outlook

In this chapter, we have formally defined the operational and declarative semantics for XChange^{EQ}, illustrated them on the sensor network use case and proved their equivalence. Both semantics are generic and easily transferable to an arbitrary EQL. On the basis of the approach described here, some other points essential for CEP can be immediately implemented. Two of them, garbage collection and query optimization, are addressed in this section.

As mentioned above, evaluation of complex event queries over time involves storing events in materialization points. Naive query evaluation simply stores all events forever. Since the event stream is not bounded into the future every naive query evaluation engine can run out of memory sooner or later. Therefore there is a need for garbage collection of irrelevant events, i.e., events which cannot contribute to the derivation of new events (any more). We refer the reader to [7] for an approach to static determination of temporal relevance for incremental evaluation of complex event queries. Temporal relevance is particularly suitable for garbage collection because one of the main principles of a reasonable CEP engine is that no rule must wait for an event for ever.

The so-called general relevance of events including temporal, causal, structural relevance as well as relevance with regards to event data, addresses query optimization. In addition to the consideration of general relevance, automatic query optimization is possible by means of application specific knowledge formalized as a model. With the help of the model one can, e.g., recognize the unsatisfiability of a (sub-) query in order to suspend or delete it and to avoid the storage of irrelevant events.

Acknowledgements

This research has been founded in part by the European Commission within the the project "EMILI — Emergency Management in Large Infrastructures" under grant agreement number 242438 and by the German Research Foundation (Deutsche Forschungsgemeinschaft) within the project "QONCEPT — Query Optimization in Complex Event Processing Technologies" under reference number BR 2355/1-1.

References

1. Abiteboul, S., Hull, R., Vianu, V.: Foundations of Databases. Addison-Wesley, Reading (1995)
2. Apt, K.R., Blair, H.A., Walker, A.: Towards a theory of declarative knowledge. In: Foundations of Deductive Databases and Logic Programming, pp. 89–148. Morgan Kaufmann, San Francisco (1988)
3. Arasu, A., Babu, S., Widom, J.: The CQL continuous query language: Semantic foundations and query execution. The VLDB Journal 15(2), 121–142 (2006)
4. Babu, S., Srivastava, U., Widom, J.: Exploiting k-constraints to reduce memory overhead in continuous queries over data streams. ACM Transactions on Database Systems 29(3), 545–580 (2004)
5. Bry, F., Eckert, M.: Rule-Based Composite Event Queries: The Language XChange^{EQ} and Its Semantics. In: Marchiori, M., Pan, J.Z., de Sainte Marie, C. (eds.) RR 2007. LNCS, vol. 4524, pp. 16–30. Springer, Heidelberg (2007)
6. Bry, F., Eckert, M.: Temporal order optimizations of incremental joins for composite event detection. In: Proc. Int. Conf. on Distributed Event-Based Systems, ACM, New York (2007)
7. Bry, F., Eckert, M.: On static determination of temporal relevance for incremental evaluation of complex event queries. In: Proc. Int. Conf. on Distributed Event-Based Systems, pp. 289–300. ACM, New York (2008)
8. Bry, F., Eisinger, N., Eiter, T., Furche, T., Gottlob, G., Ley, C., Linse, B., Pichler, R., Wei, F.: Foundations of rule-based query answering. In: Antoniou, G., Aßmann, U., Baroglio, C., Decker, S., Henze, N., Patranjan, P.-L., Tolksdorf, R. (eds.) Reasoning Web. LNCS, vol. 4636, pp. 1–153. Springer, Heidelberg (2007)
9. Bry, F., Schaffert, S.: Towards a declarative query and transformation language for XML and semistructured data: Simulation unification. In: Stuckey, P.J. (ed.) ICLP 2002. LNCS, vol. 2401, p. 255. Springer, Heidelberg (2002)
10. Ding, L., Chen, S., Rundensteiner, E.A., Tatemura, J., Hsiung, W.-P., Candan, K.S.: Runtime semantic query optimization for event stream processing. In: Proc. Int. Conf. on Data Engineering, pp. 676–685. IEEE Computer Society, Los Alamitos (2008)
11. Eckert, M.: Complex Event Processing with XChange^{EQ}: Language Design, Formal Semantics and Incremental Evaluation for Querying Events. PhD thesis, Institute for Informatics, University of Munich (2008)
12. EsperTech Inc. Event stream intelligence: Esper & NEsper, http://esper.codehaus.org
13. Furche, T.: Implementation of Web Query Languages Reconsidered: Beyond Tree and Single-Language Algebras at (Almost) No Cost. PhD thesis, Institute for Informatics. University of Munich (2008)
14. Garcia-Molina, H., Ullman, J., Widom, J.: Database Systems: The Complete Book. Prentice-Hall, Englewood Cliffs (2001)
15. Lloyd, J.W.: Foundations of Logic Programming. Springer, Heidelberg (1993)
16. Morrell, J., Vidich, S.D.: Complex Event Processing with Coral8. White Paper (2007), http://www.coral8.com/system/files/assets/pdf/Complex_Event_Processing_with_Coral8.pdf
17. Schaffert, S.: Xcerpt: A Rule-Based Query and Transformation Language for the Web. PhD thesis, Institute for Informatics. University of Munich (2004)
18. Tucker, P.A., Maier, D., Sheard, T., Stephens, P.: Using production schemas to characterize strategies for querying over data streams. IEEE Transactions on Knowledge and Data Engineering 19(9), 1227–1240 (2007)

ETALIS: Rule-Based Reasoning in Event Processing

Darko Anicic, Paul Fodor, Sebastian Rudolph, Roland Stühmer,
Nenad Stojanovic, and Rudi Studer

Abstract. Complex Event Processing (CEP) is concerned with timely detection of complex events within multiple streams of atomic occurrences, and has useful applications in areas including financial services, mobile and sensor devices, click stream analysis and so forth. In this chapter, we present ETALIS Language for Events. It is an expressive language for specifying and combining complex events. For this language we provide both a syntax as well as a clear declarative formal semantics. The execution model of the language is based on a compilation strategy into Prolog. We provide an implementation of the language, and present experimental results of our running prototype. Further on, we show how our logic rule-based approach compares with a non-logic approach in respect of performance.

1 Introduction

There has recently been a significant paradigm shift toward *real-time* computing. Databases and data warehouses are concerned with looking at what happened in the past. On the other hand, Complex Event Processing (CEP) is concerned with processing real-time events, i.e., CEP is concerned with what has just happened or what is about to happen in the future.

An *event* represents something that occurs, happens or changes the current state of affairs. For example, an event may signify a problem or an impending problem, a

Darko Anicic · Roland Stühmer · Nenad Stojanovic · Rudi Studer
FZI Forschungszentrum Informatik, Germany
e-mail: {darko.anicic,stuehmer,nenad.stojanovic,studer}@fzi.de

Paul Fodor
Stony Brook University, Stony Brook, NY 11794, U.S.A.
e-mail: pfodor@cs.sunysb.edu

Sebastian Rudolph
Karlsruhe Institute of Technology, Germany
e-mail: rudolph@kit.edu

S. Helmer et al.: Reasoning in Event-Based Distributed Systems, SCI 347, pp. 99–124.
springerlink.com   © Springer-Verlag Berlin Heidelberg 2011

threshold, an opportunity, some information becoming available, a deviation and so forth. An event can also represent something that did not happen (i.e., an absence of an event in certain time).

We distinguish between *atomic* and *complex* events. An atomic event is defined as an instantaneous occurrence of interest at a point in time. In order to describe more complex dynamic matters that involve several atomic events, formalisms have been created which allow for combining atomic into *complex events*, using different event operators as well as temporal, causal and semantic relationships. The field of CEP has the task of processing streams of atomic events with the goal of detecting complex events according to meaningful event patterns.[1]

Our own work in the CEP area is based on *logic programming* and a *rule-based* approach for event processing. There has been recently renewed interest in using logic programming in many areas outside the traditional knowledge representation area. Examples include work on cloud computing [5], declarative networking systems [10], natural language processing [12] and so forth.

We have observed that logic programming can be useful with respect to many concepts of CEP. First, a rule-based formalism (like the one we present in this chapter) is expressive enough and convenient to represent diverse complex event patterns. Also declarative rules are free of side-effects. Second, integration of query processing, that is essential in many event-based applications, with event processing is easy and natural. Third, our experience with use of logic programing in implementation of the main constructs in CEP as well as in providing extensibility of a CEP system is very positive and encouraging (e.g., number of code lines in logic programming is significantly smaller than in procedural programming). Ultimately, a logic-based event model enables *reasoning* over events, their relationships, entire state, and possible contextual *knowledge*. This knowledge captures the *domain* of interest, or *context* related to business critical actions and decisions (that are triggered in real-time by complex events). Its purpose is to be evaluated during detection of complex events in order to *enrich* recorded events with background information; to detect more complex *situations*; to propose certain intelligent *recommendations* in real-time; or to accomplish complex event *classification, clustering, filtering* and so forth. CEP, integrated with real-time knowledge evaluation, can potentially enable a new generation of programmers to innovate on novel event-driven applications in Artificial Intelligence (AI).

In this chapter, we propose a new event processing approach in which complex events are *deduced* or derived from simpler events. Complex events are defined as *deductive rules*, and events are represented as *facts*. Every time an atomic event (relevant with respect to the set of monitored complex events) occurs, the system updates the knowledge base, i.e., it adds a respective fact to the internal state of complex events. Essentially, this internal state encodes what atomic events have already happened and what are still missing for the completion of a certain complex event. Complex events are detected as soon as the last event required for their detection has occurred. Descriptions telling which occurrence of an event furthers the

---

[1] Apart from this task (also known as pattern matching), CEP further addresses other issues like event filtering, routing, transformation and so forth.

detection of complex events (including the relationships between complex events, events they consist of, or additional domain knowledge) are given by deductive rules and facts. Consequently, detection of complex events then amounts to an *inferencing* problem.

The contribution of this chapter consists of a novel event-driven approach for Complex Event Processing based on logic programming. We define an expressive complex event description language with a rule-based syntax and a clear declarative formal semantics. The language is found on a new execution model that compiles complex event patterns into logic rules and enables timely, event-driven detection of complex events. We also describe an implementation of the language, and show experimental results of our evaluation with respect to a non logic programming CEP system.

The chapter is organized as follows. Section 2 specifies the problem that we address in this chapter. In Section 3, we introduce a new language for event processing and define its syntax. Section 4 defines the declarative semantics of the language. We discuss implementation details of our formalism, and give performance results of a prototype implementation in Section 7. Section 8 reviews existing work in this area, and compares it to ours. Finally, Section 9 summarizes the chapter, and give an outline of the future work.

2 Problem Statement

The general task of Complex Event Processing can be described as follows. Within some dynamic setting, events take place. Those *atomic events* are instantaneous (i.e., they happen at one specific point in time and have a duration of zero). Notifications about these event occurrences together with their timestamps and possibly further associated data (such as involved entities, numerical parameters of the event, or provenance data) enter the CEP system in the order of their occurrence.[2]

The CEP system further features a set of *complex event descriptions*, by means of which *complex events* can be specified as temporal constellations of atomic events. The complex events thus defined can in turn be used to compose even more complex events and so forth. As opposed to atomic events, those complex events are not considered instantaneous but are endowed with a time *interval* denoting when the event started and when it ended.

The purpose of the CEP system is now to detect complex events within this input stream of atomic events. That is, the system needs to notify that the occurrence of a certain complex event has been detected, as soon as the system is notified of an atomic event that completes a sequence which makes up the complex event according to the complex event description. This notification may be accompanied by additional information composed from the atomic events' data. As a consequence of this detection (and depending on the associated data), corresponding actions can be taken, yet this is outside the scope of this chapter.

---

[2] The phenomenon of *out-of-order events* meaning delayed notification about events that have happened earlier, is outside the focus of this chapter.

In summary, the problem we address in our approach is to detect complex events (specified in an appropriate formal language) within a stream of atomic events. Therefore we assume that the timeliness of this detection is crucial and algorithmically optimize our method towards a fast response behavior.

3 Syntax

In this section we present the formal syntax of the *ETALIS Language for Events*, while in the remaining sections of the chapter we will gradually introduce other aspects of the language (i.e., the declarative and operational semantics as well as the performance of a prototype based on the language[3]).

The syntax of ETALIS Language for Events allows for the description of *time* and *events*. We represent time instants as well as durations as nonnegative rational numbers $q \in \mathbb{Q}^+$. Events can be atomic or complex. An *atomic event* refers to an instantaneous occurrence of interest. Atomic events are expressed as ground atoms (i.e., predicates followed by arguments which are terms not containing variables). Intuitively, the arguments of a ground atom describing an atomic event denote information items (i.e., event data) that provide additional information about that event.

Atomic events can be composed to form *complex events* via *event patterns*. We use event patterns to describe how events can (or have to) be temporally compared to other events or absolute time points. The language P of event patterns is formally defined by

$$P ::= \text{PR}(t_1,\ldots,t_n) \mid P \text{ WHERE } t \mid q \mid (P).q$$
$$\mid P \text{ BIN } P \mid \text{NOT}(P).[P,P]$$

Here PR is a predicate name with arity n, t_i denote terms, t is a term of type boolean, q is a nonnegative rational number, and BIN is one of the binary operators SEQ, AND, PAR, OR, DURING, MEETS, EQUALS, STARTS, or FINISHES. As a side condition, in every expression P WHERE t, all variables occurring in t must also occur in the pattern P.

Finally, an *event rule* is defined as a formula of the form $\text{PR}(t_1,\ldots,t_n) \leftarrow P$ where P is an event pattern containing all variables occurring in $\text{PR}(t_1,\ldots,t_n)$.

After introducing the formal syntax of our formalism, we will give some examples to provide some intuitive understanding before proceeding with the formal semantics in the next section. According to a stock market scenario, one instantaneous event (not requiring further specification) might be Market_closes(). Other events with additional information associated via arguments would be Bankrupt(*lehman*) or Buys(*citigroup, wachovia*). Within patterns, variables instead of constants may occur as arguments, whence we can write Bankrupt(X) as a pattern matching all bankruptcy events irrespective of the victim.

---

[3] Our prototype, ETALIS, is an open source project, available at:
http://code.google.com/p/etalis

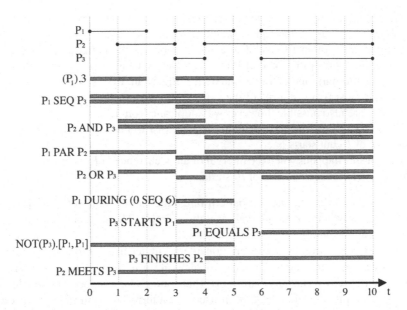

Fig. 1 Language for Event Processing - Composition Operators

Figure 1 demonstrates the various ways of constructing complex event descriptions from simpler ones in ETALIS Language for Events. Moreover, the figure informally introduces the semantics of the language, which will further be defined in Section 4.

Let us assume that instances of three complex events, P_1, P_2, P_3, are occurring in time intervals as shown in Figure 1. Vertical dashed lines depict different time units, while the horizontal bars represent detected complex events for the given patterns. In the following, we give the intuitive meaning for all patterns from the figure:

- $(P_1).3$ detects an occurrence of P_1 if it happens within an interval of length 3, i.e., 3 represents the (maximum) time window.
- P_1 SEQ P_3 represents a sequence of two events, i.e., an occurrence of P_1 is followed by an occurrence of P_3; here P_1 must end before P_3 starts.
- P_2 AND P_3 is a pattern that is detected when instances of both P_2 and P_3 occur no matter in which order.
- P_1 PAR P_2 occurs when instances of both P_2 and P_3 happen, provided that their intervals have a non-zero overlap.
- P_2 OR P_3 is triggered for every instance of P_2 or P_3.
- P_1 DURING $(0$ SEQ $6)$ happens when an instance of P_1 occurs during an interval; in this case, the interval is built using a sequence of two atomic time-point events (one with $q = 0$ and another with $q = 6$, see the syntax above).
- P_3 STARTS P_1 is detected when an instance of P_3 starts at the same time as an instance of P_1 but ends earlier.
- P_1 EQUALS P_3 is triggered when the two events occur exactly at the same time interval.

- NOT$(P_3).[P_1,P_1]$ represents a negated pattern. It is defined by a sequence of events (delimiting events) in the square brackets where there is no occurrence of P_3 in the interval. In order to invalidate an occurrence of the pattern, an instance of P_3 must happen in the interval formed by the end time of the first delimiting event and the start time of the second delimiting event. In this example delimiting events are just two instances of the same event, i.e., P_1. Different treatments of negation are also possible, however we adopt one from [2].
- P_3 FINISHES P_2 is detected when an instance of P_3 ends at the same time as an instance of P_2 but starts later.
- P_2 MEETS P_3 happens when the interval of an occurrence of P_2 ends exactly when the interval of an occurrence of P_3 starts.

It is worth noting that the defined pattern language captures the set of all possible 13 relations on two temporal intervals as defined in [4]. The set can also be used for rich temporal reasoning.

In this example, event patterns are considered under the *unrestricted policy*. In event processing, consumption policies deal with an issue of *selecting* particular events occurrences when there are more than one event instance applicable and *consuming* events after they have been used in patterns. We will discuss different consumption policies and their implementation in ETALIS Language for Events in Section 6.

It is worthwhile to briefly consider the modeling capabilities of the presented pattern language. For example, one might be interested in defining an event matching stock market working days:

```
WorkingDay() ←
  NOT(MarketCloses())[MarketOpens(),MarketCloses()].
```

Moreover, we might be interested in detecting the event of two bankruptcies happening on the same market working day:

```
DieTogether(X,Y) ←
  (Bankrupt(X) SEQ Bankrupt(Y)) DURING WorkingDay().
```

This event rule also shows, how event information (about involved institutions, provenance, and so forth) can be "passed" on to the defined complex events by using variables. Furthermore, variables may be employed to conditionally group events into complex ones if they refer to the same entity:

```
IndirectlyAcquires(X,Y) ← Buys(Z,Y) AND Buys(X,Z)
```

Even more elaborate constraints can be put on the applicability of a pattern by endowing it with a boolean type term as filter.[4] Thus, we can detect a stock prize increase of at least 50% in a time frame of 7 days.

```
RemarkableIncrease(X) ←
  (Price(X,Y₁) SEQ Price(X,Y₂)).7 WHERE Y₂ > Y₁ · 1.5
```

---

[4] Note that also comparison operators like $=,<$ and $>$ can be seen as boolean-typed binary functions and, hence, fit well into the framework.

4 Declarative Semantics

We define the declarative formal semantics of ETALIS Language for Events in a model-theoretic way.

Note that we assume a fixed interpretation of the occurring function symbols, i.e., for every function symbol f of arity n, we presume a predefined function $f^* : Con^n \to Con$. That is, in our setting, functions are treated as built-in utilities.

As usual, a *variable assignment* is a mapping $\mu : Var \to Con$ assigning a value to every variable. We let μ^* denote the extension of μ to terms defined in the usual way:

$$\mu^* : \begin{cases} v \mapsto \mu(v) & \text{if } v \in Var, \\ c \mapsto c & \text{if } c \in Con, \\ f(t_1,\ldots,t_n) \mapsto f^*(\mu^*(t_1),\ldots,\mu^*(t_n)) & \text{otherwise.} \end{cases}$$

In addition to the set of rules $\mathscr{R}$, we define an *event stream*. The event stream is formalized as a mapping $\varepsilon : Ground \to 2^{\mathbb{Q}^+}$ from ground predicates into sets of non-negative rational numbers. It thereby indicates at what time instants what elementary events occur. As a side condition, we require ε to be free of accumulation points, i.e., for every $q \in \mathbb{Q}^+$, the set $\{q' \in \mathbb{Q}^+ \mid q' < q \text{ and } q' \in \varepsilon(g) \text{ for some } g \in Ground\}$ is finite.

Now, we define an interpretation $\mathscr{I} : Ground \to 2^{\mathbb{Q}^+ \times \mathbb{Q}^+}$ as a mapping from the ground atoms to sets of pairs of nonnegative rationals, such that $q_1 \leq q_2$ for every $\langle q_1, q_2 \rangle \in \mathscr{I}(g)$ for all $g \in Ground$.

Given an event stream ε, an interpretation $\mathscr{I}$ is called a *model* for a rule set $\mathscr{R}$ – written as $\mathscr{I} \models_\varepsilon \mathscr{R}$ – if the following conditions are satisfied:

C1 $\langle q, q \rangle \in \mathscr{I}(g)$ for every $q \in \mathbb{Q}^+$ and $g \in Ground$ with $q \in \varepsilon(g)$

C2 for every rule *Atom* $\leftarrow$ *Pattern* and every variable assignment μ we have $\mathscr{I}_\mu(Atom) \subseteq \mathscr{I}_\mu(Pattern)$ where $\mathscr{I}_\mu$ is inductively defined as displayed in Fig. 2.

Given an interpretation $\mathscr{I}$ and some $q \in \mathbb{Q}^+$, we let $\mathscr{I}|_q$ denote the interpretation defined by $\mathscr{I}|_q(g) = \mathscr{I}(g) \cap \{\langle q1, q2 \rangle \mid q2 - q1 \leq q\}$.

Given two interpretations $\mathscr{I}$ and $\mathscr{J}$, we say that $\mathscr{I}$ is *preferred* to $\mathscr{J}$ if there exists a $q \in \mathbb{Q}^+$ with $\mathscr{I}|_q \subset \mathscr{J}|_q$. In words, a model is preferred to another model if it contains less events of duration $\leq q$.

A model $\mathscr{I}$ is called *minimal* if there is no other model preferred to $\mathscr{I}$. It is easy to show that for every event stream ε and rule set $\mathscr{R}$ there is a unique minimal model $\mathscr{I}^{\varepsilon,\mathscr{R}}$. Essentially, this model can be obtained by starting from ε and applying the rules in ascending order w.r.t. the duration of the event generated by the rule.[5]

---

[5] Note that we deviate from the standard minimal-model semantics for Horn logic in order to properly handle negation introduced via NOT.

pattern	$\mathscr{I}_\mu(\text{pattern})$
$\text{PR}(t_1,\dots,t_n)$	$\mathscr{I}(\text{PR}(\mu^*(t_1),\dots,\mu^*(t_n)))$
P WHERE t	$\mathscr{I}_\mu(P)$ if $\mu^*(t) = true$ $\emptyset$ otherwise.
q	$\{\langle q,q\rangle\}$ for all $q \in \mathbb{Q}^+$
$(P).q$	$\mathscr{I}_\mu(P) \cap \{\langle q_1,q_2\rangle \mid q_2 - q_1 = q\}$
$P1$ SEQ $P2$	$\{\langle q_1,q_4\rangle \mid \langle q_1,q_2\rangle \in \mathscr{I}_\mu(P1)$ and $\langle q_3,q_4\rangle \in \mathscr{I}_\mu(P2)$ and $q_2 < q_3\}$
$P1$ AND $P2$	$\{\langle \min(q_1,q_3),\max(q_2,q_4)\rangle \mid \langle q_1,q_2\rangle \in \mathscr{I}_\mu(P1)$ and $\langle q_3,q_4\rangle \in \mathscr{I}_\mu(P2)\}$
$P1$ PAR $P2$	$\{\langle \min(q_1,q_3),\max(q_2,q_4)\rangle \mid \langle q_1,q_2\rangle \in \mathscr{I}_\mu(P1)$ and $\langle q_3,q_4\rangle \in \mathscr{I}_\mu(P2)$ and $\max(q_1,q_3) < \min(q_2,q_4)\}$
$P1$ OR $P2$	$\mathscr{I}_\mu(P1) \cup \mathscr{I}_\mu(P2)$
$P1$ EQUALS $P2$	$\mathscr{I}_\mu(P1) \cap \mathscr{I}_\mu(P2)$
$P1$ MEETS $P2$	$\{\langle q_1,q_3\rangle \mid \langle q_1,q_2\rangle \in \mathscr{I}_\mu(P1)$ and $\langle q_2,q_3\rangle \in \mathscr{I}_\mu(P2)\}$
$P1$ DURING $P2$	$\{\langle q_3,q_4\rangle \mid \langle q_1,q_2\rangle \in \mathscr{I}_\mu(P1)$ and $\langle q_3,q_4\rangle \in \mathscr{I}_\mu(P2)$ and $q_3 < q_1 < q_2 < q_4\}$
$P1$ STARTS $P2$	$\{\langle q_1,q_3\rangle \mid \langle q_1,q_2\rangle \in \mathscr{I}_\mu(P1)$ and $\langle q_1,q_3\rangle \in \mathscr{I}_\mu(P2)$ and $q_2 < q_3\}$
$P1$ FINISHES $P2$	$\{\langle q_1,q_3\rangle \mid \langle q_2,q_3\rangle \in \mathscr{I}_\mu(P1)$ and $\langle q_1,q_3\rangle \in \mathscr{I}_\mu(P2)$ and $q_1 < q_2\}$
NOT$(P1).[P2,P3]$	$\mathscr{I}_\mu(P2\ \text{SEQ}\ P3) \setminus \mathscr{I}_\mu(P2\ \text{SEQ}\ P1\ \text{SEQ}\ P3)$

Fig. 2 Definition of extensional interpretation of event patterns. We use $P(x)$ for patterns, $q_{(x)}$ for rational numbers, $t_{(x)}$ for terms and PR for event predicates.

Finally, given an atom A and two rational numbers q_1,q_2, we say that the event $A^{[q_1,q_2]}$ is a *consequence* of the event stream ε and the rule base $\mathscr{R}$ (written $\varepsilon,\mathscr{R} \models A^{[q_1,q_2]}$), if $\langle q_1,q_2\rangle \in \mathscr{I}_\mu^{\varepsilon,\mathscr{R}}(A)$ for some variable assignment μ.

It can be easily verified that the behavior of the event stream ε beyond the time point q_2 is irrelevant for determining whether $\varepsilon,\mathscr{R} \models A^{[q_1,q_2]}$ is the case.[6] This justifies taking the perspective of ε being only partially known (and continuously unveiled along a time line) while the task is to detect event-consequences as soon as possible.

5 Operational Semantics

In Section 4 we have defined complex event patterns formally. This section describes how complex events, described in ETALIS Language for Events, can be detected at run-time (following the semantics of the language). Our approach is established on *goal-directed, event-driven* rules and decomposition of complex event patterns into *two-input intermediate events* (i.e., *goals*). Goals are automatically asserted by rules as relevant events occur. They can persist over a period of time "waiting" to support detection of a more complex goal. This process of asserting more and

---

[6] More formally, for any two event streams ε_1 and ε_2 coinciding up to timepoint q_2 (i.e., $\varepsilon_1(g) \cap \{\langle q,q'\rangle \mid q' \le q_2\} = \varepsilon_2(g) \cap \{\langle q,q'\rangle \mid q' \le q_2\}$ for all $g \in Ground$) we have that $\varepsilon_1,\mathscr{R} \models A^{[q_1,q_2]}$ exactly if $\varepsilon_2,\mathscr{R} \models A^{[q_1,q_2]}$.

more complex goals shows the progress towards detection of a complex event. In the following subsection, we give more details about a *goal-directed, event-driven* mechanism with respect to event pattern operators (formally defined in Section 4).

Sequence of Events. Let us consider a sequence of events represented by rule (1), i.e., E is detected when an event A^7. is followed by B, and followed by C. We can always represent the above pattern as $E \leftarrow ((A \text{ SEQ } B) \text{ SEQ } C)$. In general, rules (2) represent two equivalent rules.[8]

$$E \leftarrow A \text{ SEQ } B \text{ SEQ } C. \tag{1}$$

$$E \leftarrow P1 \text{ BIN } P2 \text{ BIN } P3... \text{ BIN } Pn.$$
$$E \leftarrow (((P1 \text{ BIN } P2) \text{ BIN } P3)... \text{ BIN } Pn). \tag{2}$$

We refer to this kind of "events coupling" as *binarization* of events. Effectively, in binarization we introduce *two-input* intermediate events (goals). For example, now we can rewrite rule (1) as $IE_1 \leftarrow A \text{ SEQ } B$, and the $E \leftarrow IE_1 \text{ SEQ } C$. Every monitored event (either atomic or complex), including intermediate events, will be assigned with one or more *logic rules*, fired whenever that event occurs. Using the binarization, it is more convenient to construct *event-driven* rules for three reasons. First, it is easier to implement an event operator when events are considered on a "two by two" basis. Second, the binarization increases the possibility for *sharing* among events and intermediate events, when the granularity of intermediate patterns is reduced. Third, the binarization eases the *management* of rules. As we will see later in this section, each new use of an event (in a pattern) amounts to appending one or more rules to the existing rule set. However it is important that for the management of rules, we do not need to *modify* existing rules when adding new ones[9].

In the following, we give more details about assigning rules to each monitored event. We also provide an algorithm (using Prolog syntax) for detecting a sequence of events.

Algorithm 5.1 accepts as input a rule referring to a binary sequence $IE_i \leftarrow A \text{ SEQ } B$, and produces Event-Driven Backward Chaining Rules (EDBCR), i.e., executable rules for the sequence pattern. The binarization step must precede the rule transformation. Rules, produced by Algorithm 5.1, belong to one of two different classes of rules. We refer to the first class as *goal inserting rules*. The second class corresponds to *checking rules*. For example, rule (4) belongs to the first class as it inserts $goal(B(\_,\_),A(T_1,T_2),IE_1(\_,\_)$. The rule will fire when A occurs, and the meaning of the goal it inserts is as follows: "an event A has occurred

---

[7] More precisely, by "an event A" is meant an *instance* of the event A

[8] If no parentheses are given, we assume all operators to be left-associative. While in some cases, like SEQ sequences, this is irrelevant, other operators such as PAR are not associative, whence the precedence matters.

[9] This holds even if patterns with negated events are added.

at $[T_1, T_2]$,[10] and we are waiting for B to happen in order to detect IE_1". The goal does not carry information about times for B and IE_1, as we do not know when they will occur. In general, the *second* event in a goal always denotes the event that has just occurred. The role of the *first* event is to specify what we are waiting for to detect an event that is on the *third* position.

Algorithm 5.1 Sequence.
Input: event binary goal $IE_1 \leftarrow A$ SEQ B.
Output: event-driven backward chaining rules for SEQ operator.
Each event binary goal $IE_1 \leftarrow A$ SEQ B is converted into: {
$\quad A(T_1, T_2) : -for\_each(A, 1, [T_1, T_2])$.
$\quad A(1, T_1, T_2) : -assert(goal(B(\_, \_), A(T_1, T_2), IE_1(\_, \_)))$.
$\quad B(T_3, T_4) : -for\_each(B, 1, [T_3, T_4])$.
$\quad B(1, T_3, T_4) : -goal(B(T_3, T_4), A(T_1, T_2), IE_1), T_2 < T_3,$
$\quad\quad retract(goal(B(T_3, T_4), A(T_1, T_2), IE_1(\_, \_))), IE_1(T_1, T_4)$.
}

$$for\_each(Pred, N, L) : -((FullPred = ..[Pred, N, L]),$$
$$event\_trigger(FullPred), (N1 is N + 1), for\_each(Pred, N1, L)) \lor true. \quad (3)$$

$$A(1, T_1, T_2) : -assert(goal(B(\_, \_), A(T_1, T_2), IE_1(\_, \_))). \quad (4)$$

$$B(1, T_3, T_4) : -goal(B(T_3, T_4), A(T_1, T_2), IE_1), T_2 < T_3,$$
$$retract(goal(B(T_3, T_4), A(T_1, T_2), IE_1(\_, \_))), IE_1(T_1, T_4). \quad (5)$$

Rule (5) belongs to the second class being a *checking rule*. It checks whether certain prerequisite goals already exist in the database, in which case it triggers the more complex event. For example, rule (5) will fire whenever B occurs. The rule checks whether $goal(B(T_3, T_4), A(T_1, T_2), IE_1)$ already exists (i.e., A has previously happened), in which case the rule triggers IE_1 by calling $IE_1(T_1, T_4)$. The time occurrence of IE_1 (i.e., T_1, T_4) is defined based on the occurrence of constituting events (i.e., $A(T_1, T_2)$, and $B(T_3, T_4)$, see Section 4). Calling $IE_1(T_1, T_4)$, this event is effectively propagated either upward (if it is an intermediate event) or triggered as a finished complex event.

We see that our *backward* chaining rules compute goals in a *forward* chaining manner. The goals are crucial for computation of complex events. They show the current state of progress toward matching an event pattern. Moreover, they allow for determining the "completion state" of any complex event, at any time. For instance, we can query the current state and get information how much of a certain pattern is currently fulfilled (e.g., what is the current status of certain pattern, or notify me if the pattern is 90% completed). Further, goals can enable *reasoning* over events (e.g., answering which event occurred before some other event, although we do not

---

[10] Apart from the timestamp, an event may carry other data parameters. They are omitted here for the sake of readability.

know a priori what are explicit relationships between these two; correlating complex events to each other; establishing more complex constraints between them and so forth). Goals can persist over a period of time. It is worth noting that *checking rules* can also delete goals. Once a goal is "consumed", it is removed from the database[11]. In this way, goals are kept persistent as long as (but not longer) than needed. In Section 6, we will return to different policies for removing goals from the database.

Finally, in Algorithm 5.1 there exist more rules than the two mentioned types (i.e., rules inserting goals and checking rules). We see that for each different event type (i.e., A and B in our case) we have one rule with a *for_each* predicate. It is defined by rule (3). Effectively, it implements a loop, which for any occurrence of an event goes through each rule specified for that event (predicate) and fires it. For example, when A occurs, the first rule in the set of rules from Algorithm 5.1 will fire. This first rule will then loop, invoking all other rules specified for A (those having A in the rule head). In our case, there is only one such a rule, namely rule (4). However, in general, there may be as many of these rules as usages of a particular event in an event program. Let us observe a situation in which we want to extend our event pattern set with an additional pattern that contains the event A (i.e., additional usage of A). In this case, the rule set representing a set of event patterns needs to be updated with new rules. This can be done even at runtime. Let us assume the additional pattern to be monitored is $IE_j \leftarrow K$ SEQ A. Then the only change we need to make is to add one rule to insert a goal and one checking rule (in the existing rule set). The change is sketched as an update of Algorithm 5.1 below[12].

Updating rules from Algorithm 5.1 to accommodate an additional usage of the event A.

$A(2,T_1,T_2) : -assert(goal(B(\_,\_),A(T_1,T_2),IE_1(\_,\_)))$.
$A(3,T_1,T_2) : -goal(A(\_,\_),K(T_3,T_4),IE_j(\_,\_)),T_4 < T_1,$
$\quad retract(goal(A(\_,\_),K(T_3,T_4),IE_j(\_,\_))),IE_j(T_3,T_2)$.

So far, we have described in detail a mechanism for event processing with *data or event-driven backward chaining rules* (EDBCR). We have also described the transformation of event pattern rules into rules for real-time events detection using the *sequence* operator. In general, for a given set of rules (defining complex patterns) there will be as many transformed rules as there are usages of distinct atomic events. Some rules however may be *shared* among different patterns. As said, the binarization brakes up patterns into binary sub-patterns (intermediate events). If two or more patterns share the same sub-patterns, they will also share the same set of EDBCR. That is, during the transformation, only one set of EDBCR will be produced for a distinct event binary goal (no matter how many times the goal is used in the whole event program). In large programs (e.g., where event patterns are built incrementally, i.e., one pattern upon another one) such a sharing may improve the overall system performance as the execution of redundant rules is avoided.

---

[11] Removal of "consumed" goals is typically needed for space reasons but might be omitted if events are required in a log for further processing or analyzing.

[12] Note that an *id* of rules is incremented for each next rule being added (i.e., 2,3...).

The set of transformed rules is further accompanied by rules to implement loops (as many as there are distinct atomic events). The same procedure is repeated for intermediate events (for example, IE_1, IE_2). The complete transformation is proportional to the number and length of user defined event pattern rules, hence such a transformation is linear, and moreover is performed at design time.

Conceptually, our backward chaining rules for the sequence operator look very similar to rules for other operators. In the remaining part of this section we show the algorithms for other event operators, and briefly describe them.

Conjunction of Events. Conjunction is another typical operator in event processing. An event pattern based on conjunction occurs when all events which comprise that conjunction occur. Unlike the sequence operator, here the constitutive events can happen at times with no particular order between them. For example, $IE_1 \leftarrow A$ AND B defines IE_1 as conjunction of A and B.

Algorithm 5.2 Conjunction.
Input: event binary goal $IE_1 \leftarrow A$ AND B.
Output: event-driven backward chaining rules for AND operator.
Each event binary goal $IE_1 \leftarrow A$ AND B is converted into: {

$A(T_1, T_2) : -for\_each(A, 1, [T_1, T_2])$.
$A(1, T_3, T_4) : -goal(A(\_,\_), B(T_1, T_2), IE_1(\_,\_)), retract(goal(A(\_,\_), B(T_1, T_2), IE_1(\_,\_)))$,
$T_5 = min\{T_1, T_3\}, T_6 = max\{T_2, T_4\}, IE_1(T_5, T_6)$.
$A(2, T_1, T_2) : -\neg(goal(A(\_,\_), B(T_1, T_2), IE_1(\_,\_))), assert(goal(B(\_,\_), A(T_1, T_2), IE_1(\_,\_)))$.
$B(T_3, T_4) : -for\_each(B, 1, [T_3, T_4])$.
$B(1, T_3, T_4) : -goal(B(\_,\_), A(T_1, T_2), IE_1(\_,\_)), retract(goal(B(\_,\_),$
$A(T_1, T_2), IE_1(\_,\_))), T_5 = min\{T_1, T_3\}, T_6 = max\{T_2, T_4\}, IE_1(T_5, T_6)$.
$B(2, T_1, T_2) : -\neg(goal(B(\_,\_), A(T_1, T_2), IE_1(\_,\_))), assert(goal(A(\_,\_), B(T_1, T_2), IE_1(\_,\_)))$.
}

Algorithm 5.2 shows the output of a transformation of *conjunction* event patterns into EDBCR (for conjunction). The procedure for dividing complex event rules into *binary event goals* is the same as in Algorithm 5.1. However, rules for *inserting* and *checking* goals are different. Both classes of rules are specific to conjunction. We have a pair of these rules created for both an event A as well as for B. Whenever A occurs (denoted as some interval (T_1, T_2)) the algorithm checks whether an instance of B has already happened (see rule (6) from Algorithm 5.2). An instance of B has already happened if the current database state contains $goal(A(\_,\_), B(T_1, T_2), IE_1(\_,\_))$. In this case the event $IE_1(T_5, T_6)$ is triggered (i.e., a call for $IE_1(T_5, T_6)$ is issued). Otherwise, a goal which states that an instance of A has occurred, is inserted (i.e., $assert(goal(B(\_,\_), A(T_1, T_2), IE_1(\_,\_)))$ is executed by rule (7)). Now if an instance of B happens later (at some $(T_3, T4)$), rule (8) will succeed (if A has previously happened). Otherwise rule (9) will insert $goal(A(\_,\_), B(T_1, T_2), IE_1(\_,\_))$.

$$A(1, T_3, T_4) : -goal(A(\_,\_), B(T_1, T_2), IE_1(\_,\_)), retract(goal(A(\_,\_),$$
$$B(T_1, T_2), IE_1(\_,\_))), T_5 = min\{T_1, T_3\}, T_6 = max\{T_2, T_4\}, IE_1(T_5, T_6). \quad (6)$$

$$A(2,T_1,T_2) : -\neg(goal(A(\_,\_),B(T_1,T_2),IE_1(\_,\_))),$$
$$assert(goal(B(\_,\_),A(T_1,T_2),IE_1(\_,\_))). \tag{7}$$

$$B(1,T_3,T_4) : -goal(B(\_,\_),A(T_1,T_2),IE_1(\_,\_)),retract(goal(B(\_,\_),$$
$$A(T_1,T_2),IE_1(\_,\_))),T_5 = min\{T_1,T_3\},T_6 = max\{T_2,T_4\},IE_1(T_5,T_6). \tag{8}$$

$$B(2,T_1,T_2) : -\neg(goal(B(\_,\_),A(T_1,T_2),IE_1(\_,\_))),$$
$$assert(goal(A(\_,\_),B(T_1,T_2),IE_1(\_,\_))). \tag{9}$$

Concurrency. A concurrent or parallel composition of two events ($IE_1 \leftarrow A$ PAR B) is detected when events A and B both occur, and their intervals overlap (i.e., we also say they happen *synchronously*).

Algorithm 5.3 shows what is an output of automated transformation of a *concurrent* event pattern into rules which serve a *data-driven backward chaining* event computation. The procedure for dividing complex event rules into *binary event goals* is the same (as already described), and takes place prior to the transformation. Rules for *inserting* and *checking* goals are similar to those in Algorithm 5.2. The only change in Algorithm 5.3 is a *sufficient* condition, ensuring the interval overlap (i.e., $T_3 < T_2$).

Algorithm 5.3 Concurrency.
Input: event binary goal $IE_1 \leftarrow A$ PAR B.
Output: event-driven backward chaining rules for PAR operator.
Each event binary goal $IE_1 \leftarrow A$ PAR B is converted into: {
$A(T_3,T_4) : -for\_each(A,1,[T_3,T_4])$.
$A(1,T_3,T_4) : -goal(A(\_,\_),B(T_1,T_2),IE_1(\_,\_)),T_3 < T_2,retract(goal(A(\_,\_),B(T_1,T_2),IE_1(\_,\_))),$
$T_5 = min\{T_1,T_3\},T_6 = max\{T_2,T_4\},IE_1(T_5,T_6)$.
$A(2,T_3,T_4) : -\neg(goal(A(\_,\_),B(T_1,T_2),IE_1(\_,\_))),T_3 < T_2,$
$assert(goal(B(\_,\_),A(T_3,T_4),IE_1(\_,\_)))$.
$B(T_3,T_4) : -for\_each(B,1,[T_3,T_4])$.
$B(1,T_3,T_4) : -goal(B(\_,\_),A(T_1,T_2),IE_1(\_,\_)),T_3 < T_2,retract(goal(B(\_,\_),A(T_1,T_2),IE_1(\_,\_))),$
$T_5 = min\{T_1,T_3\},T_6 = max\{T_2,T_4\},IE_1(T_5,T_6)$.
$B(2,T_3,T_4) : -\neg(goal(B(\_,\_),A(T_1,T_2),IE_1(\_,\_))),T_3 < T_2,$
$assert(goal(A(\_,\_),B(T_3,T_4),IE_1(\_,\_)))$.
}

Disjunction. An algorithm for detecting *disjunction* (i.e., OR) of events is trivial. The disjunction operator divides rules into separate disjuncts, where each disjunct triggers the parent (complex) event. Therefore we omit presentation of the algorithm here, but later in Section 7 we present experimental results also using an implementation of this operator.

Negation. Negation in event processing is typically understood as *absence* of an event that is negated. In order to create a time interval in which we are interested to detect absence of an event, we define a negated event in the scope of other complex events. Algorithm 5.4 describes how to handle negation in the scope of a sequence. It is also possible to detect negation in an arbitrarily defined time interval.

Algorithm 5.4 Negation.
Input: event pattern $IE_1 \leftarrow \text{NOT}(C).[A,B]$.
Output: event-driven backward chaining rules for negation.
Each event binary goal $IE_1 \leftarrow \text{NOT}(C).[A,B]$ is converted into: {
$\quad A(T_1,T_2) : -for\_each(A,1,[T_1,T_2])$.
$\quad A(1,T_1,T_2) : -assert(goal(B(\_,\_),A(T_1,T_2),IE_1(\_,\_)))$.
$\quad B(T_3,T_4) : -for\_each(B,1,[T_3,T_4])$.
$\quad B(1,T_5,T_6) : -goal(B(\_,\_),A(T_1,T_2),IE_1(\_,\_)), \neg(goal(\_,C(T_3,T_4),\_))$,
$\quad T_2 < T_5, T_2 < T_3, T_4 < T_5, retract(goal(B(\_,\_),A(T_1,T_2),IE_1(\_,\_))),IE_1(T_1,T_6)))$.
$\quad C(T_1,T_2) : -for\_each(C,1,[T_1,T_2])$.
$\quad C(1,T_1,T_2) : -assert(goal(\_,C(T_1,T_2),\_))$.
}

Rules for detection of negation are similar to rules from Algorithm 5.1. We need to detect a sequence (i.e., A SEQ B), and additionally to check whether an occurrence of C happened in-between the event A and B. That is why a rule $B(1,T_5,T_6)$ needs to check whether $\neg(goal(\_,C(T_3,T_4),\_))$ is true. If yes, this means that an C has not happened during a detected sequence (i.e., $A(T_1,T_2)$ SEQ $B(T_5,T_6)$), and $IE_1(T_1,T_6)$ will be triggered. It is worth noting that a non-occurrence of C is monitored from the time when A has been detected until the beginning of an interval which the event B is detected on.

In the following part of this section we provide brief descriptions for the remaining relations between two intervals. Each relation is easily checkable with one rule.

Duration. An event happens during (i.e., DURING) another event if the interval of the first is contained in the interval of the second. Rule (10) takes two intervals as parameters[13]. First, it checks whether all parameters are intervals using rule (11). Then it compares whether the start of the second interval (TI_2_S) is less than the start of the first interval (TI_1_S). Additionally it checks whether the end of the first interval (TI_1_E) is less than the end of the second interval (TI_2_E).

$$
\begin{aligned}
&duration(TI_1,TI_2) : - \\
&TI_1 = [TI_1\_S,TI_1\_E], validTimeInterval(TI_1), \\
&TI_2 = [TI_2\_S,TI_2\_E], validTimeInterval(TI_2), \\
&TI_2\_S @ < TI_1\_S, TI_1\_E @ < TI_2\_E.
\end{aligned}
\tag{10}
$$

$$
validTimeInterval(TI) : -TI = [TI\_S,TI\_E], TI\_S @ < TI\_E. \tag{11}
$$

Start Relation. We say that an event starts another if an instance of the first event starts at the same time as an instance of the second event, but ends earlier. Therefore

---

[13] Symbol '@' is used in Prolog built-in predicates ($>, <, \geq$ etc.) to compare terms alphabetically or numerically. When this symbol is omitted, terms are compared arithmetically.

rule (12) checks whether the start of both intervals are equal and whether the end of the first event is smaller than the end of the second one.

$$starts(TI_1, TI_2) : -$$
$$TI_1 = [TI_1\_S, TI_1\_E], validTimeInterval(TI_1),$$
$$TI_2 = [TI_2\_S, TI_2\_E], validTimeInterval(TI_2), \quad (12)$$
$$TI_1\_S = TI_2\_S, TI_1\_E @ < TI_2\_E.$$

Equal Relation. Two events are equal if they happen right at the same time. Rule (13) implements this relation.

$$equals(TI_1, TI_2) : -$$
$$TI_1 = [TI_1\_S, TI_1\_E], validTimeInterval(TI_1),$$
$$TI_2 = [TI_2\_S, TI_2\_E], validTimeInterval(TI_2), \quad (13)$$
$$TI_1\_S = TI_2\_S, TI_1\_E = TI_2\_E.$$

Finish Relation. One event finishes another one if an occurrence of the first ends at the same time as an occurrence of the second event, but starts later. Rule (14) check this condition.

$$finishes(TI_1, TI_2) : -$$
$$TI_1 = [TI_1\_S, TI_1\_E], validTimeInterval(TI_1),$$
$$TI_2 = [TI_2\_S, TI_2\_E], validTimeInterval(TI_2), \quad (14)$$
$$TI_2\_S @ < TI_1\_S, TI_1\_E = TI_2\_E.$$

Meet Relation. Two events meet each other when the interval of the first ends exactly when the interval of the second event starts. Hence, the condition $TI_1\_E = TI_2\_S$ in rule (15) is sufficient to detect this relation.

$$meets(TI_1, TI_2) : -$$
$$TI_1 = [TI_1\_S, TI_1\_E], validTimeInterval(TI_1),$$
$$TI_2 = [TI_2\_S, TI_2\_E], validTimeInterval(TI_2), \quad (15)$$
$$TI_1\_E = TI_2\_S.$$

6 Event Consumption Policies

When detecting a complex event, there may be several event occurrences (of the same type), that could be used to form that complex event. *Consumption policies* (or event contexts) deal with the issue of selecting particular occurrence(s), which will be used in the detection of a complex event. For example, let us consider rule (1) from Section 5, and a sequence of atomic events that happened in the following order: $A(1), A(2), A(3), B(4), B(5), C(6)$ (where an event attribute denotes a time point when an event instance has occurred). We expect that, when an event of type B occurs, an intermediate event IE_1 must be triggered. However, the question is, which occurrence of A will be selected to build that event, $A(1)$, $A(2)$ or $A(3)$? Different consumption policies define different strategies. Here, we illustrate three widely used consumption policies: *recent*, *chronological*, and *unrestricted* policy [8, 23], and show how they can be naturally implemented by rules in our framework.

6.1 Consumption Policies Defined on Time Points

In the above example, we assumed that the stream of events $A(1), A(2), A(3), B(4),$ $B(5), C(6)$ contains only atomic events.

Recent Policy. The most recent event of its type is considered to construct complex events. In our example, when $B(4)$ occurs, $A(3)$ will be selected to compose $IE_1(3,4)$. After a more recent occurrence $B(5)$ occurs, older (which are less recent) occurrences of B are deleted (i.e., they are no longer eligible for further compositions). The next pair, $A(3), B(5)$, is selected to form $IE_1(3,5)$. It replaces the less recent occurrence IE_1. Finally, when $C(6)$ occurs, it will trigger $E(3,6)$ (using $IE_1(3,5)$ as the more recent occurrence of IE_1 in comparison to $IE_1(3,4)$).

The recent policy can be easily implemented in our framework. Let us consider Algorithm 5.1, particularly the rule which inserts a goal (in our example, $goal(B, A, IE_1)$). Whenever an instance of A occurs, there will be a new goal inserted with a corresponding timestamp. E.g., for $A(1)$, $goal(B(\_), A(1), IE_1(\_,\_))$ is added; for $A(2)$, $goal(B(\_), A(2), IE_1(\_,\_))$, and so forth). If we insert these goals into the database using LIFO (Last In First Out) structure, we obtain the *recent policy*. In our prototype implementation, this is done with a rule of the following form:

$$assert(goal(X)) : -assertA(goal(X)). \tag{16}$$

asserta is a standard Prolog built-in that adds a term to the *beginning* of the database. Whenever a goal is inserted to the database, it is put on the top of a relation. Hence whenever we read a goal, the one inserted last will be returned.

Chronological Policy. This policy "consumes" the events in chronological order. In our example, this means that $A(1)$ and $B(4)$ will form $IE_1(1,4)$, and further $A(2)$ followed by $B(5)$ will trigger $IE_1(2,5)$. When $C(6)$ happens, it will trigger $E(1,6)$.

It is straightforward to implement the chronological policy, too. Now, the goals in Algorithm 5.1 are inserted in a FIFO (First In First Out) mode. Equivalently, we use the following rule to realize the chronological policy:

$$assert(goal(X)) : -assertz(goal(X)). \tag{17}$$

assertz is a standard Prolog built-in that adds a term to the *end* of the database. Whenever a goal is inserted to the database, it is put at the end of a relation. Consequently, whenever we read a goal, the first inserted goal will be returned first.

Unrestricted Policy. In this policy, all occurrences are valid. Consequently, no event is consumed (and no event is deleted), which makes this policy not suitable for practical use. Going back to our example, this implies that we detect the following instances of IE_1: $IE_1(1,4)$, $IE_1(2,4)$, $IE_1(3,4)$, $IE_1(1,5)$, $IE_1(2,5)$, $IE_1(3,5)$. The event E will be triggered just as many times, that is: $E(1,6)$, $E(2,6)$, $E(3,6)$...

We obtain the unrestricted policy simply by not using the construct for deleting goals (i.e., *retract*) from the database. If we replace the rule for $B(1)$ in Algorithm 5.1 with rule (18), even consumed goals will not be deleted from the database. Hence they will be available for future compositions.

$$B(1) : -goal(B,A,IE_1) \text{ SEQ } IE_1. \tag{18}$$

Consumption policies are an important part of an event processing framework. We notice that different policies change the semantics of event operators. For example, with the same operator we have detected different complex events (the recent policy detects $E(2,6)$, while the chronological policy detects $E(1,6)$).

6.2   Consumption Policies Defined on Time Intervals

We have so far discussed consumption policies assuming that we consider atomic events (in an input stream). As atomic events happen in time points, it is possible to establish a *total order* of their occurrences. Consequently it is easy to answer which event instance, out of two, happened more recently. When we deal with complex events ($T_1 \neq T_2$), a total order is not always possible. This subsection provides possible options in defining consumption policies in such a case.

Recent Policy. Let us consider the following sequence of input events: $A(1,30)$, $A(15,30)$, $B(35,50)$. In our example rule (1) (from Section 5), the question now is which instance of A is more *recent*, $A(1,30)$ or $A(15,30)$? In our opinion, this question depends on the particular application domain. There are three possible options. First, an event detected on a *longer event duration* is selected to be the recent one (i.e., $A(1,30)$). This option is suitable when aggregation functions (for example, sum, average and so forth) are applied along time windows. Hence, events detected on longer durations possibly reflect more accurate results. The second option is to choose an event with a *shorter duration* (i.e., $A(15,30)$). This preference is, for example, suitable when indeed more recent events are desired. For example, we are interested in data (carried by events) that are as up to date as possible. Finally, the third possibility is to pick up an event instance based on *data value selection* i.e., non-temporal properties. For instance, events ending at the same time, $A(1,30,X,Vol = 1000)$ and $A(15,30,X,Vol = 10000)$, are selected based on an attribute value (for example, greater $Volume$).

We implement these three cases with rules (19)-(21). When an A occurs, there is a policy check performed. In rule (19), for two events with the same ending (i.e., $A(T_1,T_3)$ and $A(T_2,T_3)$) we make sure that one with a longer path ($T_1 > T_2$) is selected. In rule (20), we replace goals if the time condition is opposite ($T_1 < T_2$). Finally, in data value (or attribute value) selection, we distinguish based on a chosen attribute (e.g., $Vol_1 > Vol_2$).

$event\_trigger(A(T_1, T_3, Vol_1)) : -$
$goal(\_, A(T_2, T_3, \_, \_), \_)$ SEQ $T_1 > T_2$ SEQ $assert(goal(\_, A(T3, T4, \_, \_), \_))$ (19)
SEQ $assert(goal(\_, A(T1, T3, Vol_1), \_))$.

$event\_trigger(A(T_1, T_3, Vol_1)) : -$
$goal(\_, A(T_2, T_3, \_, \_), \_)$ SEQ $T_1 < T_2$ SEQ $retract(goal(\_, A(T3, T4, \_, \_), \_))$ (20)
SEQ $assert(goal(\_, A(T1, T3, Vol_1), \_))$.

$event\_trigger(A(T_1, T_3, Vol_1)) : -$
$goal(\_, A(T_2, T_3, Vol_2), \_)$ SEQ $Vol_1 > Vol_2$ SEQ $retract(goal(\_, A(T3, T4, Vol_2), \_))$
SEQ $assert(goal(\_, A(T1, T3, Vol_1), \_))$.

$$(21)$$

Policy rules (19)-(21) are fired before inserting a new goal. It is worth noting that such an update of a goal is performed incrementally. We pay an additional price for forcing a particular consumption policy. However, the policy rules (19)-(21) are rather simple rules. In return, they ensure that no more than one goal with the same timestamp (with respect to certain policy) is kept in memory (during processing). Hence the rules enable a better *memory management* in our framework.

Chronological Policy. The main principle in the implementation of this policy is the same as in the recent policy. The only difference is that now we consider the same start and the different ending in multiple event occurrences $(A(T1, T2), A(T1, T3))$. To implement this policy, rule (19) will now contain the time condition from rule (20), and vice versa. Rule (21) remains unchanged, as well as *unrestricted policy* (which is the same as for the case with atomic events, see Subsection 6.1).

7 Implementation and Experimental Results

The execution model of ETALIS Language for Events is established on *goal-directed* EDBCR and decomposition of complex event patterns into *two-input intermediate events* (i.e., *goals*). Goals are automatically asserted by rules as relevant events occur. They can persist over a period of time "waiting" to support detection of a more complex goal. This process of asserting more and more complex goals shows the progress towards detection of a complex event. Important characteristics of these goals are that they are asserted only if they are used later on (to support a more complex goal or an event pattern), goals are all unique, and persist as long as they remain relevant (after that they can be deleted). Goals are asserted by rules which are executed in the backward chaining mode. The notable property of these rules is that they are event-driven. Hence, although the rules are executed backwards, overall they exhibit a forward chaining behavior.

As a proof of concept, we have provided a prototype implementation of the language. In this section, we present experimental results of our logic programming based implementation in comparison to Esper 3.3.0[14]. Esper is an engine primarily

---

[14] Esper: http://esper.codehaus.org

relying on state machines, i.e., a different paradigm that is today widely used in CEP systems [11, 3].

The test cases presented here were carried out on a workstation with Intel Core Quad CPU Q9400 2,66GHz, 8GB of RAM, running Windows Vista x64. Since our prototype automatically compiles the user-defined complex event descriptions into Prolog rules, we used SWI Prolog version 5.6.64[15] and YAP Prolog version 5.1.3[16]. All tested engines ran in a single dedicated CPU core.

To run tests, we have implemented an event stream generator, which creates time series data with probabilistic values. Event streams are generated so that every event in a stream is used for detection of one or more complex events (except the test defined by rule (22)). Such streams put the maximum workload on tested engines. The whole output generated from all tests is validated, so we have made sure that all tested systems produce the same, correct, results.

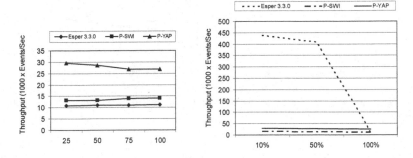

Fig. 3 Experiments for sequence operator - (a) Throughput (b) Throughput vs. Predicate Selectivity

$$C(Id, X, Y) : -A(Id, X) \text{ SEQ } B(Id, Y) \text{ WHERE } (Y < K). \qquad (22)$$

Figure 3 shows experimental results we obtained for the *sequence* operator (SEQ). In particular, Figure 3(a) shows the throughput measurements for a pattern that exhibits a sequence of three events and the join operation on their *Id* attribute, see rule (1) from Section 5. The Y-axis shows the event throughput achieved by the three different CEP systems: Esper 3.3.0, and our prototype (P) running on SWI and YAP Prolog, denoted as P-SWI and P-YAP respectively). The X-axis shows different sizes of event streams, used for detection of complex events, defined by rule (1). In this test, our system outperforms Esper. The throughput achieved by the YAP engine is more than twice as big as the one produced by Esper. Also comparing YAP and SWI, our implementation is significantly faster on YAP. This happens because YAP implements several optimizations to improve indexing. In Figure 3(b) we have

---

[15] SWI Prolog http://www.swi-prolog.org/.

[16] YAP Prolog: http://www.dcc.fc.up.pt/~vsc/Yap/. Our prototype ran by YAP was using Windows x32, as we could not find YAP version x64 available. Other two systems (Esper and SWI) were running on Windows x64.

evaluated the patterns which (apart from the join operation) also contain a selection parameter K (see rule (22)). K varies the selectivity of the Y attribute, ranging from 10% till 100%. When 10%-50% of the input events are selected, Esper shows significant advantage over our system. Hence in the future we need to review our implementation so to select events as early as possible. When all events are taken into account (100% selectivity), our system running on YAP is slightly better than Esper. We did this test on a stream of 25K events.

In Figure 4(a) we extended the tests (for 100% selectivity) to check out whether the system throughput will remain constant for bigger streams (for example, 50K-100K).

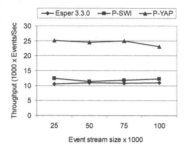 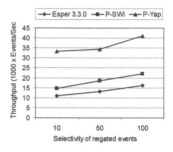

Fig. 4 (a) Sequence - Throughput vs. Workload Change (b) Negation - Throughput vs. Selectivity

Figure 4(b) presents experimental results for *negation* (NOT). The figure shows results obtained by evaluating a negated pattern from rule (23). The pattern is detected when an instance of A is followed by an occurrence of B with no C in between the two events. We have generated input event streams with different percentage of occurrences of events of type C (that is, 10%-100%). We see that our prototype (either run by SWI or YAP) dominates over Esper. We also notice that the throughput increases as the percentage of C occurrences increases. This is happening as the number of detected complex events decreases by increasing the frequency of occurrences of C. The test is computed on a stream of 25K.

$$D(Id,X,Y) : -\text{NOT}(C(Id,Z)).[A(Id,X),B(Id,Y)]. \tag{23}$$

Figure 5(a) shows that the throughput does not go down even though we increased the stream size (for example, 50K-100K).

We have tested the conjunction operator too. The pattern is specified by rule (24), and results are presented in Figure 5(b). Esper was faster in this test. Our algorithm for handling conjunction contains twice as many rules as the algorithm for sequence (that is, two events in a conjunct may occur in any order). As a future work, we will try to improve the implementation of conjunction by simplifying the event-driven backward chaining rules in this algorithm.

$$D(Id,X,Y) : -A(Id,X) \text{ AND } B(Id,Y) \text{ AND } C(Id,Z). \tag{24}$$

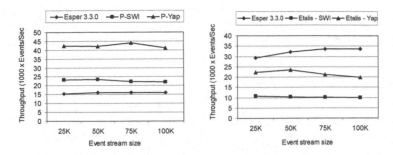

Fig. 5 (a) Negation - Throughput vs. Workload Change (b) Conjunction - Throughput.

Figure 6(a) shows results for *disjunction*, and evaluation of rule (25). In this test our system running on YAP was the most effective. The throughput for this test is similar to results for sequence (Figure 3(a)); this means that the presence of a disjunct does not affect the performance of the sequence. We have also tested computation of the transitive closure (see rule (26)). The throughput change for different sizes of event streams are presented in Figure 6(b). Evaluation results were obtained under chronological consumption policy. Our system on YAP was the fastest, however the difference between evaluations running on YAP and SWI was huge (as discussed earlier, due to better optimizations for indexing in YAP).

Finally, Figure 7 compares the tested systems with respect to event plan sharing. We have run an event program containing the same pattern (similar to rule (1) from Section 5) multiplying the pattern 1, 8, and 16 times. The focus was on examining how well the systems can exhibit computation sharing among patterns. In our prototype, we have implemented a plan sharing by decoupling events in a complex event pattern. A pattern is represented as a set of binary events, and each couple can be shared among multiply complex event patterns. Despite this feature, our system run by YAP was not resistant to increase of pattern rules. However our prototype executed on SWI was still faster than Esper, see Figure 7.

$$D(Id,X,Y) : -A(Id,X) \text{ SEQ } (B(Id,Y) \text{ OR } C(Id,Y)). \quad (25)$$

$$TC(X,Y) : -A(X,Y).$$
$$TC(X,Y) : -TC(X,Z) \text{ SEQ } A(Z,Y). \quad (26)$$

At the end, let us mention that the cost of compilation of an event program (written in the proposed language) into Prolog rules is minor. Typically a program is compiled in few micro seconds. Hence the compilation phase does not cause a significant overhead.

In this section, we have provided measurement results of our running CEP engine. Even though there is a lot of room for improvements, preliminary results show that logic-based event processing has the capability to achieve significant performance. Working one 18 months on this project, we have managed to develop a CEP

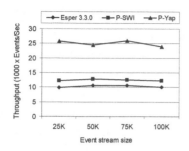
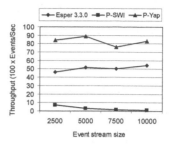

Fig. 6 (a) Experiments for Disjunction Operator - Throughput (b) Evaluation of Transitive Closure.

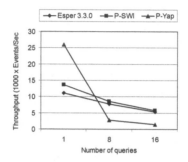

Fig. 7 Experiment for Testing Computation Sharing for Sequence Operator

language and a corresponding system that is competitive to mature CEP engines such as Esper 3.3.0. Taking inference capability into account, logic-based CEP goes beyond the state-of-the art in providing a powerful combination of *deductive* capabilities and *temporal* features[17], while at the same time exhibiting competitive run-time characteristics.

8 Related Work

In order to capture relevant changes in a system and respond to those changes adequately, a number of formal reactive frameworks have been proposed. Work on modeling *behavioral* aspects of an application (using various forms of reactive rules) started in the Active Database community a long time ago. Different aspects have been studied extensively, ranging from modeling and execution of rules to discussing architectural issues [21]. However, what is missing in this work is a clean integration of active behavior with pure *deductive* and *temporal* capabilities.

A lot of work [19, 17, 20, 7] in the area of rule-based CEP has been carried out, proposing various kinds of *logic rule*-based approaches to process complex events.

---

[17] We have skipped comparative tests requiring interval-based operators (for example, PAR , DURING and so forth), as Esper language semantics is based on time points.

As pointed out in [7], rules can be effectively used for describing so-called "virtual" event patterns. There exist a number of other reasons to use rules: Rules serve as an abstraction mechanism and offer a higher-level event description. Also, rules allow for an easy extraction of different views of the same reactive system. Rules are suitable to mediate between the same events differently represented in various interacting reactive systems. Finally, rules can be used for reasoning about causal relationships between events.

To achieve the aforementioned aims, these approaches all represent complex events as rules (or queries). Rules can then be processed either in a bottom-up manner [22], a top-down manner [9, 1], or in a manner that combines both [6]. However, all these evaluation strategies have not particularly been designed for event-driven computation. They are rather suited for a *request-response* paradigm. That is, given (and triggered by) a request, an inference engine will search for and respond with an answer. This means that, for a given event pattern, an event inference engine needs to check if this pattern has been satisfied or not. The check is performed at the time when such a request is posed. If satisfied by the time when the request is processed, a complex event will be reported. If not, the pattern is not detected until the next time the same request is processed (though it can become satisfied in between the two checks, being undetected for the time being). For instance, [20] follows the mentioned request-response (or so called *query-driven*[18]) approach. It proposes to define queries that are processed repetitively at given intervals, e.g., every 10 seconds, trying to discover new events. However, generally events are not periodic or if so might have differing periods, and nevertheless complex events should be detected as soon as they occur (not in a predefined time window). This holds in particular for time-critical scenarios such as monitoring stock markets or nuclear power plants.

To overcome this issue, in [7], an incremental evaluation was proposed. The approach is aimed at avoiding redundant computations (particularly re-computation of joins) every time a new event arrives. The authors suggest utilizing relational algebra evaluation techniques such as incremental maintenance of materialized views [14].

Prova [16] is close to our approach in sense that it supports declarative rules. On the other hand it is a reactive system, supporting agent programming. Complex event patterns can be created in Prova as Event Condition Action (ECA) rules. The Prova language however does not provide event operators (e.g., SEQ, AND, OR etc.); they rather need to be encoded as ECA rules. ETALIS is a dedicated CEP system where complex event patterns are defined as rules. ETALIS Language for Events defines a set of operators and enables the specification of complex events from other atomic or complex events. ETALIS is grounded on EDBCR, and is a logic-programming system. Prova combines imperative, declarative and functional programming styles, and unlike ETALIS (which is a Prolog-based system), Prova is implemented in Java.

A big portion of related work in the area of rule-based CEP is grounded on the Rete algorithm [13]. Rete is an efficient pattern matching algorithm, and it has been the basis for many production rule systems. The algorithm creates a decision tree that combines the patterns in all the rules of the knowledge base. Rete was intended

---

[18] If a request is represented as a query (what is a usual case).

to improve the speed of forward chained production rule systems at the cost of space for storing intermediate results. Production rules can be utilized to form complex event patterns, in which case a Rete-based production rule system is used as a CEP engine. Thanks to forward chaining of rules, Rete is also event-driven (data-driven).

Close to our approach is [15]. It is an attempt to implement business rules also with a Rete-like algorithm. However, the work proposes the use of subgoals and data-driven backward chaining rules. It has deductive capabilities, and detects satisfied conditions in business rules (using backward chaining), as soon as relevant facts become available. In our work, we focus rather on complex event detection, and enable a framework for event processing in pure Logic Programming style [18]. Our framework can accommodate not only events but conditions and actions (i.e., reactions on events), too. As this is not a topic of this chapter, an interested reader is referred to our previous work for details.

Concluding this section, many mentioned studies aim to use more formal semantics in event processing. Our approach based on ETALIS Language for Events may also be seen as an attempt towards that goal. It features data-driven computation of complex events as well as rich deductive capabilities.

9   Conclusions and Future Work

We have proposed a language for Complex Event Processing based on deductive rules. The language comes with a clear declarative, formal semantics for complex event patterns. Further, our contribution includes an execution model which detects complex events in a data-driven fashion (based on *goal-directed event-driven rules*). We have also provided a prototype implementation of our approach, which allows for specification of complex events and their detection at occurrence time. The approach substantiates existing event-driven systems with declarative semantics, and extends them with the power of deductive *reasoning*. A logic-based CEP enables reasoning over events, their relationships, and possible contextual knowledge available for a particular domain of interest. Although, so far, this feature has not been utilized enough, potentially it can enable a new generation of programmers to innovate in novel event-driven applications in AI. We believe that the proposed rule-based approach is also more pragmatic from the implementation and optimization point of view (as many techniques from logic programming and deductive databases are also applicable here). Extensibility of deductive rule systems is also higher than for systems based on imperative programming. Our experimental results presented in the chapter are encouraging.

As the next steps, we plan to investigate how our approach may show clear advantages over non-logic-based CEP. In particular, we plan to investigate how a rule representation of complex events (in large pattern bases) may help in *verification* of event patterns (e.g., discovering patterns that can never be detected according to *inconsistency* problems). Further, event *revision* is another area where logic reasoning may help in discovering consequences when certain events are *retracted*. *Out-of-order* events can also be handled in a logic CEP framework. Event retraction and out-of-order events can be seen as facts being retracted or added late to an event processing knowledge base, respectively. Hence an inference system can be deployed to *reason* about logical consequences of retracted or events added late on the whole pattern detection process. *Dynamic* event pattern management (i.e., patterns are created or discarded on-the-fly when certain situations are detected) is another interesting topic where the logic approach may help to control such an event-driven computation.

Acknowledgments

This work was supported by European Commission funded project SYNERGY (FP7-216089). We thank Ahmed Khalil Hafsi and Jia Ding for their help in the implementation and testing of ETALIS prototype.

References

1. Abiteboul, S., Hull, R., Vianu, V.: Foundations of Databases. Addison-Wesley, Reading (1995)
2. Adaikkalavan, R., Chakravarthy, S.: Snoopib: Interval-based event specification and detection for active databases. In: Data Knowledge Engineering. Elsevier Science Publishers B. V., Amsterdam (2006)
3. Agrawal, J., Diao, Y., Gyllstrom, D., Immerman, N.: Efficient pattern matching over event streams. In: SIGMOD (2008)
4. Allen, J.F.: Maintaining knowledge about temporal intervals. Communications of the ACM 26(11), 832–843 (1983)
5. Alvaro, P., Condie, T., Conway, N., Elmeleegy, K., Hellerstein, J.M., Sears, R.C.: Boom: Data-centric programming in the datacenter. Technical Report UCB/EECS-2009-113, EECS Department. University of California, Berkeley (2009)
6. Bancilhon, F., Maier, D., Sagiv, Y., Ullman, J.D.: Magic sets and other strange ways to implement logic programs. In: PODS 1986, Massachusetts, United States, ACM, New York (1986)
7. Bry, F., Eckert, M.: Rule-based composite event queries: The language xChange^{EQ} and its semantics. In: Marchiori, M., Pan, J.Z., de Sainte Marie, C. (eds.) RR 2007. LNCS, vol. 4524, pp. 16–30. Springer, Heidelberg (2007)
8. Chakravarthy, S., Mishra, D.: Snoop: an expressive event specification language for active databases. In: Data Knowledge Engineering, Elsevier Science Publishers B. V., Amsterdam (1994)
9. Chen, W., Warren, D.S.: Tabled evaluation with delaying for general logic programs. Journal of the ACM (1996)
10. Condie, T., Chu, D., Hellerstein, J.M., Maniatis, P.: Evita raced: metacompilation for declarative networks. In: Proc. VLDB Endow (2008)
11. Demers, A.J., Gehrke, J., et al.: Cayuga: A general purpose event monitoring system. In: CIDR. Stanford University, USA (2007)
12. Eisner, J., Goldlust, E., Smith, N.A.: Compiling comp ling: Weighted dynamic programming and the dyna language. In: Proc. HLT-EMNLP (2005)
13. Forgy, C.L.: Rete: A fast algorithm for the many pattern/ many object pattern match problem. Artificial Intelligence (1982)

14. Gupta, A., Mumick, I.S.: Maintenance of materialized views: Problems, techniques and applications. IEEE Data Engineering Bulletin (1995)
15. Haley, P.: Data-driven backward chaining. In: International Joint Conferences on Artificial Intelligence, Milan, Italy (1987)
16. Kozlenkov, A., Penaloza, R., Nigam, V., Royer, L., Dawelbait, G., Schröder, M.: Prova: Rule-based java scripting for distributed web applications: A case study in bioinformatics. In: Grust, T., Höpfner, H., Illarramendi, A., Jablonski, S., Fischer, F., Müller, S., Patranjan, P.-L., Sattler, K.-U., Spiliopoulou, M., Wijsen, J. (eds.) EDBT 2006. LNCS, vol. 4254, pp. 899–908. Springer, Heidelberg (2006)
17. Lausen, G., Ludäscher, B., May, W.: On active deductive databases: The statelog approach. In: Kifer, M., Voronkov, A., Freitag, B., Decker, H. (eds.) Dagstuhl Seminar 1997, DYNAMICS 1997, and ILPS-WS 1997. LNCS, vol. 1472, p. 69. Springer, Heidelberg (1998)
18. Lloyd, J.W.: Foundations of Logic Programming. Computer Science Press, Rockville (1989)
19. Motakis, I., Zaniolo, C.: Composite temporal events in active database rules: A logic-oriented approach. In: Ling, T.-W., Vieille, L., Mendelzon, A.O. (eds.) DOOD 1995. LNCS, vol. 1013, Springer, Heidelberg (1995)
20. Paschke, A., Kozlenkov, A., Boley, H.: A homogenous reaction rules language for complex event processing. In: EDA-PS. ACM, New York (2007)
21. Paton, N.W., Díaz, O.: Active database systems. ACM Comput. Surv. (1999)
22. Ullman, J.D.: Principles of Database and Knowledge-Base Systems, 2nd edn., vol. I and II. W. H. Freeman & Co., New York (1990)
23. Yoneki, E., Bacon, J.M.: Unified semantics for event correlation over time and space in hybrid network environments. In: Chung, S. (ed.) OTM 2005. LNCS, vol. 3760, pp. 366–384. Springer, Heidelberg (2005)

Acronyms

GINSENG Data Processing Framework

Zbigniew Jerzak, Anja Klein, and Gregor Hackenbroich

Abstract. For many applications guided by sensor networks, such as production automation and health monitoring, an efficient data processing with performance assurance is crucial, especially for metrics such as delay and reliability. Our study of current middleware approaches showed that they do not allow a sophisticated complex event processing, neither the performance monitoring. In this chapter we present the GINSENG middleware architecture that provides a 3-tier data processing framework to exploit the benefits of basic publish/subscribe systems, traditional event stream processing and complex business rule processing. Furthermore, the GINSENG middleware architecture provides performance control mechanisms, i.e., monitoring metrics and improvement methods, both of the underlying sensor network and the middleware itself. Finally, it supports the constraints of industrial environments by allowing for the distributed middleware deployment and data processing.

1 Introduction and Motivation

The overall goal of the GINSENG project is to develop a Wireless Sensor Network (WSN) that meets application-specific performance targets and integrates existing industry resource management systems. In order to achieve this goal, the GINSENG project focuses not only on the development of the physical sensor network but also on the development of a middleware platform which gathers and processes information coming from wireless sensors and connects them to ERP systems.

The target application of the GINSENG project is pipe and oil tank monitoring in a refinery environment. Here, pressure, volume flow, and tank level sensors are applied to control the status and enable the predictive maintenance of pipelines and

Anja Klein · Zbigniew Jerzak · Gregor Hackenbroich
SAP Research Dresden, Chemnitzer Straße 48, 01187 Dresden, Germany
e-mail: {zbigniew.jerzak,anja.klein,gregor.hackenbroich}@sap.com

S. Helmer et al.: Reasoning in Event-Based Distributed Systems, SCI 347, pp. 125–150.
springerlink.com   © Springer-Verlag Berlin Heidelberg 2011

tanks. Moreover, refinery employees may be equipped with mobile sensors to measure gas leakages and raise warnings in case of hazardous situations. In these scenarios a continuous performance control of all involved systems is crucial in order to prevent undetected emergencies. Such emergencies can be a result of simple delays concerning data transfer or data processing. Further performance-critical application areas relevant for the GINSENG project include: fire detection, energy management in manufacturing plants or health care, where status of patients is monitored at intensive care units.

Beyond the continuous performance control, these scenarios require also the analysis of temporal relationships between incoming events as well as continuous data stream processing. Moreover, all information has to be processed as close to the source as possible. This reduces the transferred data volume and minimizes the congestion probability for the limited bandwidth wireless sensor networks. Finally, the physical distribution of sensors requires a distributed middleware deployment.

To meet these constraints, the GINSENG project is developing a modular, hierarchical middleware architecture. The GINSENG middleware can be deployed in a distributed manner on three different levels (sensor node, gateway and central server) providing a flexible platform for distributed data processing – the GINSENG Data Processing Framework.

The remainder of this chapter is structured as follows. In Section 2 we present the overall architecture of the GINSENG middleware. Subsequently, in Section 3, we focus on the GINSENG Data Processing Framework: a 3-tier architecture for distributed data processing. In Section 4 we describe how the GINSENG Data Processing Framework is extended to monitor the performance of the wireless sensor network and the GINSENG middleware. Finally, in Section 5, we summarize the related work in the field of middleware technologies, publish/subscribe mechanisms, event stream and business rule processing. We conclude with a summary of this chapter and an outlook on future work in Section 6.

2   System Architecture

This section gives a brief overview of the GINSENG middleware architecture with specific focus on the event processing. The GINSENG middleware architecture is driven by two important factors: (i) it provides an event-based middleware which (ii) decouples GINSENG components from each other.

The event-based middleware of GINSENG equips the application programmer with an extensive functionality for the creation of distributed systems. Through its components the GINSENG middleware allows the programmer to collect, process and reason on the data as it moves through the system. Moreover, the GINSENG middleware supports a many-to-many interaction scheme overcoming the shortcomings (tight coupling, inflexibility) of the traditional request-reply schemes [39].

The GINSENG middleware is an event-based system which provides a strong notion of decoupling which applies to both internal middleware components and external components which communicate using the GINSENG middleware. The

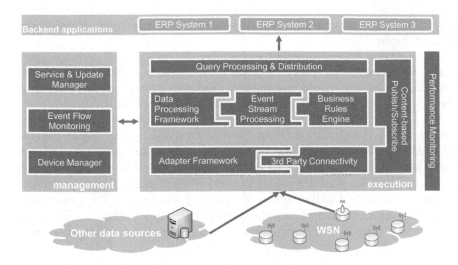

Fig. 1 The architecture of the GINSENG middleware – centralized deployment

external components include Wireless Sensor Networks (or other data sources) and Business Applications (or visualization components) – see Figure 1. Internal components include query processing and distribution, and the adapter framework.

The decoupling of GINSENG components is provided by the underlying communication scheme: the publish/subscribe system. Using the publish/subscribe scheme, components can register their interest for particular data and receive asynchronous notifications about events matching their interest. This type of communication is realized by all components within the GINSENG middleware. This approach is also aligned with the asynchronous (push-based) creation of events by wireless sensors.

Figure 1 shows the details of the GINSENG middleware architecture. The GINSENG middleware is split in two parts: (i) design-time components for all management functionality and (ii) runtime components that are relevant during the execution. The design time components provide user interfaces for all administrative tasks and thus allow the monitoring and configuration of the runtime components.

2.1 Core Components

All core components of the GINSENG middleware are connected using the *Content-based Publish/Subscribe* system. The publish/subscribe system plays a crucial role in the GINSENG middleware as it not only delivers events, but also takes part in the filtering of the information, allowing movement of the computation to within the proximity of the data sources.

The *Adapter Framework* is a pluggable infrastructure which allows the GINSENG middleware to connect to arbitrary data (event) sources. Any required 3rd party driver or service enabling the connectivity can be dynamically plugged into the adapter framework as an independent module. Such modules provide connectivity not only to SQL-databases, the SAP Business Suite (e.g., an ERP system) or arbitrary web services, but also to smart items such as wireless sensor nodes or smart meters.

The *Data Processing Framework* acts as a foundation which allows for plugging of arbitrary Event Stream Processing (ESP) and Business Rules (BR) engines to benefit from existing well-known and well-tested complex event processing techniques and to support declarative as well as rule-based event processing.

Finally, the *Query Processing and Distribution* controls the ESP and BR engine(s). It is plays a significant role in distributed deployments (see Figure 2) of the GINSENG middleware as it allows optimization of the distributed query execution. It also provides a single, generic query language to the user or backend application to encapsulate the applied ESP engine and performance-related meta-information.

The *Performance Monitoring* is a cross-cutting component. The Performance Monitoring receives relevant meta-information (e.g., latency, packet loss, processing time) from all middleware components, provides them to the interested user and triggers activities for performance improvement.

2.2 Core Technology

The component infrastructure of the GINSENG middleware is developed based on the OSGi Service Platform [59] – a dynamic module system for Java. Each GINSENG middleware component is developed as two OSGi bundles[1] – one for the design-time configuration and one for the runtime execution. The design-time bundle exposes the administration interface by component-specific User Interfaces which are incorporated into the Management Console – see Figure 2. The Management Console connects to the Central Instance which is a single stop for master copies of all configuration data and the runtime component code.

The runtime system consists of one or more middleware nodes. Each node provides a middleware runtime environment, where components' runtime bundles responsible for device connectivity, system integration, data processing, data querying, performance monitoring and performance improvement are deployed and executed. The runtime bundle of a component is a piece of Java code that communicates with a real world entity or external application via the content-based publish/subscribe system.

Each middleware node may run on a regular personal computer, on an embedded system, or within a virtual machine. The middleware runtime environment is built on Java technology and therefore platform-independent. All agents deployed within a node run in an OSGi environment (Eclipse Equinox [35]) which enables dynamic remote code modifications, without requiring a reboot. This OSGi functionality enables a minimal footprint for the runtime. Only bundles required for the specific application scenario are deployed and executed. Nevertheless, it allows this footprint

---

[1] An OSGi bundle consists of Java classes and other resources that deliver functions of a specific application (component) to application users, as well as providing services and packages to other bundles.

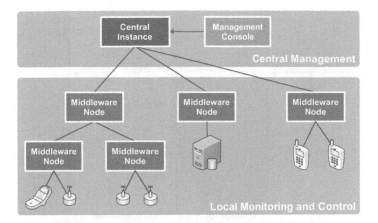

Fig. 2 The distributed deployment of the GINSENG middleware

to grow according to the changing requirements. The OSGi environment allows the deployment of additional bundles, e.g., adapters to connect new data sources, during runtime. Of course, the lightweight implementation of runtime bundles is a key to optimal middleware footprint, that developers need to keep in mind.

3 Data Processing Framework

The GINSENG data processing framework advances the current state of the art in that it combines three technologies to build a unified data processing framework. The three technologies are a content-based publish/subscribe communication system, Event Stream Processing (ESP) engine and Business Rules Engine (BRE). To the best of our knowledge it is the first approach which combines these technologies to build a unified, event-driven data processing framework – see Figure 3.

Events created by wireless sensors are transformed using the Adapter Framework of the GINSENG middleware into the internal GINSENG middleware event format, which is used by all middleware components. Events in this format are subsequently passed to the data processing framework which is responsible for stateful processing of events according to the specified rules. The result of the stateful event processing in the data processing framework (see Figure 3) is a set of complex (business) events which are consumed by the backend applications and/or management components.

The data processing framework in the GINSENG middleware consists of two main parts: the stateless and the stateful part. The input to the data processing framework consists of simple events produced by the adapter framework. Within the GINSENG middleware the content-based publish/subscribe system is considered as the part of the data processing framework. However, since the publish/subscribe system is also used to handle events leaving the data processing framework we indicate this fact by placing it as a separate component in Figure 1.

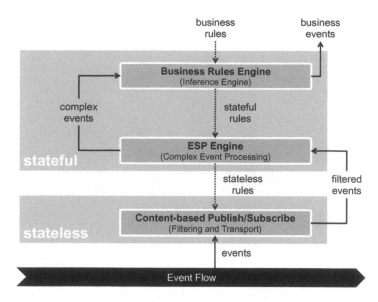

Fig. 3 Data processing framework in the GINSENG middleware

The data processing framework follows the principle of upstream evaluation and downstream replication, which is a well established concept in the literature [10]. Following this principle the GINSENG middleware routes an event in one copy as far as possible and replicates events only downstream. This means events are replicated as close as possible to the interested components (downstream replication). Filters and rules are applied, and patterns are assembled upstream, i.e., as close as possible to the sources of events (upstream evaluation).

The upstream evaluation and downstream replication principle has the following impact on the GINSENG middleware: stateful patterns and business rules entering the GINSENG middleware are evaluated as close to the source of relevant events as possible. However, since publish/subscribe systems are not well suited for handling of stateful rules, the GINSENG middleware decomposes the business rules into stateless and stateful parts, pushing the processing of the stateless rules within the proximity of the data producers.

The stateless event processing in the data processing framework is therefore handled by the content-based publish/subscribe system. Publish/subscribe systems use stateless filters to decide upon the destination of events. In case no destination matches a given event it is dropped – thus reducing the workload on the stateful parts of the data processing framework.

The stateful part of the data processing framework consists of two processing engines: the Event Stream Processing engine (based on the PIPES [33] system) and the Business Rules Processing engine (based on the JBoss Drools [27]). This GINSENG approach is aligned with the vision of ESP and BRE approaches merging into unified CEP platform [7]. The GINSENG vision is driven by the fact that

Business Rules Engines expose a more mature interface for the non-technical users while Event Stream Processing engines provide better performance for operations on multiple sources of homogeneous events.

In addition to the above, the GINSENG middleware extends this approach by asserting that the future data processing frameworks will be based on an asynchronous, data oriented communication protocol: the content-based publish/subscribe system. The three technologies: ESP, BRE and publish/subscribe share a set of similarities which further underline the applicability the GINSENG approach. All three technologies are event-based. They are inherently asynchronous and very well suited for the processing of large quantities of events. In what follows we describe in detail our effort of merging the three data driven techniques.

3.1 Business Rules Engine

Business rules engines rely on the Rete algorithm [22] to process incoming events against a set of user-defined rules – see Section 5.4 for details. In the GINSENG middleware we apply the Business Rules Management System (BRMS) JBoss Drools [27], where events are represented by Java classes. Every event in JBoss Drools can be equipped with meta-data which can state the role of the event, a timestamp (time-point), duration or an expiration time. JBoss Drools processes events as they arrive, and due to the ability to perform temporal reasoning it has a built-in mechanism for garbage collection of events that can no longer match any existing rule.

The heart of the JBoss Drools system is the fast ReteOO algorithm, which provides support for sliding windows and temporal operators (before, after, coincide, during, finishes, finished by, includes, meets, met by, overlaps, overlapped by, starts, started by) for temporal reasoning. Rules can be specified either using the Drools

Listing 1 Two example JBoss Drools rules

```
1  define rule1:
2      if s1.val>5
3          && s2.val<8
4          && s3.tmp=5
5          && s1.loc=s2.loc
6          && s2.loc=s3.loc
7      then alarm()
8  end

10 define rule2:
11     if s1.val>5
12         && s2.val<8
13         && s1.qos=s2.qos
14     then alarm()
15 end
```

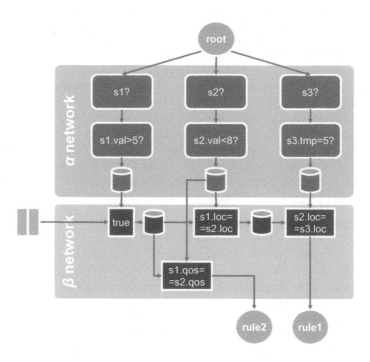

Fig. 4 The Rete network containing rules depicted in Listing 1

native procedural rule language, or via the use of a custom defined domain specific language (DSL). Within the GINSENG project we are developing a GINSENG domain specific language, tailored for use within a refinery environment, with specific focus on the oil tank and pipe monitoring. The use of the GINSENG DSL allows building of an interface between the non-technical refinery personnel and the rule engine.

The basic idea of the Rete algorithm is to create a directed acyclic graph of rule conditions, a so-called Rete network. Nodes in the graph represent rule conditions, e.g., a node can realize a selection operator that filters data based on certain constraints. Whenever a new event appears or the state of the network changes a representation of the event, a so-called working memory element (WME) is created. The WME is then propagated through the Rete network. This is performed in a forward-chaining fashion from the root to the leaf nodes of the network. During this process, every node in the network checks conditions or performs joins and only matching WMEs are passed on to child nodes. Every WME or tuple of WMEs that reaches a leaf node represents a match and results in an activation of the corresponding rule. A rule firing can influence the working memory, i.e., it can change events. If this is the case, the system again creates WMEs from these events and propagates them through the network.

Let us consider the set of rules specified in Listing 1. The rule rule1 states that if an event of type s1 and field val greater than 5 and an event of type s2 and field

Table 1 Filters and events in predicate-based semantics

Node	Description
Root	Starts each Rete network (`root`). It has no ancestor nodes.
Type	Distinguishes between different event types (e.g. `s1?`). Type node has only one input, and acts as a filter by passing only events matching the type of the node. The number of type nodes is equal to the number of event types occurring in rules.
Alpha (α)	Performs stateless filtering similar to a selection in relational algebra (e.g. `s1.val>5?`).
Beta (β)	Combines two different types of WMEs to produce a joined result (e.g. `s1.loc==s2.loc`). Beta nodes usually perform joins, however, extensions to realize a universal quantifier, an existential quantifier, a negation and different aggregation functions are available.
Terminal	A leaf node in the Rete network (e.g. `rule2`). If an event reaches the terminal node, this represents the fulfillment of the corresponding rule. Therefore, the number of terminal nodes is equal to the total number of rules.

`val` less than 8 and an event of type `s3` and field `tmp` equal 5 all have the field `loc` set to the same value than the `alarm()` function should be called. Similarly, the rule `rule2` states that if an event of type `s1` and field `val` greater than 5 and an event of type `s2` and field `val` less than 8 have the same value of field `qos` than the `alarm()` function should be called as well.

The set of rules specified in the Listing 1 after loading into the working memory of the Rete algorithm is presented in Figure 4. The description of each of the node types which are illustrated in the Figure 4 are presented in Table 1.

3.2 BRM and Publish/Subscribe

For the evaluation of the Business Rules Engine we have used the Linear Road Benchmark [5] – see Figure 5. The Linear Road Benchmark simulates a tolling system on a fictional expressway, where every car is equipped with a transponder which every 30 seconds emits a car's position. Position reports are used to generate traffic statistics which in turn determine the toll charges. Our evaluation has indicated that the GINSENG Data Processing Framework requires additional mechanisms to lower the memory consumption of the rules processing engine. We have developed a two-stage strategy for coping with this issue. The first stage encompasses the use of the publish/subscribe layer while the second (currently in development) extends this approach to embrace event stream processing engines.

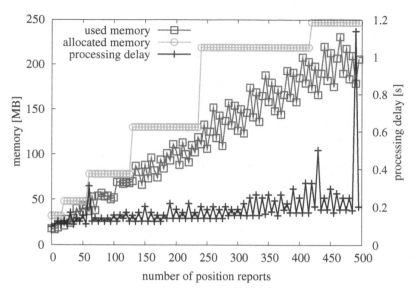

Fig. 5 Linear Road Benchmark – memory consumption using the Drools business rules engine

For the first stage – using the publish/subscribe layer – we exploit the nodes available in the α network in order to drive the processing in the publish/subscribe layer. Specifically, we extract the stateless conditions contained within α network and re-use them in the publish/subscribe layer. We extract both type and α nodes and use their conditions as topics and filters within the publish/subscribe system, respectively.

For example, let us consider the leftmost branch of the α network depicted in the Figure 4. This branch will be converted to an OSGi filter with following properties (props):

Listing 2 OSGi filter

```
1 Properties p = new Properties();
2 p.put(EventConstants.EVENT_TOPIC, "ginseng/events/s1");
3 p.put(EventConstants.EVENT_FILTER, "s1.val>5");
4 context.registerService(EventHandler.class.getName(), this, p);
```

The important aspect of the extraction process is that the filters are not removed from the α network. Instead, filter copies are extracted and following the upstream evaluation principle moved to within the proximity of the data producer. We have decided to use this approach for pre-processing of the business rules as it allows us to exploit the rule parsing and merging mechanism within the Rete network.

3.3 BRE and ESP

The second stage aiming at the optimization of the event processing within the GINSENG Data Processing Framework involves event stream processing engines. Our initial experiments have shown that performing operations like joins on simple events consumes less memory if done within the Event Stream Processing network than within the Business Rules Engine. Therefore, we the goal of the second stage is to migrate parts of the β network into the event stream processing system.

The migration of the β network nodes imposes several implications for the BRE. The most important issue is the need to change the corresponding parts of the Rete network. The reason for the change is the fact that execution of the operators within the β network results in the creation of new event types. This in turn, invalidates both the α and the β network constructs as new events appear at the input of the BRE. For example, let us assume that β node `s1.loc==s2.loc` visible in Figure 4 has been migrated to an ESP system. This implies that ESP system produces a new type of event which is a join between events of type `s1` and `s2`. This in turn requires the creation of new α node in the Rete network which would detect such a new compound even type.

We are currently evaluating the set of operators and corresponding event types which pose the best candidates for the migration from the BRM into the ESP system. In order to provide a general framework we will perform tests using not only the GINSENG specific data, but also the general benchmarking systems [5].

3.4 Domain Specific Languages for Rule Definitions

In order to shield users from changes in the underlying system GINSENG provides a domain specific language (DSL) which allows non-technical users to specify new and modify existing rules. The GINSENG DSL is based on an existing set of tested rules, which constitute the GINSENG knowledge base. Based on these rules a set of DSL queries is developed. As an example let us consider the following rule expressed using the Drools native procedural rule language:

```
1  If WSNMessage(TankLevel<20, GenTimestamp>152467362)
2      then System.out.println("Oil tank is almost empty.");
```

The above rule monitors the level of a tank in the refinery. It states that messages of type `WSNMessage` contain information with respect to the generation time stamp (`GenTimestamp`) and current tank level – `TankLevel`. The above rule fires, i.e., produces a message `Oil tank is almost empty`, whenever the tank level is below 20 and the generation time stamp is newer than 152467362. It can be translated into a DSL rule using the following specification:

```
1 [when]  If a message indicates that=WSNMessage()
2 [when]  - oil tank level is lower than
3            {EnteredLevel}=TankLevel>{EnteredLevel}
4 [when]  - and it was generated no later than at
5            {EnteredTimestamp}=GenTimestamp<{EnteredTimestamp}
6 [then]  then log
7            "{Message}"=System.out.println("{Message}");
```

In the above specification first a test for a correct message type (line 1) is executed. Subsequently (lines 2–3), it is tested whether the user entered tank level (EnteredLevel) is lower than the one contained in the message and (lines 4–5) and whether the user entered time stamp (EnteredTimestamp) is newer than the one contained in the message. Finally (lines 6–7) a user entered message (Message) is displayed whenever the previous conditions are met. The final rule written in the GINSENG Domain Specific Language can take the following – easy to write and understand – form:

```
1 If a message indicates that
2 - oil tank level is lower than 20
3 - and it was generated no later than at 152467362
4 then log "Oil tank is almost empty."
```

The DSL definition of the rule acts like a template for the technical definitions, which allows the business user not only to understand the rule meaning, but also frees him from the underlying implementation details, simultaneously providing the ability to modify the DSL rules. In the example above users can select single constraints in DSL rules, allowing them to, e.g., test only for the tank level without performing the test for generation time.

4   Performance Control in Data Processing Framework

The first step towards a comprehensive performance monitoring is the collection and definition of available and required performance parameters. To evaluate the performance and quality of the event streams, we identified three classes of performance and data quality indicators: (i) event latency, (ii) event loss (reliability), and (iii) event content quality.

The metadata dimensions detailing these classes depend on (i) application requirements and (ii) used sensor nodes and their capabilities of metadata provisioning. To allow the comprehensive evaluation of the data quality of sensor measurement streams, we propose a set of 13 data quality (DQ) and performance dimensions derived from the DQ categories provided by [52]. We show these in Tables 2, 3, and 4. The source of the respective metadata item is either the sensor node (S) or the middleware (MW) itself. Further metadata dimensions can be calculated (C) based on other performance information, allowing easy extension of the provided list. This calculation is performed by the GINSENG middleware, when the respective meta-information is required. This list can be easily extended by deriving further dimensions.

Table 2 List of performance dimensions for event latency

Dimension	Description	Source
GenerationTimestamp	Timestamp of event message generation, e.g., timestamp of sensor measurement	S
MwArrivalTimestamp	Timestamp, indicating the event's arrival at the middleware	MW
MwLeavingTimestamp	Timestamp, indicating the event is leaving the middleware layer towards an application	MW
NetworkLatency	Time interval required for transferring this specific event message within the WSN	C
NodeLatency	Average time interval required for transferring events from the source mote of this event	C
MWLatency	Time interval required for transferring and processing this event in the middleware	C

Table 3 List of performance dimensions for event loss (reliability)

Dimension	Description	Source
PacketLossPerMote	Number of message packets lost during data transmission in the WSN per mote (calculated based on the MoteID and MessageID)	C
PacketLossAvg	Average packet loss over all sensor motes	C

Table 4 List of performance dimensions for event content quality

Dimension	Description	Source
Timeliness	Age of this event message since its generation, calculated as difference of current system time and generationTimestamp	C
Completeness	Fraction of original sensor values	C
Accuracy	Maximal systematic numeric error of a sensor measurement	MW
Confidence	Maximal statistical error of a sensor measurement	C
DataBasis	Amount of raw data underlying a data processing result or complex event	C

4.1 Performance Monitoring Infrastructure

To record and manage the above listed parameters within the WSN and the middle-ware, event messages have to be enriched with performance information. However, the metadata dimensions listed above would significantly increase the data volume. Thus, within the GINSENG middleware we apply the window-based approach (first proposed in [30]) for the data quality management in data streams. To allow for efficient data quality management, the event stream D, comprising a continuous stream of m events consisting of n attribute values $A_i (1 \leq i \leq n)$, is partitioned into κ consecutive, non-overlapping data quality windows $w_i(k)(1 \leq k \leq \kappa)$, each of which is identified by its starting point t_b, its end point t_e, the window size ω and the corresponding attribute A_i. In addition to the event data $e_i(j)(t_b \leq j \leq t_e)$, the window contains a set of d performance and data quality information items, each describing one performance dimension. Each window-wise performance information item is calculated as an average of the original event-wise meta-information items. For example, Figure 6 shows the window's network latency $l_{WSN,w}(k)$ the window accuracy $a_w(k)$ and the window completeness $c_w(k)$ with $\omega = 5$.

The window size ω can be defined independently for each event attribute and/or window. Small jumping windows result in high-granularity performance informa-tion at the expense of a higher data overhead. A wider window definition guarantees the important resource savings that are essential for data stream environments; this happens by risking information with lower granularity and decreased correctness due to error deviations introduced by the window-wise metadata aggregation.

4.2 Performance and Data Quality Algebra

To compute the performance and quality of event steam processing results, the tradi-tional stream operators have to be extended as illustrated in Figure 7. For each data processing function F consisting of operators $o \in O$, a metadata function F^M has to be composed of the data quality operators $o^M \in O^M$ to compute the metadata M^Y, describing the derived knowledge $Y = F(X)$.

Table 5 lists the operators o extracted from traditional event stream processing engines and the GINSENG application scenarios, for which metadata operators o^M have to be described. The data quality algebra defines how operators influence each DQ dimension. For a more detailed description the reader is referred to [32].

Generation Timestamp	...	210	211	212	213	214	215	216	217	218	219	220	221	222	223	224	225	226	227	228	229
Pressure	...	180	178	177	175	176	181	189	201	204	190	194	192	189	183	215	210	211	199	187	184
Network Latency	...					12,92					13,34					12,1					12,12
Completeness	...					0,9					0,8					0,9					0,99
Accuracy	...					3					3,9					2,78					2,86

Fig. 6 Window-based approach for an event stream sample

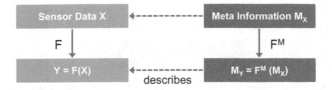

Fig. 7 Generic metadata algebra

Table 5 Data quality algebra operators

Type	Operators
Numeric Operators	Unary operators (e.g., square root); Binary operators (e.g., addition)
Signal Analysis	Sampling; Interpolation; Frequency analysis; Frequency filtering
Relational Algebra	Projection; Selection; Aggregation; Join
Rule-based Operators	Set operators (e.g., union); Boolean operators; Threshold comparison

4.3 Performance Improvement

In [31], we present the quality-driven optimization of stream processing that improves the resulting quality of data and service. We identify the targeted quality-driven process optimization as multi-objective, non-linear, continuous optimization problem with side conditions. We define the optimization objectives (for each data quality dimension) and optimization parameters (that configure the required stream processing operators). In the following, we briefly describe the developed generic optimization framework and discuss major evaluation results.

4.3.1 Optimization Framework

The optimization framework is illustrated in Figure 8. The data quality-driven optimization is executed continuously to tune the data stream processing during system runtime. As soon as an optimal parameter set is found and deployed, it has to be checked against the currently processed data stream. The online tuning allows the seamless adaptation to varying stream rates, measurement values and data quality requirements.

First, the system evaluates by means of static information like maximal sensor stream rate or sensor precision, if the user-defined quality requirements can be accomplished or if conflicts exclude a realization of all sub-objectives. In the latter case, the conflict is reported to the user. To check the satisfiability of DQ requirements, no access to streaming data is needed.

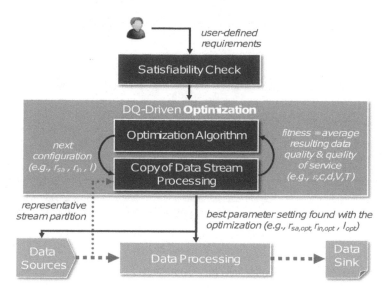

Fig. 8 Optimization framework

As the optimization must not interfere with the ongoing data stream processing, it is separated into an independent system component. To execute the optimization in parallel with the traditional data stream processing, the processing path with all its operators is copied into the optimization component. To execute the optimization, the framework applies a heuristic optimization algorithm (see Section 4.3.2 for comparison of different algorithms). In each algorithm iteration, solution individuals, defining the stream operators' configuration, have to be applied, evaluated and compared. The solution individuals are ranked according to the achieved fitness, that summarizes all sub-objectives of the optimization problem, and the best solutions are used to start the next iteration. Step by step, the resulting fitness and, thus, the configuration of the stream operators is improved.

Each solution is evaluated by directing a representative data stream partition through the copied processing path in the optimization component. This stream partition for optimization constitutes a data stream window of n data tuples. It may either be selected in batch-mode at the beginning of each optimization run and used in each iteration without changes. On the other hand, the partition can be updated for each algorithm iteration with current tuples from the original stream, to reflect the dynamic progression of the data stream and allow the continuous optimization. After the stream partition is completely processed, the fitness of the tested configuration is calculated.

As soon as the fitness accomplishes the user requirements, the optimization problem is solved. The new parameter setting is applied to the original processing path. The sampling operators are updated with the optimized sampling rates $r_{sa,opt}$. The frequency analyses and aggregations are updated with the determined group sizes

l_{opt}. Finally, the data quality initialization at the sensor nodes is re-configured with the new window sizes ω_{opt}.

The logical distinction between optimization and processing enables also their physical separation, for example on a distinct server node. Thus, the optimization task has no negative impact on the performance of the traditional data stream processing.

4.3.2 Evaluation Results

The evaluation of the optimization framework showed that the processing time rises nearly linearly for small to medium partition lengths of 100 to 1000 stream tuples. Only for very large stream partitions does the iteration duration exhibit an exponential character. Further, we can prove linear scalability for increasing processing complexity. The time required for one algorithm iteration rises linearly with increasing numbers of sensors as data sources and operators in the processing path.

Besides the evaluation of the generic framework, we compared different optimization algorithms. The Monte-Carlo-Search (MC) performs a random search over the problem domain and serves as reference value defining the lower performance bound [38]. The single-objective optimization using the heuristic Evolution Strategy [36] is executed with randomly chosen weights (SO-R) as well as optimal weights (SO-O), which determine a well-balanced objective compromise. Finally, the multi-objective optimization (MO) approximates the Pareto front of all optimal compromises.

We evaluated the overall time performance of single- and multi-objective optimization with respect to the achieved quality improvement for 16 sensor data sources and 10 randomly inserted aggregations. The quality improvement is expressed as percentage value $(q_{before} - q_{after})/q_{before}$. The Monte-Carlo-Search performs worst followed by the randomly initiated single-objective optimization, requiring 5.2 and 2.9 seconds, respectively, for a DQ improvement of 10%. The single-objective optimization executed with well-balanced weights performs best (1.6s for 10%). However, the definition of these weights requires multiple optimization runs and has to be adapted as soon as stream characteristics or user requirements change. The multi-objective optimization (MO) is a little slower (1.9s) due to the complex computation of the Pareto front. However, the result comprises the complete set of all optimal compromises and no pre-processing to determine optimal weights is necessary.

The evaluation showed that the designed quality-driven optimization provides good scalability with regard to the applied stream partition length as well as increasing complexity of the data stream processing. Data quality and quality of service could be improved within a few seconds. Further, we deduce that the single-objective optimization in batch mode is the best choice for constant user requirements and steady data streams. If streaming data values present high fluctuations or user requirements are often adjusted, the multi-objective optimization constitutes the better option.

4.3.3   Open Issues

The research work presented above provides a good starting point toward performance control in the GINSENG middleware. We need to extend this quality-driven
optimization framework towards a performance-driven optimization. The optimization objectives have to be extended to cover the performance AND data quality
indicators defined above. Further, the applied optimization algorithms, e.g., the evolution strategy, has to be adapted to embrace these novel objectives and the related
optimization parameters. After that, the optimization framework can be applied
without modifications to enable the performance maximization of event stream processing in the GINSENG middleware.

5   Related Work

This section provides an overview of the state-of-the art concerning middleware
developments. Further, we discuss publish/subscribe systems and event- and rule-
based processing engines applicable to our processing framework.

5.1   Middleware Technology

In this section we analyze a representative set of existing middleware approaches
with regard to performance control and data processing supported. The project on
Wireless Accessible Sensor Populations (WASP) addresses the energy-efficient and
secure integration of Wireless Sensor Networks (WSN) into applications like traffic
and herd control and integrates data cleansing techniques for data quality improvement [58]. WASP fulfills business applications requirements related to communication (e.g. on demand, event based sensor data acquisition, WSN discovery), but does
not support time synchronization. With respect to sensor data processing, WASP is
restricted to a basic operator set and does not support complex event processing, nor
performance or quality-guided stream processing.

The middleware developed within the project on PROduct lifecycle Management
and Information tracking using Smart Embedded systems (PROMISE) allows managers, designers, and operators to track, manage and control product information at
any phase of its lifecycle (design, manufacturing, maintenance, recycling), at any
time and anywhere in the world [40]. However, PROMISE does not provide any
tools or components for performance or quality management and monitoring.

The Collaborative Business Items (CoBIs) project [14] provides a service-oriented
approach to support business processes that involve physical entities (goods, tools,
etc.) in large-scale enterprise environments in a transparent way. The CoBIs project
primarily focuses on the design of a service-oriented middleware for deploying, running, and querying services on sensor nodes and does not support data and event
processing as proposed by the GINSENG approach.

The SAP Auto-ID Infrastructure (AII) enables the integration of all automated
communication and sensing devices (e.g., RFID, Bluetooth and bar-code devices)

as well as intelligent programmable language controls [45]. SAP AII provides functionalities for the data transfer from sensor motes to backend applications, but does not include any data processing and performance monitoring technology.

Beyond this small excerpt of existing middleware systems, our comprehensive analysis showed that middleware approaches for connecting field devices with backend systems only support basic pre-processing of sensor data and do not provide any quality and/or performance control mechanism required by the industrial applications, where unreliable or deferred data may risk a system breakdown or even personal injuries. The GINSENG middleware addresses the performance assurances (reliability, timeliness and precision) and control related issues. Moreover, it provides a) an abstraction to conceal the heterogeneity of underlying sensor motes, b) a complex data stream and event processing engine, c) a robust monitoring and management of application logic.

5.2 Publish/Subscribe Communication

The GINSENG middleware uses the OSGi-based publish/subscribe paradigm as its underlying communication infrastructure. In what follows we give a brief overview of the existing approaches towards the design and implementation of publish/subscribe systems.

Publish/subscribe is the first communication paradigm to unify three important decoupling properties: space, time and synchronization decoupling [21] which allows for a flexible communication between content producers (publishers) and content consumers (subscribers). The decoupling properties ensure that the communication is anonymous (space decoupling), asynchronous (synchronization decoupling) and communicating parties do not need to be active at the same time (time decoupling).

The above properties position pub/sub paradigm as a very attractive interaction scheme for building loosely coupled, event driven applications. Moreover, due to the decoupling properties publish/subscribe can act as a enabling technology for higher level dynamic and distributed event-driven services, like Complex Event Processing or Business Rules Processing. It is strictly for this reason that we design the GINSENG middleware to rely on the publish/subscribe paradigm as its basic communication primitive.

The basic interaction scheme of the publish/subscribe system is based on the well known observer pattern [24]. Data consumers express their interest using subscriptions. The first publish/subscribe systems started appearing over two decades ago. One of the first publish/subscribe systems was Information Bus [37] which is similar in concept to the generative communication model of tuple spaces [9]. The Information Bus implements a topic-based publish/subscribe paradigm, i.e., filters in subscription events took a form of fixed topics. The Information Bus and other topic-based publish/subscribe systems [49, 6] allow content consumers to dynamically specify topics which segment the information into channels. Publishers can specify to which channel the produced information belong, thus allowing interested subscribers to receive it.

The increasing need for heterogeneity and expressiveness among publish/subscribe systems has lead to the development of the content-based [41] and type-based [20] systems. Type-based systems provide type safety and encapsulation by using the hierarchy of the class structure to filter events. Content-based systems provide a much greater degree of flexibility by relying on arbitrary filter expressions (often in form of conjunctions of predicate functions [10, 29]) in order to select events of interest to the subscribers.

Parallel to the academic development of the publish/subscribe communication systems, the industry has started the adoption and standardization of the publish/-subscribe communication. One of the first widely recognized and available specifications was the Java Message Service [26] (JMS) which has been implemented in multiple commercial, e.g., SonicMQ [51], and open source products, e.g., HornetQ [28]. Other publish/subscribe specifications are WS-Eventing [17] and WS-Notification [12, 25]. WS-Eventing and WS-Notification standards are not limited to topic-based subscriptions (like JMS), allowing the definition of arbitrary filters in form of XPath [18] queries in case of WS-Eventing or user defined queries (including, e.g., XPath) in case of WS-Notification.

The OSGi publish/subscribe system has been proposed by the OSGi Alliance in the Service Platform Specification [53]. The publish/subscribe communication scheme implemented in the GINSENG middleware is defined within the Event Admin service specification [54] and provides a topic-based publish/subscribe system with additional ability to increase the filter selectivity (content-based publish/subscribe) by using a LDAP-style filter specification [50].

5.3 Event Stream Processing

The author in [34] states that CEP systems process incoming raw events from multiple data sources in real-time using algorithms and rules to determine correlations, trends and patterns expressed in outgoing complex events. This goal is achieved by handling the correlation of temporal as well as other events which occur simultaneously and in high volumes [7]. While the Event Stream Processing (ESP) part of the CEP technology is relatively new to the market with first commercial offerings appearing in 2004 [46] the Business Rules part of the CEP technology is already familiar to most organizations [44].

The event stream processing systems can perform filtering, correlation, transformation and aggregation operations on multiple event streams. There exist a number of academic prototypes which have also been partially transformed into commercial offerings. Examples include Aurora/Borealis [1, 2] (commercialized by Coral8 [15]), TelegraphCQ [11] (commercialized by Truviso [56]) and PIPES [33] (commercialized by RTM Realtime Monitoring GmbH [42]). The processing of data within ESP engines is driven by queries which can take be either declarative, e.g., CQL: The Continuous Query Language [4], or procedural, e.g., SQuAl [2]. Declarative queries resemble the standard SQL syntax with event stream specific extensions (e.g. support for windows), while procedural approaches allow for composition of queries out of well known building blocks (operators).

In recent years due to the distributed nature of data sources and the increasing availability of the on-demand resources [16] the distributed ESP approaches have been the focus of academic research. Recent developments include systems like Borealis [1], System S [3] and NextCEP [47]. The focus of the distributed ESP research lies on the scalability (via load and query distribution across multiple system nodes) and fault tolerance (in most cases via state-machine replication [13, 48]) issues.

In what follows we give a brief descriptions of two open source ESP systems which are used within the GINSENG middleware. The first system we use is PIPES [33]. PIPES is a flexible and extensible infrastructure providing fundamental building blocks to implement a data stream management system. It constructs directed acyclic query graphs based on a publish/subscribe mechanism which is integrated into the graph nodes. PIPES is based on XXL [8] – a Java library that contains a rich infrastructure for implementing advanced query processing functionality.

The second ESP system which is used in the GINSENG middleware is Esper [19]. Esper implements an Event Query Language (EQL) which allows for registering of queries in the engine. A listener class is called by the engine when the EQL condition is matched as events flow in. The EQL enables the expression of complex matching conditions that include temporal windows, joining of different event streams, as well as filtering, aggregation, and sorting. Esper statements can also be combined together with "followed by" conditions thus deriving complex events from more simple events. Events can be represented as Java classes, XML documents or `java.util.Map`, which promotes reuse of existing systems acting as messages publishers. Esper also includes a historical data access layer to connect to databases, combining historical data and real time data in one single query.

5.4   Business Rules Engines

Business Rules Management Systems (BRMS) provide the ability to easily express the rules in a simple and understandable way by using abstractions, such as flowcharts, decision trees and decision tables as well as scoring models and textual if-then rules [7]. Therefore, the benefit of BRMS lies in the fact that a change in the rules can be easily reflected in the system by a non-technical user. BRMS, in contrast to CEP solutions, are a mature offering with good market penetration.

Business rules engines (BRE), which constitute the core of the BRMS offerings, are in most cases designed to accept discrete events which typically have a complex payload (multiple elements) [7]. There exist currently a number of commercial offerings for business rules platforms, including, but not limited to, Tibco Business Events [55], UC4 Automation Engine [57] and ruleCore CEP Server [43]. There exist also non-commercial or open source systems, examples being JBoss Drools [27] and Jess [23]. The common denominator for most of the business rules engines is the use of the Rete algorithm [22] to process incoming events against the stored rules. The use of the Rete algorithm allows reuse of the common parts of rules and thus reducing the number of operations which are necessary to match incoming events.

6 Summary

In this chapter, we presented the GINSENG middleware which closes the gap between (wireless) sensors networks and arbitrary backend applications, such as the monitoring tool for a manufacturing site, the health care system at an intensive care unit in a hospital or a zoological warehouse to track the routes of endangered animals.

After a short description of the overall GINSENG architecture, the main part of this chapter focused on the distributed event stream processing. With the presented 3-tier data processing framework composed of a low-level publish/subscribe system, an exchangeable event stream processing engine and the high-level business rule processing engine, we enabled the seamless integration of raw sensor events into the complex business rule evaluation.

As GINSENG targets performance-critical application scenarios where long data transfer delays, and missing or incorrect events may be hazardous, we proposed mechanisms for the on-the-fly performance monitoring. Besides metrics for the basic performance measurement, a performance algebra to compute the accuracy of event processing results and methods for the eventual performance improvement were illustrated. Finally, we gave an overview of related work concerning middleware approaches, as well as publish/subscribe, event and rule-based processing engines applicable to our processing framework.

In further work, we will investigate and refine the integration between the event processing engine and the business rules system. Moreover, the data quality driven optimization of the event stream processing will be extended to cover all dimensions listed in Tables 2, 3, and 4 as well as the GINSENG domain specific query language. Finally, we will evaluate the GINSENG middleware against existing middleware, event stream and business rules approaches to demonstrate the advantages of the 3-tier processing architecture over stand-alone processing engines.

Acknowledgments

The research leading to these results has received funding from the European Community's Seventh Framework Program (FP7/2007-2013) under grant agreement No. 224282. We would also like to thank Sebastian Weng for his support with the execution of the Linear Road benchmark.

References

1. Abadi, D.J., Ahmad, Y., Balazinska, M., Cetintemel, U., Cherniack, M., Hwang, J.-H., Lindner, W., Maskey, A.S., Rasin, A., Ryvkina, E., Tatbul, N., Xing, Y., Zdonik, S.: The design of the borealis stream processing engine. In: CIDR 2005: Second Biennial Conference on Innovative Data Systems Research, pp. 277–289 (2005)
2. Abadi, D.J., Carney, D., Çetintemel, U., Cherniack, M., Convey, C., Lee, S., Stonebraker, M., Tatbul, N., Zdonik, S.B.: Aurora: a new model and architecture for data stream management. VLDB J. 12(2), 120–139 (2003)
3. Amini, L., Jain, N., Sehgal, A., Silber, J., Verscheure, O.: Adaptive control of extreme-scale stream processing systems. In: ICDCS 2006: 26th IEEE International Conference on Distributed Computing Systems, Lisboa, Portugal, July 2008, p. 71. IEEE Computer Society, Los Alamitos (2006)
4. Arasu, A., Babu, S., Widom, J.: The cql continuous query language: semantic foundations and query execution. The VLDB Journal 15(2), 121–142 (2006)
5. Arasu, A., Cherniack, M., Galvez, E.F., Maier, D., Maskey, A., Ryvkina, E., Stonebraker, M., Tibbetts, R.: Linear road: A stream data management benchmark. In: Nascimento, M.A., Özsu, M.T., Kossmann, D., Miller, R.J., Blakeley, J.A., Schiefer, K.B. (eds.) VLDB, pp. 480–491. Morgan Kaufmann, San Francisco (2004)
6. Baldoni, R., Beraldi, R., Quéma, V., Querzoni, L., Piergiovanni, S.T.: Tera: topic-based event routing for peer-to-peer architectures. In: Jacobsen, H.-A., Mühl, G., Jaeger, M.A. (eds.) DEBS 2008: Proceedings of the 2007 Inaugural International Conference on Distributed Event-Based Systems, Toronto, Ontario, Canada, June 2007. ACM International Conference Proceeding Series, vol. 233, pp. 2–13. ACM, New York (2007)
7. Brett, C., Gualtieri, M.: Must you choose between business rules and complex event processing platforms? Forrester Research (January 2009)
8. Cammert, M., Heinz, C., Krämer, J., Schneider, M., Seeger, B.: A status report on xxl - a software infrastructure for efficient query processing. IEEE Data Eng. Bull. 26(2), 12–18 (2003)
9. Carriero, N., Gelernter, D.: Linda in context. Commun. ACM 32(4), 444–458 (1989)
10. Carzaniga, A., Rosenblum, D.S., Wolf, A.L.: Design and evaluation of a wide-area event notification service. ACM Trans. Comput. Syst. 19(3), 332–383 (2001)
11. Chandrasekaran, S., Cooper, O., Deshpande, A., Franklin, M.J., Hellerstein, J.M., Hong, W., Krishnamurthy, S., Madden, S., Raman, V., Reiss, F., Shah, M.A.: Telegraphcq: Continuous dataflow processing for an uncertain world. In: CIDR 2003: First Biennial Conference on Innovative Data Systems Research (2003)
12. Chappell, D., Liu, L.: Web Services Brokered Notification. 1.3 (2006), http://docs.oasis-open.org/wsn/wsn-ws_brokered_notification-1.3-spec-os.htm
13. Clement, A., Kapritsos, M., Lee, S., Wang, Y., Alvisi, L., Dahlin, M., Riche, T.: UpRight cluster services. In: Proceedings of the 22 nd ACM Symposium on Operating Systems Principles (SOSP), pp. 277–290 (2009)
14. CoBIs. Collaborative Business Items, http://www.cobis-online.de/
15. Coral8 Inc. Complex event processing with coral8, http://download.microsoft.com/.../complex_event_processing_with_coral8_final.pdf
16. Creeger, M.: Cloud computing: An overview. Queue 7(5), 3–4 (2009)
17. Davis, D., Malhotra, A., Warr, K., Chou, W.: Web service eventing, w3c working draft (2009), http://www.w3.org/tr/2009/wd-ws-eventing-20090317/

18. DeRose, J.C.S.: Xml path language, xpath (1999),
    http://www.w3.org/tr/xpath
19. EsperTech. Esper reference documentation (1999),
    http://esper.codehaus.org/esper-3.3.0/doc/reference/en/
    pdf/esper_reference.pdf
20. Eugster, P.: Type-based publish/subscribe: Concepts and experiences. ACM Transactions on Programming Languages and Systems 29(1), 1–50 (2007)
21. Eugster, P.T., Felber, P.A., Guerraoui, R., Kermarrec, A.-M.: The many faces of publish/-subscribe. ACM Comput. Surv. 35(2), 114–131 (2003)
22. Forgy, C.: Rete: A fast algorithm for the many patterns/many objects match problem. Artif. Intell. 19(1), 17–37 (1982)
23. Friedman-Hill, E.: Jess, http://www.jessrules.com/
24. Gamma, E., Helm, R., Johnson, R., Vlissides, J.: Design Patterns. Elements of Reusable Object-Orineted Software. Addison-Wesley Professional, Reading (1995)
25. Graham, S., Hull, D., Murray, B.: Web Services Brokered Notification. 1.3. Web Services Base Notification. 1.3 (2006), http://docs.oasis-open.org/wsn/
    wsn-ws_base_notification-1.3-spec-os.htm
26. Hapner, M., Burridge, R., Sharma, R., Fialli, J., Stout, K.: Java message service (April 2002), http://java.sun.com/products/jms/
27. JBOSS. Drools, http://labs.jboss.com/drools
28. JBoss. Hornetq, http://www.jboss.org/hornetq
29. Jerzak, Z., Fetzer, C.: Bloom filter based routing for content-based publish/subscribe. In: DEBS 2008: Proceedings of the second international conference on Distributed event-based systems, Rome, Italy, July 2008, pp. 71–81. ACM, New York (2008)
30. Klein, A.: Incorporating quality aspects in sensor data streams. In: Proceedings of the 1st ACM Ph.D. Workshop in CIKM (PIKM), pp. 77–84 (2007)
31. Klein, A., Lehner, W.: How to optimize the quality of sensor data streams. In: ICCGI 2009: Proceedings of the 2009 Fourth International Multi-Conference on Computing in the Global Information Technology, pp. 13–19. IEEE Computer Society, Los Alamitos (2009)
32. Klein, A., Lehner, W.: Representing data quality in sensor data streaming environments. J. Data and Information Quality 1(2), 1–28 (2009)
33. Kraemer, J., Seeger, B.: Pipes - a public infrastructure for processing and exploring streams. In: Weikum, G., Koenig, A.C., Deßloch, S. (eds.) Proceedings of the 9th ACM SIGMOD International Conference on Management of Data, pp. 925–926. ACM, New York (2004)
34. Leavitt, N.: Complex-event processing poised for growth. Computer 42(4), 17–20 (2009)
35. McAffer, J., VanderLei, P., Archer, S.: OSGi and Equinox: Creating Highly Modular Java Systems. Addison-Wesley Professional, Reading (2010)
36. Michalewicz, Z.: Genetic Algorithms Plus Data Structures Equals Evolution Programs. Springer, Heidelberg (1994)
37. Oki, B.M., Pflügl, M., Siegel, A., Skeen, D.: The information bus – an architecture for extensible distributed systems. In: Liskov, B. (ed.) Proceedings of the 14th Symposium on the Operating Systems Principles, pp. 58–68. ACM Press, New York (1993)
38. Patel, N.R., Smith, R.L., Zabinsky, Z.B.: Pure adaptive search in monte carlo optimization. Mathematical Programing 43(3), 317–328 (1989)
39. Pietzuch, P.R.: Hermes: A Scalable Event-Based Middleware. PhD thesis, Computer Laboratory, Queens' College. University of Cambridge (February 2004)

40. PROMISE. PROduct lifecycle Management and Information tracking using Smart Embedded system, http://www.promise.no/
41. Rosenblum, D.S., Wolf, A.L.: A design framework for internet-scale event observation and notification. SIGSOFT Softw. Eng. Notes 22(6), 344–360 (1997)
42. RTM Realtime Monitoring GmbH, http://www.realtime-monitoring.de/
43. ruleCore. Cep server, http://rulecore.com/
44. Rymer, J.R., Gualtieri, M., Brown, M., Salzinger, C.: The forrester wave: Business rules platforms, q2 2008 (April 2008)
45. SAP AG. SAP Auto-ID Infrastructure, http://www.sap.com/platform/netweaver/autoidinfrastructure.epx
46. Schulte, W., Blechar, M., Jones, T., Sholler, D., Thompson, J., Malinverno, P., Gassman, B.: The growing impact of commercial complex-event processing products. Gartner Research (October 2009)
47. Schultz-Moller, N.P., Migliavacca, M., Pietzuch, P.: Distributed complex event processing with query rewriting. In: DEBS 2009: Proceedings of the 2009 International Conference on Distributed Event-Based Systems, pp. 1–12 (2009)
48. Singh, A., Fonseca, P., Kuznetsov, P., Rodrigues, R., Maniatis, P.: Zeno: eventually consistent byzantine-fault tolerance. In: NSDI 2009: Proceedings of the 6th USENIX Symposium on Networked Systems Design and Implementation, pp. 169–184. USENIX Association, Berkeley (2009)
49. Skeen, M.D., Bowles, M.: Apparatus and method for providing decoupling of data exchange details for providing high performance communication between software processes. U.S. Patent No. 5,557,798 (July 1989)
50. Smith, M., Howes, T.: Lightweight directory access protocol (ldap): String representation of search filters. Request for Comments: 4515 (2006)
51. SonicMQ, http://web.progress.com/en/sonic/
52. Strong, D.M., Lee, Y.W., Wang, R.Y.: Data quality in context. Communications of the ACM 40(5), 103–110 (1997)
53. The OSGi Alliance. Osgi service platform - core specification (2009), http://www.osgi.org/
54. The OSGi Alliance. Osgi service platform - service compendium (2009), http://www.osgi.org/
55. TIBCO. Businessevents, http://www.tibco.com/software/complex-event-processing/businessevents
56. Truviso. Web analytics software, http://www.truviso.com/
57. UC4. Automation engine, http://www.uc4.com/products-solutions/automation-engine.html
58. WASP. Wireless Accessible Sensor Populations., http://www.wasp-project.org/
59. Wütherich, G., Hartmann, N., Kolb, B., Lübken, M.: Die OSGi Service Platform: Eine Einführung mit Eclipse Equinox. dpunkt, Heidelberg (2008)

Glossary

GINSENG   The goal of the EU-project GINSENG is the development of a performance-controlled wireless sensor network.

WSN   A **W**ireless **S**ensor **N**etwork is a network of sensor nodes that communicate wirelessly.

Middleware   The middleware is a computer software that connects software components or applications.

Publish/-   Publish/Subscribe (or pub/sub) is a messaging paradigm where senders
Subscribe   (publishers) broadcast messages that are only received by Subscribers who defined their interest in advance.

BR   **B**usiness **R**ules are application- or domain-specific rules defining the selection of alternative execution paths in complex business processes.

BRP   The **B**usiness **R**ule **P**rocessing evaluates incoming business and/or event data against business rules to guide business processes.

BRM   The **B**usiness **R**ule **M**anagement includes the definition, management and processing of business rules.

ESP   The **E**vent **S**tream **P**rocessing embraces all techniques and methods for the real-time processing of continuous or discrete event data.

CEP   The **C**omplex **E**vent **P**rocessing combines and evaluates incoming raw events against rules or patterns to create outgoing complex events.

DSL   A **D**omain **S**pecific **L**anguage is a programming or specification language dedicated to a particular problem domain, a particular problem representation technique, and/or a particular solution technique.

DQ   The **D**ata **Q**uality defines the appropriateness of a given data item for a specific task, expressed e.g., as accuracy or completeness.

Performance   The performance of a system describes its non-functional ability to solve a specific task, expressed e.g., as latency or reliability.

Performance   The performance control includes the monitoring of the current system per-
Control   formance as well as methods for the performance improvement.

MO   The **M**ulti-objective **O**ptimization targets for the Pareto front of optimal compromises between all involved sub-objectives.

SO   The **S**ingle-objective **O**ptimization summarizes all sub-objectives in one objective function, which is optimized afterwards.

MC   The **M**onte-**C**arlo-**S**earch performs a random search over all possible solutions to find the optimal one.

Security Policy and Information Sharing in Distributed Event-Based Systems

Brian Shand, Peter Pietzuch, Ioannis Papagiannis, Ken Moody,
Matteo Migliavacca, David M. Eyers, and Jean Bacon

Abstract. Linking security policy into event-based systems allows formal reasoning about information security. In the applications we address, highly confidential data must be shared both dynamically and for historical analysis. Principals with rights to access the data may be widely distributed, existing in a federation of independent administrative domains. Domain managers are responsible for the data held within domains and transmitted from them; security policy must be specified and enforced in order to meet these obligations. We motivate the event-driven paradigm and take healthcare as a running example, because the confidentiality of healthcare data must be guaranteed over many years. We first consider how to enforce authorisation policy at the client level through parametrised role-based access control (RBAC), taking context into account. We then discuss the additional requirements for secure information flow through the infrastructure components that contribute to communication within and between distributed domains. Finally, we show how this approach supports reasoning about event security in large-scale distributed systems.

Brian Shand
CBCU / Eastern Cancer Registry and Information Centre, National Health Service,
Unit C – Magog Court, Shelford Bottom, Hinton Way, Cambridge CB22 3AD, UK
e-mail: `Brian.Shand@cbcu.nhs.uk`

Peter Pietzuch · Ioannis Papagiannis · Matteo Migliavacca
Department of Computing, Imperial College London, 180 Queen's Gate,
London SW7 2AZ, UK
e-mail: `{prp,ip108,migliava}@doc.ic.ac.uk`

Ken Moody · David M. Eyers · Jean Bacon
Computer Laboratory, University of Cambridge, JJ Thomson Avenue,
Cambridge CB3 0FD, UK
e-mail: `first.last@cl.cam.ac.uk`

S. Helmer et al.: Reasoning in Event-Based Distributed Systems, SCI 347, pp. 151–172.
springerlink.com © Springer-Verlag Berlin Heidelberg 2011

1　Introduction

Large event-based computer systems must protect the confidentiality and integrity of data as it passes through them. This task is too large and dynamic to be done in an ad hoc or centralised way: instead, access must be controlled through explicit, distributed security policy. At the same time, policy enables formal reasoning about the security of information within the event system.

In this chapter, we show how security policy support and enforcement can be embedded into event-based systems. By adding fine-grained Distributed Information Flow Control restrictions, we enhance end-to-end security; this allows security analysis to extend beyond the boundaries of the event-based middleware. Reasoning about event security can then support system-wide security audit and information governance, and protect sensitive data in large-scale distributed systems.

Many large-scale distributed applications are best modelled as a federation of domains. A domain is defined as an independently administered unit in which a domain manager has, or may delegate, responsibility for naming and policy specification. The naming of individual users, groups and roles forms the basis of authentication and authorisation.

For example, a national health service comprises many independently administered hospitals, clinics, primary-care practices, etc. The care of a patient may move between domains, from primary care to treatment in hospital. Specialists may be associated with more than one domain, such as hospitals and clinics. Researchers may need to perform statistical analysis on health data. Records may need to be transmitted to auditors and universally accessible Electronic Health Record services. Patient treatment may necessitate the ordering of medication or services from separate domains. Communication within and between domains must therefore be supported, but because of the sensitivity of the data must be strictly controlled.

In such applications the need to communicate is driven by the actions of people (administration of treatment, taking medication), observations (derived automatically from sensors, laboratory test results), and changes in patient state or context (that might indicate emergency situations). Communication is naturally *event-driven*, requiring the asynchronous transmission of data that captures the nature of some occurrence according to application-specific event-naming, specification and management. In some domains patient treatment records are held in databases and the entry of a record may trigger communication. We shall not focus on this particular source of communication; details of how to integrate databases with event-based communication are given in [1, 2, 3].

Security models must reflect the structure and requirements of the application environment. When security policy is enforced within a federation of domains it must be clear where responsibility for each data item lies, and domain managers must be accountable for the information they generate, hold and are obliged to share. General rules may hold about the data that can be sent from the domain responsible for it to other specific domains. Complete data may be sent to a patient's primary care practice and current hospital clinician, but certain attributes may be withheld from other domains, such as pharmacy and audit. Medical data to be used for research purposes must be de-identified, but it must be possible to relate the different records of a given person.

Traditional access control mechanisms tend to focus on authentication and authorisation of individual clients. In highly dynamic, distributed applications, context becomes increasingly important; and the circumstances in which access is appropriate may also need to be captured. Context may include absolute and relative times, prerequisite actions and procedures, credential checks, the relationship between the principal accessing the data and the principal to whom the data relates, any individual exclusions that may have been placed on the data and whether an emergency situation exists.

Often data is highly sensitive, and must be protected, yet must also be delivered in a timely manner to those parties that have a need and right to know, for example, clinicians carrying out emergency procedures. Security policy enforcement must not unduly delay the transmission of data. Heavyweight security procedures tend to be bypassed by human users when there is an option.

Data must be protected, not only from inappropriate clients end-to-end, but also during transmission or processing by intermediate infrastructure components. Data may remain confidential for as long as a human lifetime, or longer.

In summary, policy specification and enforcement must therefore capture: (1) Data is communicated asynchronously in the form of messages that embody events. (2) Security policy must ensure that event type names and specifications are unique within a system and that their evolution is controlled. (3) When events are structured, separate components of events may need different protection. (4) Security policy must specify the principals that are authorised to access data. In an event-based system, control is enforced in terms of who can send and receive events. (5) Parametrised RBAC has the required properties of simple administration and subtle expression of policy. In a multi-domain system, inter-domain negotiation establishes access rights in terms of roles within domains. (6) It must be possible to qualify access rights by context. (7) Domain managers must be supported in their obligations to protect data. General domain-level rules must be established and enforced on transmitted data. (8) Mechanisms must be in place to secure the flow of data through infrastructure components. (9) Security policy enforcement must not delay communication unduly. Real-time communication may be required by some clients. (10) Security policy should support analysis and reasoning about event security.

Section 2 of this chapter presents our model of event-based communication, showing how access control policy can be incorporated. In Section 3, we address the challenges of secure event type management for Internet-scale environments. Section 4 extends event access controls, to give end-to-end event security protection, even inside event broker nodes and event recipients. Section 5 demonstrates how security policy can allow formal reasoning about event security, and shows how end-to-end security enhances this. We conclude in Section 6.

This chapter focuses on policy expression and enforcement. As we introduce topics, we outline briefly the assumptions we make about the system architecture and implementation within which policy must be enforced. Other papers on the architecture and engineering of distributed event-based systems give further detail [4, 5, 6].

2   Integrating Access Control into Event-Based Systems

In this section we discuss both client-level and communication security. First we cover how the event data, that is to be communicated and accessed, is defined, named uniquely, and its evolution controlled.

2.1   *Event Type Specification and Ownership in a Single Domain*

Before events can be advertised, published or subscribed-to they must be defined and their specification made available system-wide. We assume that each event type has an owner who registers the type definition with the local domain's secure server. We assume a secure server per domain capable of supporting a Public Key Infrastructure (PKI). When a type is registered, a public/private key-pair is associated with it and the public key is bound into its definition; the type owner holds the private key. The management of the type, for example any changes that need to be made to its specification, are controlled within this PKI.

We model events as structures consisting of multiple named parts (or attributes). This fine-grained structure is useful for characterising event filters, and for restricting which subscribers may receive which parts of an event.

In Section 3.1 we discuss how events may be made known system-wide in a multi-domain system. We can then assume that event type specifications are available to all application domains, statically through a registration service, and dynamically when publishers advertise that they are live and ready to publish. First, we describe the basic communication mechanism and access control policy and mechanisms.

2.2   *Event Communication: Advertise, Publish and Subscribe*

Having defined event types, communication can then take place. We assume a publish/subscribe paradigm for event-based communication, [7]. Much of our research has been based on a large-scale pub/sub architecture Hermes, [8, 9]. The assumption is that event transmission is carried out by a shared event broker network, independent of, and shared by, the client domains that use it. An overview of the approach to secure transmission in Hermes is given in [10].

Here, we modify our assumptions about the broker network for applications with long-term confidentiality requirements as described in Section 1. We do not envisage a large-scale broker network independent of the application domains; each broker belongs to an application domain, from which its security properties, including trust, are inherited. A client-hosting broker acts on behalf of publishers and subscribers in its domain and enforces the security policy of the domain. For load balancing, more than one broker per domain can be used.

An event publisher, that becomes ready to publish, multicasts an advertisement system-wide via its domain-local event broker. The broker checks and enforces authorisation policy relating to the advertisement, see Section 2.5. Initial authentication of publishers, subscribers and brokers is done using public key pairs bound to identity certificates (e.g. X.509) [5]. There can be any number of publishers per event type and their advertisements make them known to the event-brokers.

Principals with an interest in receiving events of an advertised type (subscribers) issue a subscription, together with a subscription filter, via their domain-local event broker. Again the broker checks that the subscriber is authorised to make the subscription, see Section 2.5.

Subscriptions are routed to the relevant publishers' brokers. When an event is published, it is matched against the subscription filters and sent to those subscribers whose filters match the event attributes.

For example, a pathology laboratory would advertise its intention to publish pathology reports to its event broker, and would then publish each pathology report as an event. Authorised doctors and clinical researchers would subscribe to pathology events, with subscription filters for their patients or research studies.

We now describe how this procedure is qualified by security policy enforcement, controlling who can publish and subscribe and the data that can leave a domain and be sent to each destination domain.

2.3 Role-Based Access Control

Role-Based Access Control (RBAC) [11] is a standard technique for simplifying scalable security administration by introducing *roles* as a linking concept between *principals* (i.e. users and their agents) and *privileges*. Privileges, such as the right to use a service or to access an object managed by a service, are assigned to roles. Separately, principals are associated with roles. This separates the administration of people, and their association with roles, from the control of privileges for the use of services (including service-managed data). The motivation is that users join, leave and change roles in an organization frequently, and the policy of services is independent of such changes. Service developers need only be concerned with specifying access policy in terms of roles, and not with individual users.

RBAC is particularly suitable for securing event-based systems because the process of agreeing the notions of role between decoupled participants within an event-based system closely parallels the process by which those same decoupled participants must agree on how to interpret events. Both are well suited to operation in widely distributed, multi-domain systems. Here we focus on securing access to the communication service using RBAC.

Authentication into roles must be securely enforced to control the use of all protected services and access to the data they manage [12]. Domain managers, or their delegates, specify communication policy in terms of message types and roles; that is, which roles may create, advertise, send and receive which types of message, see Section 2.2.

Inter-domain communication is achieved through negotiated agreements, expressed as access control policy, on which roles of one domain may receive (which attributes of) which types of message of another. This negotiation must also take into account any general domain-level policies regarding data transfer, see Section 3.2.

Standard RBAC causes principals to be anonymous (i.e. the privileges available to role holders do not depend on their identity), whereas parametrised RBAC gives the option of anonymity or identification, for example, treating-doctor(*hospital-ID, doctor-ID, patient-ID*). By means of parametrised RBAC it is possible to capture relationships and patient-specified exclusions, as may be required by law. The use of parametrised roles can also help to avoid an explosion in the number of roles required when RBAC is used in large systems. For the communication service, RBAC policy indicates the visibility (to roles, intra- and inter-domain) of specified attributes of message types.

The fact that advertisement is required before messages can be published, and that both are RBAC-controlled, prevents the spam that pervades email communication between humans. Without such control, denial-of-service through publication or subscription flooding could degrade large-scale inter-software communication in the same way that it consumes resources in email management. With our approach, a spammer could only be an authorised, authenticated member of a role and therefore could be held accountable.

2.4   OASIS Role-Based Access Control

In this section we provide a brief introduction to the Open Architecture for Secure Interworking Services (OASIS) [12], developed at Cambridge and used as the basis for our research. OASIS provides a comprehensive rule-based means to check that users can only acquire the privileges that authorise them to use services by activating appropriate roles. Although we use OASIS policy in subsequent sections, we aim to highlight fundamental principles that must guide any policy implementation.

A role activation policy comprises a set of rules, where a role activation rule for a (target) role r_t takes the form

$$r_1, .., r_n, a_1, .., a_m, e_1, .., e_l \vdash r_t$$

where r_i are prerequisite roles, a_i are appointment certificates (most often persistent credentials) and e_i are environmental constraints. The latter allow restrictions to be imposed on when and where roles can be activated (and privileges exercised), for example at restricted times or from a restricted set of computers. Any predicate that must remain true for the principal to remain active in the role is tagged as a *role membership condition*. Such predicates are monitored, and their violation triggers revocation of the role and related privileges from the principal.

For example, a role activation rule could support patient referral between consultants, by allowing dynamic assignment of a treating-doctor target role, given an appropriate consultant-referral appointment certificate.

An authorisation rule for some privilege p takes the form

$$r, e_1, .., e_l \vdash p$$

An authorisation policy comprises a set of such rules. OASIS has no negative rules, and satisfying any one rule indicates success.

In our example, role authorisation rules could allow treating doctors and clinical researchers respectively the privilege priv-clinical of reading clinical data for the appropriate patients.

OASIS roles and rules are parametrised. This allows fine-grained policy requirements to be expressed and enforced, such as exclusion of individuals and relationships between principals, for example treating-doctor(*hospital-ID, doctor-ID, patient-ID*), as outlined in Section 2.3.

2.5 Access Control Policy for Publish/Subscribe Clients

The most general access control requirement in pub/sub relates to how clients connect to the pub/sub service (e.g. a local broker of a distributed broker network), and make requests using its API. This implements security at the pub/sub network edge. Many pub/sub systems [7, 13] include the following service methods in one form or another:

define(*message-type*)
advertise(*message-type*)
publish(*message-type, attribute-values*)
subscribe(*message-type, filter-expression-on-attributes*)

Some policy languages will only be able to provide a coarse specification of the client privileges required to use the API. In OASIS RBAC, the authorisation policy for any service specifies how it can be used in terms of roles and environmental constraints, and parametrised roles can be used to effect fine-grained control.

OASIS policy indicates, for each method, the role credentials, each with associated environmental constraints, that authorise invocation. OASIS role parameters can be used to limit privileges to particular *message-type*s. The define method is used to register a message type with the service and specify its security requirements at the granularity of attributes. On advertise, publish and subscribe, these requirements are enforced. We can therefore support secure publish/subscribe within a domain in which roles are named, activated and administered.

3   Multi-domain Security Architecture

Multi-domain systems can be structured as a federation of autonomous domains. This distributed approach requires a model for secure event type definition and management, and an approach to inter-domain communication control. Before covering these issues, we first clarify our domain model.

A domain-structured OASIS system is engineered with a per-domain, secure OASIS server, see [4] for details. Also, each domain has a policy store containing all the role activation and service-specific authorisation policies. This avoids the need for every service to independently perform authentication and secure role activation, allowing simple service implementation. The domain's OASIS server carries out all per-domain role activation and monitors the role membership rule conditions while the roles are active. This effectively concentrates role dependency maintenance within a single server and provides a single, per-domain, secure service for managing inter-domain authorisation policy specification and enforcement. For robustness, this would typically be a replicated service, with fail-over to a hot standby server.

3.1   *Management of Event Names, Types and Policies*

As introduced in Section 2.1, a mechanism is needed to agree on the naming of event types, when constructing policy-secured, pub/sub systems. Extending for multiple domains, we assume that domains are allocated unique names within the system as a whole and that roles are named and managed within a domain. Each domain provides a management interface through which role activation policies and service authorization policies can be specified and maintained.

A group of domains may have a parent domain from which an initial set of role names and policies is obtained. For example, health service domains may start from an initial national role-set. The domain management

Table 1 Contents of a secure event type definition

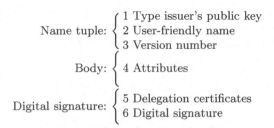

$$
\begin{array}{r l}
\text{Name tuple:} & \left\{\begin{array}{l} \text{1 Type issuer's public key} \\ \text{2 User-friendly name} \\ \text{3 Version number} \end{array}\right. \\[1em]
\text{Body:} & \left\{\begin{array}{l} \text{4 Attributes} \end{array}\right. \\[1em]
\text{Digital signature:} & \left\{\begin{array}{l} \text{5 Delegation certificates} \\ \text{6 Digital signature} \end{array}\right.
\end{array}
$$

interface allows local additions and updates, for example when national government policy needs to be customised for implementation regionally. As mentioned above, parametrised roles allow domain-specific parameters, for example treating-doctor(*hospital-ID, doctor-ID, patient-ID*). This allows relationships and exclusions to be captured as well as avoiding excessive numbers of roles in large-scale systems.

For a multi-domain pub/sub system, we introduce a format of event type definition that binds the type name and definition together in a secure manner. Public key cryptography is used to guarantee the authenticity and integrity of this type information. To achieve this we require that all participating brokers in the pub/sub system have a key-pair. We can thus require that event-type issuers incorporate this public key into the type name. This facilitates an event naming scope for each particular type issuer. Since event type names include a public key, it is intuitive that the event type definitions should be signed by the corresponding private key. This binds the type definition to the type name, and facilitates verification of event type integrity and issuer authenticity. Thus we protect the system against forged or tampered event type definitions. We also reduce the chance of accidental name collisions, and provide a unique handle through which policy can refer to the names of types and attributes. An outline is given below and further details are in [14, 15, 16, 10].

The six items that make up a secure event type definition are shown in Table 1. Items 1–3 identify the name of the type. The Attributes item 4 indicates the core event type definition, and items 5 and 6 contain a digital signature of the event type. Table 2 illustrates this for a pathology report type.

Adding a public key to the type name eliminates event name conflicts. While this is desirable, a user-friendly name is also maintained. The user-friendly name is able to encode useful aspects, such as hierarchical naming, e.g. uk.nhs.path_report. This enables administrative grouping of type definitions across multiple event type owners. Orthogonal to both of those concerns is type evolution, hence the provision of a version number. Releasing a new version of an event type definition will not conflict with previous instances still in use within the publish/subscribe system. Indeed, to avoid

Table 2 Example of a secure event type definition for pathology reports

1 Type issuer's public key
-----BEGIN PGP PUBLIC KEY BLOCK-----
Comment: Public key for NHS Information Authority ...
2 User-friendly name
uk.nhs.path_report
3 Version number
31ab47dc-4e52-4509-92ec-39f6e603d6e6
4 Attributes

patient-ID	0640d3d2-0b2b-45db-b66b-0f4200cd8358	String
patient-name	de312646-e940-466d-a71f-fa9122891d5a	String
hospital-ID	e4ab7b23-87dc-44ca-bf7c-cba5cc5ce6f7	String
lab-number	1dd5f593-1bc6-4f53-89ef-5f2a38b1d4ec	String
sample-receipt-date	a273e069-7cb6-48ca-b827-9bb11d449dbf	Date
path-report-text	b831c325-4d63-4d98-b7c0-782a1d4a1c44	Binary
SNOMED-CT-codes	3514a33d-5173-453c-8b5a-11e8ead71761	String

5 Delegation certificates
[Delegation certificate from NHS root to NHS Information Authority]
6 Digital signature
-----BEGIN PGP SIGNATURE----- ...

race conditions for version numbers when multiple type managers are releasing updated event definitions, a UUID scheme is used for version numbers. UUIDs are 128-bit values, generated by a collision-avoiding algorithm.

Item 4 in the event type definition describes the event type structure. Each attribute definition itself consists of a user-friendly name, a unique identifier (UUID), and an attribute type identifier. The set of types supported depends on the subscription filter language used. Friendly names for attributes are intended to be used by clients of the publish/subscribe system, e.g. *path-report-text*, whereas the UUID is used by intermediate brokers during distributed event routing. The friendly names only need to be unique within the context of one particular version of an event type definition. When a publisher-hosting broker receives an event to route, it looks up the friendly names used by its client using the publisher's event type definition. This allows the UUID fields to be correctly populated. The reverse of this process occurs at subscriber-hosting brokers. The UUIDs allow multiple versions of an event type to exist within the publish/subscribe system at a point in time.

The digital signature of an event type provides a guarantee of the authenticity and integrity of the type definition. The signature is calculated over items 1–4 in Table 1. It thus binds the type definition and the name tuple.

The delegation certificates (item 5) facilitate Internet-scale management of event types. Since key-pairs are involved in signing event types, without delegation certificates the type owner would need to re-sign all updates to the type. Delegation certificates facilitate a digitally-signed path of trust from the original event type owner to type managers: parties that are allowed to update event types on their behalf. In Table 2, the UK National Health Service (NHS) authorises the NHS Information Authority to securely manage certain event types. The delivery of delegation certificates to type managers can be performed out-of-band, e.g. by secure email. The delegation certificates also provide the means to specify fine-grained access rights. Our prototype implementations have typically supported rights such as addAttribute, removeAttribute, editAttributeName and editAttributeType.

3.2   Inter-domain Communication Control

When security policy is enforced within a federation of domains, the manager of each domain is responsible, and accountable, for the information held within it and propagated to other domains. Encryption can be used to protect data during communication but the confidentiality of healthcare data persists for as long as a human lifetime, or longer. On that timescale, encryption keys may be compromised by being exposed or becoming insufficiently secure to withstand attacks from the increased computing power available.

Under these requirements, some of the much-researched optimisations advocated for publish/subscribe systems become problematic. Use of a shared external broker network, with content-based routing, is ill-advised and direct communication between domains is preferable. We envisage that a given message will be sent to a handful of domains, unlike in global-scale pub/sub information dissemination.

Some data may have a short lifetime, such as that transmitted in battlefield scenarios and emergency situations. In these situations it may be appropriate to use an external broker network, perhaps assembled in an ad hoc fashion from mobile components. Messages in transit are assumed to be protected by a standard, link-level scheme such as TLS [17]. At the client level, the attributes of messages may be encrypted separately and decrypted by brokers on behalf of their authorised clients. Some brokers may be less trustworthy than others. Here, client-level encryption is appropriate, and essential, so that untrusted brokers cannot decrypt highly sensitive attributes. Details of how this can be achieved are summarised in [16].

Sensitive data should not be communicated to principals in other domains unless authorised by policy. This can be established when RBAC-based authorisation is set up by negotiation between domains. Rather than sending entire events, containing attributes that are sensitive but protected by encryption, it may be preferable to remove or to lower the precision of some of the attributes. Similarly, brokers receiving events from other domains should enforce client access rights according to RBAC policy. These issues have been explored in detail in [2, 3].

4   End-to-End Security with Information Flow Control

Thus far, we have outlined a security model for event-based systems, in which multi-domain RBAC protects access to both individual events and the event type system. This approach of integrating access control into event-based systems provides strong, consistent security at the boundaries between publishers, event broker nodes and subscribers, and between federated domains. However, it has limitations too:

- A large base of event broker code must be trusted not to leak information, e.g. accidentally disclosing events to unauthorised subscribers, or disclosing one subscriber's subscriptions to another subscriber.
- If services such as network monitoring or complex event detection are added to the event brokers, they need to be trusted too.
- Once data is released to a subscriber, information about its security is lost, limiting the potential for end-to-end security, e.g. in follow-on messages.

In this section, we show how these limitations can be overcome, using Distributed Information Flow Control (DIFC) to guard all event processing within the event broker nodes. This approach also enhances long-term data security, because unauthorised data cannot be accessed or stored within a DIFC framework, even in encrypted form.

Figures 1a and 1b illustrate this distinction, between security controls at the boundaries of the event system, and continuous, end-to-end security tracking with DIFC. In both figures, publishers and subscribers communicate via trusted event brokers in two domains. Two publishers in Domain 1 send Event A and Event B respectively. The subscriber in Domain 2 has a subscription which matches both events, but Event A is blocked for security reasons. On receiving Event B, the subscriber attempts to save it to a file, and republish it as Event C – both of which actions should be blocked for security reasons. At the same time, a monitor component Σ in Domain 2 emits an event count, which is triggered by Event A, but not Event B. In the figures, the large circles represent the boundaries of the event brokers, and a shaded background marks trusted code. Event flows with the same line style are protected with the same security restrictions. Tap icons show potential data leaks. Solid small circles are security checks.

In Figure 1a, security is enforced only at the boundaries of the event brokers. This correctly blocks Event A from the subscriber, and allows Event B through. However, the subscriber can then republish or leak to a file the confidential contents of Event B. Furthermore, there is no straightforward way to ensure that the event count contains no confidential data.

In Figure 1b, DIFC tags allow continuous end-to-end tracking of data security. These not only correctly block and allow Events A and B respectively, but also ensure that subsequent uses of Event B retain the same security properties (dashed lines), preventing republishing or saving to a file. Furthermore, the event count can be shielded from confidential data, by using a small,

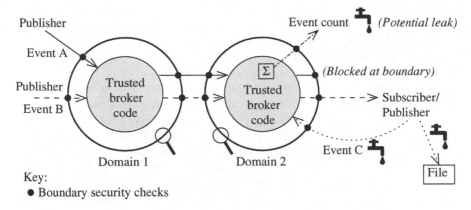

Fig. 1a Boundary security between event publishers, brokers and subscribers

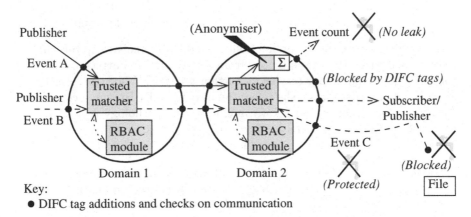

Fig. 1b End-to-end security with Distributed Information Flow Control

isolated anonymiser component before computing the sum; this reduces the need for trusted code in the event broker.

4.1 Distributed Information Flow Control

Information Flow Control [18] is a security technique that uses security labels to track and guard all information and processing. Information, that must have a security label, can flow only to processes with compatible security labels, and any resulting information is restricted by the labels of both the process and the original information – unless special privileges are exercised, such as declassification. For example, this could block confidential patient data from a process with insufficient confidentiality privileges.

Distributed Information Flow Control (DIFC) [19, 20] applies this technique to distributed systems, by allowing new security label constituents to be created dynamically by processes, instead of relying on a centrally predefined set.

We express each security label as a set of tags, each representing an atomic security concern. Each tag is used either for confidentiality or integrity purposes: data may ordinarily only increase in confidentiality, and decrease in integrity as it is processed, unless special privileges are exercised for *declassification* (lowering confidentiality) or *endorsement* (higher integrity).

Notationally, we write a DIFC security label as (S, I) with confidentiality component S and integrity component I, where S and I are each sets of tags. Information may flow from a to b with labels (S_a, I_a) and (S_b, I_b) only if $S_a \subseteq S_b$ and $I_a \supseteq I_b$. For example, this restriction would guard receipt of event data by a process. To gain access to data, the process may need to use privileges to increase its confidentiality (or secrecy), or lower its integrity – these privileges are considered weaker than their opposites, declassification and endorsement above.

To illustrate this practically, we use the following clinical example:

At a surgical outpatient appointment at Addenbrooke's Hospital, the consultant Miss Ali takes a biopsy from a patient (Mr Patterson), and sends it to the pathology service for further analysis. When she receives the results two days later, she refers the patient to Dr Brown, a consultant oncologist, for chemotherapy, including the pathology report and her own comments. The pathology report is also released to a cancer researcher, Dr Chen, for analysis in an anonymised patient cohort.

The three events in this example are: e_1: the pathology request sent in conjunction with the physical biopsy, e_2: the pathology report, a uk.nhs. path_report event as in Table 2, and e_3: the pathology report with comments.

Here, a confidentiality tag $t_{\text{identifiable}}$ could be used to restrict access to the identifiable patient data in the pathology report event e_2, such as the *patient-name* or *patient-ID* attributes. Non-identifiable data, such as *SNOMED-CT-codes* and *path-report-text* could be restricted with other confidentiality tags. Finally, the pathology laboratory would add an integrity tag t_{pathlab} to all attributes which it added, to demonstrate the authenticity of the report.

The overhead of DIFC label checks is small [21]. Computationally, the complexity is linear in the number of labels; results can be cached for efficiency.

In a distributed environment, we treat tags as local entities, ensuring that network outages need not stall the event processing system. Tags are created within an event broker node, and the creator of a tag can delegate privileges to other processes, for adding the tag to labels or removing it from them. (In [21], we detail how we can do this in an IFC-safe way, i.e. without the privilege delegations themselves leaking information.) When event broker nodes exchange labelled data, they need to translate any tags with global meaning into the local representation, e.g. using DNS-backed URIs as global names. However, the extra communication overhead is small, and local DIFC checks can then efficiently treat all tags as local.

4.2 Event Security with DIFC

DIFC is well-suited to event processing systems: DIFC can protect all event-based communication; any remaining communication channels or storage mechanisms that persist between events should be blocked or DIFC mediated. This combines nicely with our model of multi-part events in Section 2.1, as each event part (or attribute) can have its own security label. By restricting event handling to a controlled API, this allows code within an event broker to access only those parts of an event for which it has tag privileges. The existence of any other parts will effectively be hidden.

For efficient operation, we allow processes within an event broker to specify in advance that they will always release an event (possibly modified) for other processes to operate on [21]. Without this mechanism, all event parts would need to be labelled with all of the confidentiality tags (and only a subset of the integrity tags) from the releasing process's label, resulting in unnecessary complexity of design and operation.

DIFC for events improves the security of general purpose event processing systems, including event-based middleware, offering strong protection against communication leaks, and efficient end-to-end-security. The bulk of the event processing system can consist of code with only limited privileges.

Figure 1b illustrates the effect of DIFC tracking within an event broker: the security labels of multi-part events are tracked throughout an event-based system, both within and between event brokers (shown by the different line styles). If publishers and subscribers use a DIFC-enforcing API to interact with events, then all of their subsequent operations are protected too – event publication C must respect event B's security and integrity labels, unless additional declassification or endorsement privileges are exercised.

4.3 Enforcing OASIS Security with DIFC

Event systems can enforce Role-Based Access Control using DIFC restrictions. Our approach to this supports dynamic role assignment and revocation, and parametrised roles. We illustrate this using OASIS RBAC (outlined in Section 2.4), thus providing end-to-end DIFC support within the multi-domain security architecture of Section 3.

In essence, whenever an OASIS role is activated, DIFC confidentiality tags are created for the resulting privileges, and any event-handling processes that need access to the data are granted the appropriate confidentiality tag privilege. Trusted processes, such as broker-to-broker communication or shared matching, may need to be granted declassification privileges too, depending on the event system design.

In our example from Section 4.1, Miss Ali submits a subscription for all clinical events at Addenbrooke's Hospital pertaining to patient Mr Patterson. The policy for this message-type requires that a subscriber must have privilege priv-clinical(*patient-ID*) to receive events with the corresponding *patient-ID*.

Mr Patterson's pathology report is published by the pathology lab: the identifiable patient data is labelled with confidentiality tag $t_{identifiable}$, and the clinical details have tag $t_{MrPatterson}$ corresponding to the privilege of being able to handle clinical data with Mr Patterson's *patient-ID*, i.e. priv-clinical("Mr Patterson"). Relevant parts of the pathology event are shown in Table 3. Creation of this event was OASIS-mediated; the trusted OASIS support code retains the right to grant privileges over $t_{MrPatterson}$ on the local event broker node.

Table 3 A multi-part pathology report with security labels

Label (S, I)	Event part name	Event data
$(\{t_{identifiable}\}, \{\})$	patient-ID	1234567768
$(\{t_{identifiable}\}, \{\})$	patient-name	Roger John Patterson
$(\{\}, \{\})$	hospital-ID	RGT01
$(\{\}, \{t_{pathlab}\})$	lab-number	A10/21367
$(\{\}, \{t_{pathlab}\})$	sample-receipt-date	2010-02-04
$(\{t_{MrPatterson}\}, \{t_{pathlab}\})$	path-report-text	CASE HISTORY: Large solid tumour with some papillary elements on right side of trigone. Query TCC. MACROSCOPIC: ...
$(\{t_{MrPatterson}\}, \{t_{pathlab}\})$	SNOMED-CT-codes	302512001 |Bladder|, ...

When the event broker receives the pathology report, the event subscription causes role treating-doctor("Addenbrooke's Hospital", "Miss Ali", "Mr Patterson") to be activated, yielding privilege priv-clinical("Mr Patterson"). The corresponding DIFC tag $t_{MrPatterson}$ is retrieved, and the subscription's process is granted access to it. More precisely, the process receives the privilege to add $t_{MrPatterson}$ to its confidentiality component, and effectively spawns a lightweight copy of itself, allowing it to inspect the event contents without prejudicing its ability to operate with a lower confidentiality level on subsequent events.

When Miss Ali refers the patient to Dr Brown, he too obtains the appropriate treating-doctor role, and thus the right to read the pathology report in question, as well as the patient demographics tagged $t_{identifiable}$. The researcher, Dr Chen will only be granted privilege priv-clinical("Mr Patterson") to read the pathology report, but no access to the identifiable patient data. If patient linkage is required for the research, a trusted process can anonymise the *patient-ID*, adding a new event part *anonymised-ID* that Dr Chen can read.

Dynamic privilege assignment adds an extra complexity glossed over above: instead of using a single tag t for a privilege at all points in time, a sequence of tags $t_1, t_2, t_3, \ldots$ is created as needed whenever the roles actively supporting the privilege change. This is needed to enforce OASIS privilege revocation, without mandatory revocation of DIFC tag privileges (as such a mechanism would violate DIFC restrictions). Therefore, when a role is revoked, a new DIFC tag t_{n+1} is used for subsequent data, essentially making the revokee's existing tag t_n worthless, except for processing that has already been authorised.

DIFC tags provide efficient, local enforcement of privileges. The additional overheads of tag management are minor, as they affect only the relatively expensive OASIS-mediated activities of exercising privileges and roles, and role revocation. OASIS policy also provides effective translation between local DIFC tags and domain-level privileges.

We have shown that DIFC enforcement of RBAC restrictions can provide fine-grained end-to-end security for event flows, both within event processing broker nodes, and for event publishers and subscribers subject to DIFC restrictions. This also reduces the trusted code footprint of the event processing system, supporting the principle of least privilege.

As the following section shows, OASIS privileges or DIFC tags in the application domain integrate well with other aspects of secure system design, such as physical security.

5 Reasoning about Event Security

A policy-based approach to event security enables formal reasoning about information flow in an event-based system. This reasoning can be performed either at the level of the policy language, such as OASIS rules (Section 2.4), or in terms of their projection into DIFC labels, where DIFC enforces event security (Section 4.3). Policy language reasoning is generally simpler, but DIFC reasoning offers extra flexibility, such as integration with multiple policy languages, direct links to environmental factors such as physical location, and better support for end-to-end policy enforcement.

5.1 Policy-Based Reasoning

Linking RBAC privileges to particular message-types (Section 2.5) immediately allows straightforward reasoning about certain security properties. For example, suppose the pathology message-type's policy requires privilege priv-clinical(*patient-ID*) to access the event part named "path-report-text", for events with that *patient-ID*, and the only OASIS policy rules yielding priv-clinical permissions are:

treating-doctor(*hospital-ID, doctor-ID, patient-ID*) ⊢ priv-clinical(*patient-ID*)

clinical-researcher(*study-ID, researcher-ID, patient-ID*) ⊢ priv-clinical(*patient-ID*)

Then we can reason that only treating doctors and authorised clinical researchers can access the pathology text, as long as the assignment of the treating-doctor and clinical-researcher roles is suitably restricted.

Similarly, if consultants can refer patients to each other through consultant-referral role appointment certificates, then these can be supported with the following OASIS policy rule:

treating-doctor(*referring-hospital-ID, referring-doctor-ID, patient-ID*),

consultant-doctor(*hospital-ID, doctor-ID*),

consultant-referral(*referring-doctor-ID, doctor-ID, patient-ID*)

   ⊢ treating-doctor(*hospital-ID, doctor-ID, patient-ID*)

We can then conclude that the consultant referral from Miss Ali to Dr Brown described in Section 4.1's example will enable Dr Brown to read the pathology report's text when the patient is referred – as long as he receives a copy, and his treating-doctor role is still valid.

This sort of reasoning focuses principally on confidentiality restrictions, i.e. guaranteeing that data will not be received by unauthorised recipients. This focus is appropriate for our target scenarios, such as protecting healthcare data against disclosure. However, protecting integrity and availability are equally important for safety-critical systems.

Integrity can be linked to the event type, implying that all publications of a particular message-type are of high integrity, or that only high-integrity messages will have certain attributes set. Limited integrity restrictions can also be implemented indirectly in OASIS, by treating them each as a forced low-integrity role that is held by all principals except for a few high-integrity entities, like the high-integrity pathology lab in our example. However, these approaches require event publishers to manage their own integrity tracking, to ensure that they do not inadvertently raise the integrity level of data from an external source. DIFC restrictions provide much more robust, fine-grained support for protecting both confidentiality and integrity.

OASIS policy checking can also prove a limited version of availability: it can show that no access control restrictions will prevent the delivery of a particular event to a subscriber. However, this depends on the underlying availability of the event processing system and supporting infrastructure.

5.2 Event Security Reasoning with DIFC Labels

DIFC labels allow direct reasoning about confidentiality and integrity properties. The benefits of this include: (1) DIFC restrictions protect against many security vulnerabilities directly, (2) the correctness of policy enforcement can be verified in terms of its transformation to DIFC tags, (3) fine-grained, end-to-end confidentiality/integrity reasoning is supported, (4) multiple policy languages can translate to common enforcement tags, and (5) DIFC tag analysis can facilitate reasoning about physical security.

The greatest advantage of translating policy into DIFC restrictions is that the basic Information Flow Control restrictions are not violated. This system-wide property subsumes many individual security checks, and its power should not be understated. For example, this means that checking "No unauthorised users may access patient data" translates to checking what program code can declassify tagged patient data in the event processing system, and the safeguards on corresponding role assignment.

Relying on a common enforcement mechanism allows safe, consistent policy enforcement. Correctness of the OASIS policy enforcement mechanism can be assessed simply in terms of its transformation of policy rules to DIFC tags. Provided this transformation is correct, DIFC restrictions will then enforce the OASIS RBAC guarantees. Furthermore, policy-based reasoning will apply to the resulting DIFC tags too.

End-to-end confidentiality and integrity are possible, because publishers and subscribers can process event data within a DIFC-enforcing framework. This makes many security checks on client code unnecessary, as this code is protected by the DIFC framework. For example, a subscriber cannot forward patient data to an unauthorised third party, because DIFC restrictions will automatically block access to that data. Event subscribers (and untrusted code within the event broker system such as network status monitors) are also prevented from accessing or retaining confidential data, even in encrypted form; this provides essential protection against the long-term risk of disclosure.

This end-to-end support is also potentially more fine-grained than the underlying policy may suggest. If an event monitor were introduced, to count the number of pathology reports by anatomical site (e.g. the number of pathology reports with a SNOMED-CT code of "302512001 |Bladder|"), the monitor could declassify the non-disclosive result immediately after checking the "SNOMED-CT-codes" field. Then the only program code that would need to be trusted would be the code reading that field. A more coarse-grained policy-based analysis would instead suggest the entire monitoring codebase, and all subsequent messages, as potential data leaks.

DIFC tags can also protect against cross-contamination of patient data: if a recipient uses confidentiality privileges to merge two patients' data, the result will be contaminated with tags for both priv-clinical privileges, preventing accidental release as a single patient's data.

Multiple policy languages can be supported, provided that they can all translate their requirements into DIFC tags. This may limit the potential to perform security reasoning at the policy level – unless the DIFC tags of one policy language can be translated back into another, or the tag overlap between the languages can be characterised. On the other hand, this does support safe, consistent interaction between otherwise independent policy languages.

Finally, location-based policies and reasoning are well supported by DIFC tags – based on either network or physical location. For example, a confidentiality tag can be used to keep all patient data within the health service's physically secure network. This can enhance the multi-domain security architecture of Section 3, by supporting high-level specification of the interface between domains and preventing unauthorised event bridges between them. Similarly, physical security can be linked to the DIFC system, so that data displayed or entered at a hospital terminal is automatically linked to a DIFC confidentiality level, which is in turn bound to OASIS security policy rules. This allows the security of the event system to extend beyond the event interface.

For example, a doctor might be allowed to view clinical data only on a trusted terminal, e.g. inside a hospital, by defining a secure-terminal role for this (backed by environmental constraints). This results in DIFC confidentiality labels as described above, and our earlier rule can be changed to:

$$\text{treating-doctor}(\textit{hospital-ID, doctor-ID, patient-ID}),$$
$$\text{secure-terminal} \vdash \text{priv-clinical}(\textit{patient-ID})$$

Event security reasoning scales well to large-scale distributed systems: our domain model ensures that event types are managed in a structured, domain-based way. Parametrised RBAC reduces the number of rules and the frequency of rule change. This lets policy authors and users reason about event security on the basis of relatively static, domain-local information. Where policy references roles from other domains, either external rules can be checked directly, or partial conclusions can be drawn, with open-ended external role dependencies. For example, in reasoning about the consultant referral rule, the consultant-doctor role could be assumed to be assigned correctly by the appropriate governing body (the General Medical Council in the UK), without needing complete knowledge of the rules governing this assignment. High-level interface policies can further simplify analysis, particularly in large organisations such as a health service, by allowing reasoning about long-lasting security guarantees, independently of more dynamic local rules.

Thus end-to-end DIFC security enhances the scope for reasoning about event security, enabling policy-based reasoning to be extended to include event publishers, subscribers, and the environmental context in which they operate.

6 Conclusions

The ability to reason about event security depends on the underlying security architecture. In this chapter we have outlined our approach, which integrates strong, policy-based security into event-based systems. OASIS Role-Based Access Control allows simple administration and rich distributed security policy, with parametrised roles and appointment certificates. OASIS secures not only individual events, but also the event type system, supporting evolving rules and event structures for long-lived systems and data. We have shown how to use this in developing multi-domain systems, allowing security policy to reflect complex organisational structures and interconnections.

The security of event systems is enhanced by adopting fine-grained Distributed Information Flow Control restrictions on all event data. Extending these restrictions to event publishers and subscribers allows robust end-to-end security, which is essential for protecting long-term confidential data such as healthcare records. By translating OASIS policy enforcement into DIFC restrictions, we support both policy-based reasoning about event security, and the extension of this reasoning framework beyond the event processing middleware's boundaries.

In this way, event security reasoning can play a significant part in system-wide security audit and information governance, and can protect sensitive data in large-scale distributed systems.

Acknowledgement. We thank Jem Rashbass and Jon Crowcroft for their comments on this chapter. This work has been supported by grants from the UK Engineering and Physical Sciences Research Council (EPSRC) since 2000, including OASIS (GR/M75686), Contract-driven Systems (GR/S94919), EDSAC21 (GR/T28164), CareGrid (EP/C53719X) and, most recently, SmartFlow—Extendable Event-Based Middleware (EP/F042469 & EP/F044216; http://www.smartflow.org/).

References

1. Vargas, L., Bacon, J., Moody, K.: Integrating Databases with Publish/Subscribe. In: Proceedings of the 4th International Workshop in Distributed Event-Based Systems (DEBS 2005), June 2005, pp. 392–397. IEEE Press, Los Alamitos (2005)
2. Singh, J., Vargas, L., Bacon, J., Moody, K.: Policy-based information sharing in publish/subscribe middleware. In: IEEE Workshop on Policies for Distributed Systems and Networks (Policy 2008), IBM Palisades, New York, pp. 137–144. IEEE Press, Los Alamitos (2008)
3. Singh, J., Vargas, L., Bacon, J.: A Model for Controlling Data Flow in Distributed Healthcare Environments. In: Pervasive Health 2008: Second International Conference on Pervasive Computing Technologies for Healthcare, Tampere, Finland, January 2008, pp. 188–191. IEEE Press, Los Alamitos (2008)
4. Bacon, J., Moody, K., Yao, W.: Access control and trust in the use of widely distributed services. In: Liu, H. (ed.) Middleware 2001. LNCS, vol. 2218, pp. 295–310. Springer, Heidelberg (2001)
5. Bacon, J., Eyers, D.M., Moody, K., Pesonen, L.I.W.: Securing publish/Subscribe for multi-domain systems. In: Alonso, G. (ed.) Middleware 2005. LNCS, vol. 3790, pp. 1–20. Springer, Heidelberg (2005)

6. Bacon, J., Eyers, D.M., Singh, J., Shand, B., Migliavacca, M., Pietzuch, P.: Security in multi-domain event-based systems. IT - Information Technology 51(5), 277–284 (2009), doi:10.1524/itit.2009.0552
7. Eugster, P.T., Felber, P.A., Guerraoui, R., Kermarrec, A.M.: The many faces of publish/subscribe. ACM Computing Surveys 35(2), 114–131 (2003)
8. Pietzuch, P.R., Bacon, J.M.: Hermes: A distributed event-based middleware architecture. In: 1st International Workshop on Distributed Event-Based Systems (DEBS 2002), Vienna, Austria. ICDCS, pp. 611–618. IEEE Press, Los Alamitos (2002)
9. Pietzuch, P.R., Bacon, J.M.: Peer-to-peer overlay broker networks in an event-based middleware. In: 2nd International Workshop on Distributed Event-Based Systems (DEBS 2003). ICDCS, pp. 1–8. ACM SIGMOD, New York (2003)
10. Bacon, J., Eyers, D.M., Singh, J., Pietzuch, P.R.: Access control in publish/subscribe systems. In: Proceedings of the Second International Conference on Distributed Event-Based systems (DEBS 2008), pp. 23–34. ACM, New York (2008)
11. Sandhu, R., Coyne, E., Feinstein, H.L., Youman, C.E.: Role-based access control models. IEEE Computer 29(2), 38–47 (1996)
12. Bacon, J., Moody, K., Yao, W.: A model of OASIS role-based access control and its support for active security. ACM Transactions on Information and System Security (TISSEC) 5(4), 492–540 (2002)
13. Pietzuch, P., Eyers, D., Kounev, S., Shand, B.: Towards a Common API for Publish/Subscribe. In: Proceedings of the Inaugural Conference on Distributed Event-Based Systems (DEBS 2007), pp. 152–157. ACM Press, New York (2007) (short paper)
14. Pesonen, L.I.W., Bacon, J.: Secure Event Types in Content-Based, Multi-domain Publish/Subscribe Systems. In: SEM 2005: Proceedings of the 5th International Workshop on Software Engineering and Middleware, Lisbon, Portugal, September 2005, pp. 98–105. ACM Press, New York (2005)
15. Pesonen, L.I., Eyers, D.M., Bacon, J.: Access control in decentralised publish/subscribe systems. Journal of Networks 2(2), 57–67 (2007)
16. Pesonen, L.I.W., Eyers, D.M., Bacon, J.: Encryption-Enforced Access Control in Dynamic Multi-Domain Publish/Subscribe Networks. In: Proceedings of the Inaugural Conference on Distributed Event-Based Systems (DEBS 2007), June 2007, pp. 104–115. ACM Press, New York (2007)
17. Dierks, T., Allen, C.: The TLS protocol version 1.0. IETF RFC 2246 (January 1999)
18. Bell, D.E., La Padula, L.J.: Secure computer systems: Mathematical foundations and model. Technical Report M74-244. The MITRE Corp., Bedford, MA (May 1973)
19. Myers, A., Liskov, B.: Protecting privacy using the decentralized label model. ACM Transactions on Software Engineering and Methodology 9(4), 410–442 (2000)
20. Krohn, M., Yip, A., Brodsky, M., Cliffer, N., Kaashoek, M.F., Kohler, E., Morris, R.: Information flow control for standard OS abstractions. In: Proceedings of Twenty-First ACM SIGOPS Symposium on Operating Systems Principles (SOSP 2007), pp. 321–334. ACM, New York (2007)
21. Miglivacca, M., Papagiannis, I., Eyers, D., Shand, B., Bacon, J., Pietzuch, P.: High-performance event processing with information security. In: USENIX Annual Technical Conference, Boston, MA, USA, pp. 1–15 (2010)

Generalization of Events and Rules to Support Advanced Applications*

Raman Adaikkalavan and Sharma Chakravarthy

Abstract. Event-Condition-Action (ECA) rules monitor applications and systems and react to changes. In this chapter, we discuss various extensions to ECA rules to support advanced applications. We particularly use the access control domain to drive the extensions needed for expressiveness, specification, and execution of policies using the ECA paradigm. We discuss alternative actions, generalized event specification and detection, and event detection modes. We also discuss the extensions to the event detection graphs to implement the proposed ECA rule extensions.

1 Introduction

Event-Condition-Action (ECA) [1–11] have been shown to support applications such as situation monitoring, workflow management, transaction processing, change detection, pattern detection, and others in various application domains. A ECA rule consists of three parts: event, conditions, and actions. An event (simple or complex) is defined as an occurrence of interest. When events are detected, associated rules are triggered and conditions are evaluated. When conditions evaluate to *True*, (predefined) actions are performed. There has been a lot of work in event specification and detection [1–11].

Raman Adaikkalavan
Computer Science & Informatics
Indiana University South Bend
e-mail: raman@cs.iusb.edu

Sharma Chakravarthy
Computer Science and Engineering
The University of Texas At Arlington
e-mail: sharma@cse.uta.edu

* This work was supported, in part, by the NSF grant IIS 0534611. This work was also supported, in part, by an IUSB Grant.

S. Helmer et al.: Reasoning in Event-Based Distributed Systems, SCI 347, pp. 173–193.
springerlink.com   © Springer-Verlag Berlin Heidelberg 2011

Access control [12] models and mechanisms allow subjects to access only authorized objects. There are multiple access control models – discretionary, mandatory, and role-based. Discretionary allows subjects to own objects, and grant permissions to other subjects to access those objects. It allows subjects to access objects if the authorized rules are satisfied. Mandatory allows subjects to access objects if the security axioms based on sensitivity level and category are satisfied. Role-based assigns permissions to roles that can then be given to more than one subject. A subject is then allowed to activate any of the assigned roles. This allows for better management of permissions, because if a user changes job roles then it is easier to revoke roles from or grant roles to a user, rather than an entire set of permissions for each user.

ECA rules were first developed for adding change detection and notification in the context of databases. However, their role and relevance have been recognized for a number of application areas that require monitoring of different sorts. This also meant extensions to the functionality and semantics that were needed for these new applications. In this chapter, we discuss various extensions to ECA rules that make them usable to support new application domains. We particularly use the access control domain to identify the extensions discussed in this chapter. We have extended the rules themselves, the event operator semantics, and the event detection mechanisms.

Extensions to ECA rules with alternative actions are discussed in Section 2. Generalization of event semantics with attribute based conditions is discussed in Section 3. Potential as compared with actual events are discussed in Section 4. Event detection modes are discussed in Section 5. Extensions to the event detection mechanism are discussed in Section 6. Distributed event processing and complex event reasoning are discussed in Section 7. Finally, Section 8 presents our conclusions and future work.

2 Alternative Actions

An event, defined as an occurrence of interest, can be simple or complex. Simple events occur at a point in time, and complex events occur over an interval. Complex events combine one or more events (simple or complex) using event operators [13]. Whenever an event is detected, rules associated with that event are triggered. Conditions associated with those rules are evaluated, and if satisfied, actions are performed. A simple form of ECA rule specification is shown below:

$$
\begin{aligned}
&\text{RULE } [\ R_{name} \\
&\quad \text{EVENT} \qquad E_{name} \\
&\quad \text{CONDITION} \quad C_1 \dots C_k \\
&\quad \text{ACTION} \qquad A_1 \dots A_n \qquad]
\end{aligned}
$$

- RULE R_{name} – Unique rule name.
- EVENT E_{name} – Event (simple or complex) associated with the rule. The detection of this event *triggers* the rule.

- CONDITION $C_1 \ldots C_k$ – The set of conditions to be evaluated once the rule is triggered by event E_j.
- ACTION $A_1 \ldots A_n$ – The set of actions to be triggered when conditions evaluate to *True*.

A rule may have additional specifications such as coupling modes, contexts or consumption modes, and so forth [6, 8, 13]. Rules associated with an event are triggered only after that event is detected. Actions are performed when conditions are evaluated to *True*. This in turn means, though the event has happened, no actions are performed when the conditions evaluate to *False*. These events are immediately dropped from the system. However, in several applications and domains, this behavior is not sufficient to model the required policies.

Let us consider an example from access control. *"When anyone requests to open a file between 9 am and 4 pm, allow"*. Event E_{open} and rule R_{open} defined below using the existing ECA rule specification, model this policy.

Event E_{open} = FOpen(FileName);

RULE [R_{open}
      EVENT E_{open}
      CONDITION $t_{occ} >= 9am \wedge t_{occ} =< 5pm$
      ACTION   /* Allow Access */   ]

Event E_{open} is defined on the function signature *"FOpen(FileName)"*. Assume that the function *FOpen* is invoked by the underlying system when there is a file open request. Thus, when a user tries to open a file, this event raised. The event parameters include explicit or function parameters (e.g., *FileName*) and implicit parameters (e.g., t_{occ} *(time of occurrence), user_id (login information))*.

Rule R_{open} is triggered when the function *FOpen* is invoked. Assume that a user is trying to open a file at 5 pm. This raises the event and triggers the rule. The condition evaluates to *True* and starts the action procedures. Assume that the system receives a request at *6 pm*. Since this does not satisfy the condition, no actions are taken (i.e., user is not allowed to open the file). After the associated rules are triggered, this event is dropped from the system. On the other hand, this event is critical in access control as it can indicate potential security violations.

We have extended the ECA rule specification with Alternative Actions [14]. This extension allows users to specify the list of alternative actions that need to be taken when the event is detected, but the conditions *fail*:

RULE [R_{name}
      EVENT E_{name}
      CONDITION $C_1 \ldots C_k$
      ACTION $A_1 \ldots A_n$
      ALT ACTION $AA_1 \ldots AA_n$]

- ALT ACTION $AA_1 \ldots AA_n$ – The set of actions to be triggered when conditions evaluate to *False*.

Below, we rewrite the rule R_{open} using the extended **ECAA** (Event-Condition-Action-Alternative Action) rule specification. When this rule is used, the file open request at *6 pm* is logged and access is denied.

RULE [R_{open}
        EVENT E_{open}
        CONDITION $t_{occ} >= 9am \wedge t_{occ} =< 5pm$
        ACTION   /* Allow Access */
        ALT ACTION /* Add to Log, Deny Access */   ]

This can be handled via a rule with complementing conditions and actions. For example, create a rule with 4.59 pm to 8.59 am as the condition. Though this approach seems trivial, it introduces lot of overhead and rule executing complexity as both (i.e., all associated) rules have to be executed.

3 Event Generalization

In application domains such as access control, the loose coupling between the event part and the condition-action-alternative action part is not sufficient to capture and enforce the required policies. In this section, we will first discuss interval-based semantics, and then the generalization of event specification and detection using attribute-based constraints.

3.1 *Interval-Based Semantics*

Simple events happen at a point in time (e.g., 10:00 am). Complex events happen over an interval (e.g., 10:00 am to 10:05 am).

A complex event is detected when all of its constituent events occur according to the event operator semantics. The Sequence event operator composes two events and is detected when the second event follows the first event. The sequential occurrence is evaluated based on the time of occurrence. Consider E_{Seq}, a Sequence complex event, which composes two simple events E_1 and E_2. It is defined as:

    Event $E_{Seq} = E_1$ Sequence E_2

The event E_{Seq} is detected when event E_1's time of occurrence is less than event E_2's. Assume that E_1 captures opening of a building door, and E_2 captures opening of an office door. Logically, E_1 should happen before E_2.

Assume the following event occurrences: Building door is opened at 10:00 am and event E_1 is detected at time point [10:00 am]. Office door is opened at 10:05 am and event E_2 is detected at [10:05 am]. The event E_{Seq} is detected with event occurrences E_1 [10:00 am] and E_2 [10:05 am], since the latter follows the former. It event occurs over the interval [10:00 am] and [10:05 am].

Though E_{Seq} occurs over an interval, it is detected at [10:05 am] in Point-based semantics [13]. This can lead to loss of events when complex events compose other complex events, as discussed in [4]. In order to overcome the problems discussed in [4], we extended the event detection to support interval-based semantics [1, 15–17] for all the event operators in all event consumption modes. For the above example, event E_{Seq} is detected over an interval [10:00 am, 10:05 am] in interval-based semantics.

Below, we discuss the formalization of the Sequence event operator using both Point- and Interval-based semantics.

$$(E_1 \ Sequence \ E_2)[t_2] \triangleq \exists t_1, t_2 (E_1[t_1] \wedge E_2[t_2] \wedge (t_1 < t_2));$$

$$(E_1 \ Sequence \ E_2)[t_s, t_e] \triangleq \exists t_s, t_e, t, t'(E_1[t_s, t] \wedge E_2[t', t_e] \wedge (t_s \leq t < t' \leq t_e));$$

In the above, events E_1 and E_2 can be simple or complex. The first definition is using the Point-based semantics. Event E_1 is detected at $[t_1]$ and E_2 at $[t_2]$, and the Sequence event is detected when $(t_1 < t_2)$. The second definition is using the Interval-based semantics. Event E_1 is detected over an interval $[t_s, t]$ and E_2 is detected over $[t', t_e]$. The Sequence event is detected when the end time of E_1 is less than the start time of E_2 (i.e., $t_s \leq t < t' \leq t_e$). For simple events, the start time and end time are same as they occur over a point in time.

3.2   Generalization of Events

Interval- and point-based semantics have been useful, however, both of them are based on temporal conditions and are insufficient [18].

Let us consider the same Sequence event E_{Seq} defined in the previous section. With information security, the users who are responsible for the event occurrences of E_1 and E_2 are also critical to the event composition. If user 1 opens the building door and user 2 opens the office door, both cannot be composed together to form a Sequence event. With existing ECAA rules, the condition whether events E_1 and E_2 are from the same user can only be checked in the condition part (i.e., after the events are combined and detected, and the rule is triggered). This will lead to incorrect event detection.

Assume the following event occurrences: E_1 (user 1, 10:00 am), E_2 (user 2, 10:05 am), and E_2 (user 1, 10:10 am). With existing event operator semantics, the Sequence event E_{Seq} is detected with event occurrences E_1 (user 1, 10:00 am) and E_2 (user 2, 10:05 am), since the latter follows the former. This will trigger the associated rules. But, when the condition part is evaluated, it will return *False* as both the events are from different users. Thus, predefined alternative actions are performed. On the other hand, based on the event consumption modes [13], either or both the event occurrences E_1 (user 1, 10:00 am) and E_2 (user 2, 10:05 am) can be removed from the system, as they have taken part in a complex event detection. If both the events are removed, when the next event occurrence E_2 (user 1, 10:10 am) happens, there are no other E_1 event occurrences to match, and this E_2 (user 1, 10:10 am) is also dropped. The above composition of the events based on temporal semantics leads to unintended event detection. Thus, the event operator semantics used for event composition have to be extended to include conditions based on event attributes in addition to the existing temporal conditions.

Masks proposed in ODE [19] allow simple events to be detected based on attribute conditions. For instance, an event *"buyStock(double price) &&* *price > 500"* is detected when the function "buyStock" is invoked and *price > 500*. Masks (predicates) can only be based on the formal parameters of the function or based on the state of the database object. Even though some of the conditions can be specified by existing systems via masks, not all possible conditions can be specified. Similar to simple event masks, composite events can also be associated with masks in ODE. But, masks associated with composite events can be evaluated only in terms of the current state of the database and are very restricted as they are just constant integer expressions.

We have generalized the event operator semantics to include attribute conditions [18], in addition to point-based and interval-based semantics. In event processing, each event has a well-defined set of *implicit* and *explicit* attributes or parameters. Implicit parameters contain system-defined and user-defined parameters (e.g., event name, time of occurrence). Explicit parameters are collected from the event itself (e.g., stock price, stock value). The new event specification allows users to specify conditions on explicit and implicit parameters via two different expressions. These expressions are denoted as E_{expr} and I_{expr}, where E_{expr} allows the specification of conditions based on explicit parameters and I_{expr} allows the specification of condition on implicit parameters. I_{expr} subsumes the existing point-based and interval-based semantics. Generalization of simple event specification with parameter expressions is shown below.

$$Event\ E_{name} = functionSignature \land (E_{expr} \land I_{expr});$$

Generalization of complex event specification, using the Sequence event E_{Seq} discussed at the start of the section, is shown below.

$$Event\ E_{Seq} = (E_1\ Sequence\ E_2) \land (E_1.user = E_2.user);$$

Assume the same event occurrences: E_1 (user 1, 10:00 am), E_2 (user 2, 10:05 am), and E_2 (user 1, 10:10 am). With the above generalization, the Sequence event E_{Seq} is *not* detected with event occurrences E_1 (user 1, 10:00 am) and E_2 (user 2, 10:05 am), since the explicit expression that checks for the associated users fails. When the next event occurrence E_2 (user 1, 10:10 am) happens, it is combined with E_1 (user 1, 10:00 am) as the events satisfy both the temporal and attribute-based conditions.

4   Potential vs. Actual Events

Events are detected when an occurrence of interest happens. This is useful in domains where an occurrence of interest is the actual event rather than a potential event. Let us consider two examples: Event A *controls* the opening of an office door. Event B turns the lights ON when the door is actually opened. A user is allowed to open a door only if that user has authorized access. In other words, swiping of a card results in the detection of an event that checks user authorization. Thus, swiping of the card detects event A, and if the user has the permissions then the door is opened. On the other hand, event B is triggered only if the door was opened. In the above example, the opening of the door is the actual event that should trigger event B. Swiping the card is the potential event that leads to the opening of the door, if the user has required permissions.

Below, we discuss a sample policy that requires potential and actual events. We show how it is modeled using the ECAA rules.

Policy 1: Between 00:00 hrs and 06:00 hrs, only those who have entered through the building's external door are allowed to enter an office room in that building.

This policy can be made more complex by including situations wherein two or more persons enter the building together without using their cards individually. Without the loss of generality, we just discuss the basic policy defined above and assume that each person registers his/her entry individually.

Event E_{ExtReq} (external door request) is detected when there is a *request* to open any external door between 00:00 hrs and 06:00 hrs. This triggers rule R_{ExtReq}. The condition part authenticates the user, and the action part allows the door to be opened. Access is denied, otherwise.

$$E_{ExtReq} = (doorOpen(bldgId, doorId, doorType, userId),$$
$$(((t_{occ} > 00 : 00hrs) \land (t_{occ} < 06 : 00hrs))$$
$$\land(doorType = \text{``external''})));$$

```
RULE [ R_ExtReq
        EVENT         E_ExtReq
        CONDITION     /* Authenticate User */
        ACTION        /* Open door */
        ALT ACTION    /* Deny Access */        ]
```

The external door open request is handled by event E_{ExtReq} but the actual opening of the door is handled in the action part. Thus, another event $E_{ExtOpen}$ is raised by rule R_{ExtReq} in the action part to indicate that the door was opened. The parameters of the event $E_{ExtOpen}$ are the same as E_{ExtReq}. The modified rule is shown below.

RULE [R_{ExtReq}

    EVENT E_{ExtReq}

    CONDITION   /* Authenticate User */

    ACTION   /* Open door, Raise $E_{ExtOpen}$ */

    ALT ACTION /* Deny Access */   ]

When event E_{ExtReq} is detected, rule R_{ExtReq} is triggered. Conditions are evaluated and the user is authenticated. If authenticated, the door is opened, and an internal event $E_{ExtOpen}$ is raised. This door opening event is not raised directly and is raised only from the rule R_{ExtReq}. Similarly, event E_{OffReq} is raised when someone tries to open an office door.

$$E_{OffReq} = (doorOpen(bldgId, doorId, doorType, userId),$$
$$(((t_{occ} > 00 : 00hrs) \wedge (t_{occ} < 06 : 00hrs))$$
$$\wedge(doorType = \text{``}office\text{''})));$$

Similar to rule R_{ExtReq} another rule is created to allow someone to open an office door. But the policy requirement will not be met (i.e., the user should have opened the building door first). Thus, a complex event is required to model this requirement (i.e., if E_{OffReq} happens after $E_{ExtOpen}$, trigger the rule to check for access). A generalized Sequence event $E_{OffReq2}$ and rule $R_{OffReq2}$ are created as shown below to model the required policies. The explicit expression tracks each user separately.

$$E_{OffReq2} = (E_{ExtOpen} \; Sequence \; E_{OffReq}) \wedge$$
$$((E_{ExtOpen}.userId = E_{OffReq}.userId$$
$$\wedge E_{ExtOpen}.bldgId = E_{OffReq}.bldgId));$$

RULE [$R_{OffReq2}$

    EVENT $E_{OffReq2}$

    CONDITION   /* Authenticate User */

    ACTION   /* Open door */

    ALT ACTION /* Deny Access */   ]

5   Event Detection Modes

Complex events are triggered when all the required constituent events occur according to the attribute-based semantics. For example, a Sequence event is detected when the second event follows the first event. With existing semantics, when the second event happens without the first event, it is just dropped as the occurrence of interest is the sequential occurrence. Consider the example where the building door has to be opened before the office door. What happens when the office door is opened without opening the building door? With existing systems the event that captures the opening of the office door is dropped as the first event never occurred. In access control domain, this can capture an attempt to break-in.

We have extended event processing with event detection modes [20]. Below, we discuss the event detection modes using events and rules defined in Section 4. Existing systems detect event $E_{OffReq2}$ and trigger rules when event $E_{ExtOpen}$ happens before E_{OffReq}. Possible constituent event occurrences (*cases*) of the Sequence event $E_{OffReq2}$ are as follows:

1. Both $E_{ExtOpen}$ and E_{OffReq} occur: the $E_{ExtOpen}$ event is raised when the external door is opened. When the same person requests for opening an office door, event E_{OffReq} is detected. Since this is the detector event for the complex event $E_{OffReq2}$, the complex event occurrence is completed and the rule $R_{OffReq2}$ is triggered. If the person has proper authentication the office door is opened.

2. Only $E_{ExtOpen}$ occurs: This event is raised when the external door is opened. This event is the initiator and starts the complex event $E_{OffReq2}$. The detector event is raised only when the same person tries to open an office door. In case the detector does not happen, a timeout event can be triggered based on the organization's policy.

3. Only E_{OffReq} occurs: Event E_{OffReq} is detected without $E_{ExtOpen}$ i.e., complex event $E_{OffReq2}$ had not been initiated. *What will happen if the detector event happens without any initiator?* With *existing systems*, the detector event is just *ignored* as it does not capture the occurrence of interest. In access control applications, this should trigger a violation, which is *not* possible with current systems.

Detector/terminator events play an important role in enforcing policies, but are ignored when they occur *without* an initiator event. Other operators cannot be used to model the above policy as it requires one event to follow the other. Additional rules or complex conditions/actions will still not be able to model the above discussed policy, as rules are triggered only when an event happens. Since the event occurrences should follow a sequence, only a sequence event can trigger rules. Thus, binary event operators such as Sequence need to be extended to handle the occurrence of detector/terminator events without a prior occurrence of the corresponding initiator event.

Below, we discuss the issues with ternary event operators using the following policy:

Policy 2: Alert security personnel when a shoplifting activity occurs in a RFID-based retail store [21] i.e., items that were picked at a shelf and then taken out of the shop without an entry in the point of sale system.

The above policy requires to alert on a non-occurrence event and can be modeled with the NOT event operator as shown below. A NOT event is detected, when a constituent event does not occur between the Sequence of two other constituent events. In the example shown below, E_{Chk} is detected when event E_{POS} does not occur between E_{Pick} and E_{Gate}.

$$E_{Chk} = (NOT(E_{Pick}, E_{POS}, E_{Gate}) \wedge$$
$$(E_{Pick}.itemId = E_{POS}.itemId = E_{Gate}.itemId);$$

We do not show all the constituent event definitions and rules. Event E_{Pick} represents picking the item from the shelf. Event E_{POS} represents checking out at the point of sale system. Event E_{Gate} represents item leaving the gate. $\mathcal{I}_{expr}$ allows event detection in either point-based or interval-based semantics. $\mathcal{E}_{expr}$ relates all event occurrences with the same item for controlling each item simultaneously. E_{Pick} is the initiator, E_{Gate} is the detector, and middle event E_{POS} is the non-occurrence event (i.e., the event that should not occur). Possible constituent event occurrences (*cases*) of event E_{Chk} are discussed below.

1. Both E_{Pick} and E_{Gate} occur: This detects the NOT event and an alert regarding shop lifting can be sent via the actions part of a rule. Current systems handle this correctly.
2. Only E_{Pick} occurs: An item was picked but nothing happened after that. Current systems just wait for a detector to occur. One possible solution would be to raise a timeout event (e.g., shop closing) and take further actions (e.g., re-shelf the item).
3. Only E_{POS} occurs: Someone has checked out an item without picking it from the shelf. This indicates that something is malfunctioning. Since this event is just a *constituent* event it is ignored (i.e., deleted) in the current event systems. But this cannot be ignored since a E_{GATE} event might occur in the future. One possible solution would be store this event, and wait for the E_{GATE} event or raise a timeout event (e.g., shop closing) and take further actions.
4. Both E_{Pick} and E_{POS} occur: An item has been picked up and checked out. Current systems just wait for the *detector/terminator* event E_{Gate} to occur. One possible solution would be to raise a timeout event (e.g., shop closing) and take further actions (e.g., check gate sensors).
5. Both E_{POS} and E_{Gate} occur: An item was not picked out, but it was checked out and taken to the gate. Current systems just ignore all these occurrences. This cannot be the case as it might indicate that there is some malfunctioning and it has to be reported.
6. Only E_{Gate} occurs: An item was not picked up or checked out, but has reached the gate. Current systems ignore this *detector/terminator* event and purge it from the system. This is an incorrect action as it might be a shop lifting activity.
7. All of E_{Pick}, E_{POS} and E_{Gate} occur: All the occurrences are just ignored as the event that should *not* occur has happened. This case indicates that the items were checked out properly. This event occurrence can be used to create a log for inventory maintenance.

Summary: The occurrence of an initiator event without other events starts a complex event. The only solution to detect that complex event is to trigger a timeout event. In the above example, it might just mean that the item has to be searched and reshelved. When the initiator happens alone, or when initiator and constituent events occur but not the detector/terminator event, then raising a timeout event is the only solution. In the above example, it might mean that a gate sensor is malfunctioning. When the detector/terminator happens without the initiator and constituent events it is a problem that needs immediate attention. In the above example, it may indicate a shop lifting activity. Currently, events are simply dropped in all the cases except the first case, which is insufficient. Modeling of the above discussed policy using other existing event operators is not possible as the policy requires the capturing of a non-occurrence event. On the other hand, additional rules or complex conditions/actions cannot model the policy, as rules are triggered and conditions are evaluated only after the non-occurrence event detection. This requires extensions to ternary operators such as NOT with mechanisms for handling the above discussed situations.

Simple events are detected whenever they occur in the system and these extensions do not apply to them. Extensions to both binary and ternary event operators are discussed below.

5.1 Binary Event Operator Semantics

With binary operators, two constituent events are involved and they act as initiator and detector/terminator.

Complete Event: When the *detector* event occurs, operator semantics are applied and both $\mathcal{I}_{expr}$ and $\mathcal{E}_{expr}$ are checked. If any of the conditions fail then that event is *not* raised and other constituent events are dropped. Whether an event is complete is checked *only* when the detector is raised. Current systems deal only with complete events. Thus, *a complete complex event E occurs when* i) *the initiator occurs, and* ii) *the detector occurs and completes that event.*

Partial Event: When the detector event occurs without the initiator, existing event detection semantics have to be modified to trigger ECAA rules. Extending current event detection semantics to handle situations where the detector has occurred *without* the required events to complete the detection will allow the system to take additional actions. We term these events as *partial events* and define them as: *a partial complex event E occurs when* i) *event E is not initiated, and* ii) *the detector occurs.*

Policy 1 Modeling: The three cases that were analyzed in Section 4 under Policy 1 can be handled using these extensions. Specifically, Case 1 is handled by complete events and Case 3 is handled by partial events. Case 2 can be handled by a timeout event which completes the event.

5.2 Ternary Event Operator Semantics

Similar to binary operators, ternary events have an initiator and a detector/terminator. In addition there is another event that is just a constituent event.

Complete Event: Binary complete event definition is further refined as: *A complete complex event E occurs when,* i) *initiator occurs,* ii) *all the required constituent events occur, and* iii) *the detector occurs and completes that event.*

Partial Event: Partial binary event definition is further refined as: *A partial complex event E occurs when* i) *event E is not initiated,* ii) *other constituent events can occur, and* iii) *detector occurs.*

Failed Event: In addition to the above events, we define *failed events* as shown below. For example, this event is detected when the non-occurrence has failed for a NOT operator. *A failed complex event E occurs when* i) *the initiator occurs,* ii) *other constituent events occur, and* iii) *the detector occurs and completes the event, but the event fails because some constituent event that should not have occurred has occurred.*

Policy 2 Modeling: The seven cases that were analyzed in under Policy 2 can be handled using the proposed extensions (whereas current systems handle only Case 1). Specifically, Case 1 is handled by complete events, Cases 5 & 6 are handled by partial events, and Case 7 is handled by failed events. Cases 3 & 4 are handled by partial events using timeout events. Case 2 can be handled by a timeout event which completes the event.

5.3 ECAA Rule Specification

As discussed previously events can be detected as complete, partial, or failed. These three types are termed *event detection modes.* With the event extensions proposed, ECAA rules can be triggered in all the three event detection modes. Though the specification has not been changed, detection has to be changed to trigger appropriate rules. Event detection using graphs is discussed in Section 6.

As explained in Section 2, ECAA rules consist of four major components. ECAA rules are triggered when complete events are detected. With new event detection modes, *when should ECAA rules be triggered?* Partial and failed events should also trigger associated ECAA rules. In this section, we discuss the extensions needed to support complete, partial, failed and other future detection modes in a seamless way.

Event detection modes are handled by adding an optional DMODE attribute to the existing rule specification. Currently, the values of the DMODE attribute are: COMPLETE, PARTIAL, and FAILED. If a value is not specified, the rule will be triggered when a complete event is detected, by default. Using the DMODE attribute, different sets of condition-action-alternative actions are associated to the rules. This generalization allows the specification of the proposed and future event modes and their associated condition-action-alternative actions. The Rule R1 shown below will be triggered by the event E1. The condition-action-alternative actions corresponding to the complete and partial event are specified.

RULE [R_1
    EVENT E_1
    DMODE:COMPLETE {
      CONDITION   /* Conditions */
      ACTION   /* Actions */
      ALT ACTION /* Alternative Actions */ }
    DMODE:PARTIAL {
      CONDITION   /* Conditions */
      ACTION   /* Actions */
      ALT ACTION /* Alternative Actions */ }
  ]

With complete, partial, and failed events and rules we can model and capture policies that cannot be captured using existing systems. Below, we show the extended rules corresponding to the policies discussed in Section 4.

Rules for Policy 1: Rule R_{ExtReq} need not be changed as it is associated with a simple event. By default the event is triggered as a complete event. Rule $R_{OffReq2}$ that is associated with the Sequence complex event $E_{OffReq2}$ has been modified using the generalized rule specification:

RULE [$R_{OffReq2}$
    EVENT $E_{OffReq2}$
    DMODE:COMPLETE {
      CONDITION   /* Authenticate User */
      ACTION   /* Open door */
      ALT ACTION /* Deny Access */ }
    DMODE:PARTIAL {
      CONDITION   /* *True* */
      ACTION   /* Notify Security */
   ]

In rule $R_{OffReq2}$, DMODE:COMPLETE handles authentication when the external door is opened and the office door is opened after that. Specifically, it handles Case 1 under Policy 1. DMODE:PARTIAL handles Case 3 (i.e., when the office door is opened without the external door opening). When triggered it notifies security personnel of a possible break-in.

Rules for Policy 2: Below we create a rule and associate it with the NOT complex event E_{Chk}. When events E_{Pick} and E_{Gate} occur (Case 1), it detects the non-occurrence of the checkout (E_{POS}) event. This detects the NOT event and triggers the DMODE:COMPLETE part of rule shown below:

```
RULE [ R_Chk
         EVENT   E_Chk
         DMODE:COMPLETE {
            CONDITION  /* True */
            ACTION     /* Notify Security */
         DMODE:PARTIAL {
            CONDITION  /* True */
            ACTION     /* Notify Security */
         DMODE:FAILED {
            CONDITION  /* True */
            ACTION     /* Update Log */
      ]
```

When the detector/terminator event occurs with other constituent events and no initiator, a partial event is detected and the partial rule is triggered. In our example when E_{Gate} occurs alone (Case 6), or when E_{POS} and E_{Gate} occur (Case 5), it indicates some problem and should be notified. In either case the security is notified via the DMODE:PARTIAL part of rule. In addition, Cases 3 & 4 are also handled using timeout events and the DMODE:PARTIAL part of the rule.

When all the events E_{Pick}, E_{POS} and E_{Gate} occur (Case 7), the failed event is detected. This is because event E_{Chk} is modeling the non-occurrence of E_{POS}, but it has occurred. This triggers the DMODE:FAILED part of the rule.

All other cases where there is no occurrence of a detector/terminator can be handled using a timeout event.

6   Event Detection Graph Extensions

In the previous sections we have discussed event and rule specification, and their extensions. Event detection graphs [1, 9, 13] are used to represent, process and detect simple and complex events. In this section, we discuss event detection graphs and their extensions to handle the ECA rule extensions we have presented here.

Event detection graphs are acyclic graphs. Simple events are represented using leaf nodes and complex events are represented using internal nodes. The graph shown in Figure 1 represents a Sequence event ESeq that composes two simple events E1 Sequence E2. Leaf nodes represent the simple events E1 and E2. An internal node represents the Sequence complex event. An

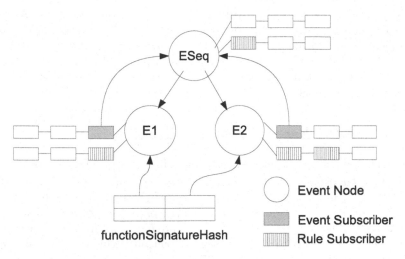

Fig. 1 Event Detection Graph

event subscriber list maintains all the events that subscribe to that event. For example, E1 is subscribed by ESeq, but ESeq is subscribed by no other event. Rule subscriber list maintains all the rules that need to be triggered when an event is detected. For example, there are two rules associated with event E2. The functionSignatureHash is used by the system to link function signatures and event nodes. When a function is invoked, appropriate event nodes are notified using this hash table.

The event detection graph shown in Figure 1 is constructed as follows. Simple events nodes are created first with appropriate rule subscribers and empty event subscribers. Complex event nodes are created with empty event subscriber lists and appropriate rule subscriber lists. Once created, child event nodes are linked with a parent node via the event subscriber lists and the parent node is linked with the child nodes. These links are necessary for propagating event occurrences from child nodes to parent nodes, and for propagating consumption mode [1, 9, 13] policies from parent nodes to child nodes. For example, if there is a rule with recent mode[1] in the parent node's rule subscriber list, this information must be propagated to the child nodes, so that the child nodes can also detect events in recent mode.

Whenever a function is invoked, the function signature hash is used by the Local Event Detector [23] system to propagate it to the appropriate event node. In the event node, all the associated event and rule subscribers are

---

[1] In the recent mode, a new occurrence of an event replaces the old occurrence. It is used by applications where events happen at a fast rate, and multiple occurrences of an event only refine the previous occurrence. Other modes are continuous, cumulative, chronicle, and recent-unique. For more details, please refer to [1, 9, 13, 22].

notified. This action is repeated till the event reaches the root node. All the rules that are in the rule subscriber list are executed by the rule processing component. The internal nodes represent complex event operators, and compose events using point- or interval-based semantics, and event consumption modes. At any point in time the event detection graph has access to the partial history of events.

6.1 Extended Event Detection Graph

An extended event detection graph is shown in Figure 2. This graph represents events E_{ExtReq}, $E_{ExtOpen}$, E_{OffReq}, and $E_{OffReq2}$ defined in Section 4, rule R_{ExtReq} defined in Section 4, and rule $R_{OffReq2}$ defined in Sections 4 and 5. The sequence composite event $E_{OffReq2}$ is represented using an internal node. The simple events E_{ExtReq} and E_{OffReq} are represented using leaf nodes. The structure of the event detection graph and the data flow architecture have not been changed. However, event detection algorithms have been modified to support the extensions. For example, current sequence event detection algorithm detects only complete events, whereas the modified algorithm detects events based on detection modes (e.g., complete and partial) and triggers appropriate rules. Below, we discuss the extensions in detail.

Alternative Actions: In order to implement the alternative actions, we did not modify the event detection graph. We have modified the rule processing component. Thus, whenever the conditions evaluate to *False* alternative actions are performed.

Attribute Conditions: In the existing event detection graphs, internal event nodes represent event operators, and compose events using interval- or point-based semantics. In order to support attribute conditions, we have extended both leaf and internal event node processing algorithms. Once an event notification arrives, the algorithm evaluates the specified implicit and explicit parameter expressions. Only if all the conditions evaluate to *True*, are events from event subscriber lists notified of this event's occurrence and are associated rules from the rule subscriber list triggered.

Potential Vs. Actual Events: All the leaf event nodes are notified via functionSignatureHash, and the internal nodes are notified using event subscriber lists. When potential events (e.g., E_{ExtReq}) are used to raise actual events (e.g., $E_{ExtOpen}$) from the action part of a rule they must be handled differently. In Figure 2, special event nodes represented with dashed circle can only be notified from a rule. The actual opening of the door captured by simple event $E_{ExtOpen}$ is represented as a special internal node which can be notified only from the rule R_{ExtReq}.

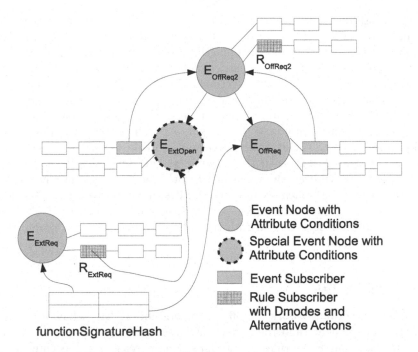

Fig. 2 Extended Event Detection Graph

Event Detection Modes: These are handled similarly to event consumption modes. In the existing graph, event consumption modes are specified along with the rule. Similarly, event detection modes are also specified as part of the rule. During the construction of the graphs and subscriber lists, event consumption modes bits are set. This allows the detection of events in various consumption modes. Event detection modes are also handled the same way.

7   Distributed Event Processing and Reasoning

In this section we discuss distributed event processing and use of rules in reasoning.

7.1   Distributed Event Processing

The need for processing event streams from devices and to compose, aggregate, and compute results in (near) real-time is ever growing due to various stream and people-centric sensing applications [24]. A number of applications involving sensors and other data that needs to be accessed are distributed in nature. Hence, there has been a lot of interest in distributed event processing [25–31]. Distributed event specification, semantics, and implementation have also been addressed in [28, 29, 32, 33]. With distributed event processing, the implementation of the local event detector has to be tailored to raise and detect events taking into account the distributed aspects such as clock synchronization, transport of events, and heterogeneity, to name a few. In general, event composition and detection and ECAA rules can be executed at the event producer, intermediate nodes, central server, or event consumers. When the filtering of events is high, processing the events and associated rules at the producer site or at intermediate nodes reduces communication and processing cost. The extensions to event and rule specification and event detection graphs discussed in this chapter are at the logical level and are applicable to distributed environments. In other words, existing distributed event processing approaches can be used without modifications, as the representation of simple and complex events using leaf and internal nodes, and the structure and data flow architecture of the event detection graph have not been modified.

7.2 Reasoning

There have been much work [34–36] on understanding the behavior of rules and their components and their effect on systems. A real-time inference algorithm based on ECA rules was proposed as a reasoning mechanism for real-time and active databases in [34]. A heuristic search approach was used, which searches the rule graph under certain timing constraints to find the set of actions corresponding to an event occurrence. On the other hand, a single event can trigger multiple rules and the systems managed using ECA rules do not provide any guarantee about the system behavior. In order to provide guarantees about the system behavior, ECA rules have been extended in [35], to contain action specifications in first-order predicate logic to reason about the enforcement order of triggering multiple rules. The use of Concurrent Transaction Logic as a underlying formalism for logic based event processing was proposed in [36] for extending rule-based reasoning capabilities. When organizations use the extended ECAA rules to enforce policies the order of triggering of rules with respect to the detection and raising of actual events, discussed in Section 4, needs to be carefully analyzed to avoid incorrect event detections. For example, assume that E_{ExtReq} occurs and E_{OffReq} happens immediately after that. In order for $E_{OffReq2}$ to be detected, the actual event $E_{ExtOpen}$ should be raised from the action part of the rule associated with event E_{ExtReq}, before detecting event E_{OffReq}. Similarly, the relationship between event detection modes specified in the ECAA rules and events, and between rules with different modes needs to be analyzed. Reasoning about the consistency, completeness, and correctness is a general problem that needs to be addressed as well. This is a critical factor in the adoption and use of ECA rules in application domains such as access control where security and privacy laws need to be adhered to and guarantees about the system behavior need to be provided.

8   Conclusions

In this chapter, we discussed various extensions to the ECA rule paradigm to support advanced applications. Alternative actions are important as they allow the capture of security violations in access control. Generalization of events and rules provides richer semantics and hence facilitates the modeling of a wider range of applications. Event detection modes are needed to capture complex policies or situations, such as security violations. Extension of rule detection modes is needed to support extended specifications to ensure correct enforcement of specified rules. We also discussed the extensions made to event detection graphs to implement the above discussed extensions. Future work includes studying the effect of these extensions on system behavior and developing techniques for propagation of events to event subscribers based on detection modes.

References

1. Adaikkalavan, R., Chakravarthy, S.: SnoopIB: Interval-Based Event Specification and Detection for Active Databases. Data and Knowledge Engineering 59(1), 139–165 (2006)
2. Carlson, J., Lisper, B.: An Interval-based Algebra for Restricted Event Detection. In: Larsen, K.G., Niebert, P. (eds.) FORMATS 2003. LNCS, vol. 2791, pp. 121–133. Springer, Heidelberg (2004)
3. Mellin, J., Adler, S.F.: A formalized schema for event composition. In: Proceedings International Conference on Real-Time Computing Systems and Applications, March 2002, pp. 201–210. IEEE Computer Society, Tokyo (2002)
4. Galton, A., Augusto, J.: Two Approaches to Event Definition. In: Hameurlain, A., Cicchetti, R., Traunmüller, R. (eds.) DEXA 2002. LNCS, vol. 2453, pp. 547–556. Springer, Heidelberg (2002)
5. Zimmer, D.: On the semantics of complex events in active database management systems. In: Proceedings International Conference on Data Engineering, p. 392. IEEE Computer Society, Washington, DC, USA (1999)
6. Paton, N.W.: Active Rules in Database Systems. Springer, New York (1999)
7. Roncancio, C.: Toward Duration-Based, Constrained and Dynamic Event Types. In: Andler, S.F., Hansson, J. (eds.) ARTDB 1997. LNCS, vol. 1553, pp. 176–193. Springer, Heidelberg (1999)
8. Widom, J., Ceri, S.: Active Database Systems: Triggers and Rules. Morgan Kaufmann Publishers, Inc., San Francisco (1996)
9. Chakravarthy, S., Mishra, D.: Snoop: An Expressive Event Specification Language for Active Databases. Data and Knowledge Engineering 14(10), 1–26 (1994)

10. Gatziu, S., Dittrich, K.R.: Events in an Object-Oriented Database System. In: Proceedings International Workshop on Rules in Database Systems (September 1993)
11. Gehani, N.H., Jagadish, H.V., Shmueli, O.: Composite Event Specification in Active Databases: Model & Implementation. In: Proceedings International Conference on Very Large Data Bases, pp. 327–338 (1992)
12. Bishop, M.: Computer Security: Art and Science. Addison-Wesley Professional, Reading (2002)
13. Chakravarthy, S., Krishnaprasad, V., Anwar, E., Kim, S.-K.: Composite Events for Active Databases: Semantics, Contexts, and Detection. In: Proceedings, International Conference on Very Large Data Bases, pp. 606–617. Morgan Kaufmann Publishers Inc., San Francisco (1994)
14. Adaikkalavan, R., Chakravarthy, S.: Active Authorization Rules for Enforcing Role-Based Access Control and its Extensions. In: Proceedings IEEE International Conference on Data Engineering (International Workshop on Privacy Data Management), Tokyo, Japan, p. 1197 (April 2005)
15. Adaikkalavan, R., Chakravarthy, S.: SnoopIB: Interval-Based Event Specification and Detection for Active Databases. In: Kalinichenko, L.A., Manthey, R., Thalheim, B., Wloka, U. (eds.) ADBIS 2003. LNCS, vol. 2798, pp. 190–204. Springer, Heidelberg (2003)
16. Adaikkalavan, R., Chakravarthy, S.: Formalization and Detection of Events Over a Sliding Window in Active Databases Using Interval-Based Semantics. In: Proceedings, East-European Conference on Advances in Databases and Information Systems, Budapest, Hungary, September 2004, pp. 241–256 (2004)
17. Adaikkalavan, R., Chakravarthy, S.: Formalization and Detection of Events Using Interval-Based Semantics. In: Proceedings International Conference on Management of Data, Goa, India, pp. 58–69 (January 2005)
18. Adaikkalavan, R., Chakravarthy, S.: Event Specification and Processing for Advanced Applications: Generalization and Formalization. In: Wagner, R., Revell, N., Pernul, G. (eds.) DEXA 2007. LNCS, vol. 4653, pp. 369–379. Springer, Heidelberg (2007)
19. Gehani, N.H., Jagadish, H.V., Shmueli, O.: Event Specification in an Object-Oriented Database. In: Proceedings International Conference on Management of Data, San Diego, CA, pp. 81–90 (June 1992)
20. Adaikkalavan, R., Chakravarthy, S.: When to Trigger Active Rules? In: Proceedings International Conference on Management of Data, Mysore, India (December 2009)
21. Gyllstrom, D., Wu, E., Jin Chae, H., Diao, Y., Stahlberg, P., Anderson, G.: SASE: Complex event processing over streams. In: Proceedings Conference on Innovative Data Systems Research (2007)
22. Elkhalifa, L., Adaikkalavan, R., Chakravarthy, S.: InfoFilter: a system for expressive pattern specification and detection over text streams. In: Proceedings Annual ACM SIG Symposium on Applied Computing, pp. 1084–1088 (2005)
23. Chakravarthy, S., Anwar, E., Maugis, L., Mishra, D.: Design of Sentinel: An Object-Oriented DBMS with Event-Based Rules. Information and Software Technology 36(9), 559–568 (1994)
24. Campbell, A.T., Eisenman, S.B., Lane, N.D., Miluzzo, E., Peterson, R.A., Lu, H., Zheng, X., Musolesi, M., Fodor, K., Ahn, G.-S.: The rise of people-centric sensing. IEEE Internet Computing 12, 12–21 (2008)

25. Schwiderski, S., Herbert, A., Moody, K.: Composite events for detecting behavior patterns in distributed environments. In: TAPOS Distributed Object Management (1995)

26. Jaeger, U., Obermaier, J.K.: Parallel Event Detection in Active Database Systems: The Heart of the Matter. In: Proceedings International Workshop on Active, Real-Time and Temporal Database Systems, pp. 159–175 (1997)

27. Luckham, D.C., Frasca, B.: Complex event processing in distributed systems. Stanford University, Tech. Rep. CSL-TR-98-754 (1998)

28. Yang, S., Chakravarthy, S.: Formal semantics of composite events for distributed environments. In: Proceedings International Conference on Data Engineering, p. 400 (1999)

29. Chakravarthy, S., Liao, H.: Asynchronous Monitoring of Events for Distributed Cooperative Environments. In: Proceedings International Symposium on Cooperative Database Systems for Advanced Applications, Beijing, China, pp. 25–32 (April 2001)

30. Pietzuch, P.R., Shand, B., Bacon, J.: Composite event detection as a generic middleware extension. IEEE Network 18(1), 44–55 (2004)

31. Akdere, M., Çetintemel, U., Tatbul, N.: Plan-based complex event detection across distributed sources. Proceedings of the VLDB Endowment 1(1), 66–77 (2008)

32. Chakravarthy, S., Tufekci, S., Honnavalli, R.: Flexible manufacturing simulation: Using an active dbms. In: Cooperative Knowledge Processing for Engineering Design, Chapman and Hill, Boca Raton (1997)

33. Tanpisuth, W.: Design and Implementation of Event-based Subscription/Notification Paradigm for Distributed Environments. Master's thesis, The University of Texas at Arlington (December 2001),
http://itlab.uta.edu/ITLABWEB/Students/sharma/theses/Tan01MS.pdf

34. Qiao, Y., Li, X., Wang, H., Zhong, K.: Real-time reasoning based on event-condition-action rules. In: Chung, S., Herrero, P. (eds.) OTM-WS 2008. LNCS, vol. 5333, pp. 1–2. Springer, Heidelberg (2008)

35. Shankar, C., Campbell, R.: Ordering Management Actions in Pervasive Systems using Specification-enhanced Policies. In: Proceedings Fourth IEEE International Conference on Pervasive Computing and Communications (March 2006)

36. Anicic, D., Stojanovic, N.: Expressive Logical Framework for Reasoning about Complex Events and Situations. In: Proceedings Intelligent Event Processing - AAAI Spring Symposium (2009)

Pattern Detection in Extremely Resource-Constrained Devices

Michael Zoumboulakis and George Roussos

Abstract. Pervasive computing anticipates a future with billions of data producing devices of varying capabilities integrated into everyday objects or deployed in the physical world. In event-based systems, such devices are required to make timely autonomous decisions in response to occurrences, situations or states. Purely decentralised pattern detection in systems that lack time synchronisation, reliable communication links and continuous power remains an active and open research area. We review challenges and solutions for pattern detection in distributed networked sensing systems without a reliable core infrastructure. Specifically, we discuss localised pattern detection in resource-constrained devices that comprise Wireless Sensor and Actuator Networks. We focus on online data mining, statistical and machine learning approaches that aim to augment decentralised pattern detection and illustrate the properties of this new computing paradigm that requires stability and robustness while accommodating severe resource limitations and frequent failures.

1 Introduction

The vast majority of research in middleware and distributed event-based systems proposes techniques that are not directly applicable to the extremely resource-constrained nodes of a Wireless Sensor Network (WSN). This is because they rely heavily on core infrastructure services such as reliable communication links, time synchronisation and a persistent event history. We target a class of pervasive computing devices with constraints in terms of power, processing, memory, bandwidth and reliability. These devices are usually found in embedded control systems, such

Michael Zoumboulakis
Birkbeck College, University of London, WC1N 3QS
e-mail: mz@dcs.bbk.ac.uk

George Roussos
Birkbeck College, University of London, WC1E 7HX
e-mail: gr@dcs.bbk.ac.uk

S. Helmer et al.: Reasoning in Event-Based Distributed Systems, SCI 347, pp. 195–216.

as Wireless Sensor Actuator Networks (WSAN) [22], with a requirement for timely response to interesting or unusual occurrences in the collected data.

A solution to this problem is to task nodes in the network to push data to base stations. The base stations are connected to powerful desktop-class machines that can perform offline processing. This data harvesting method solves the problem of event detection and pattern recognition but it incurs the significant overhead of expensive radio communication. Multihop links, non-uniform path costs and frequent route failures (cf. [26] for a detailed treatment of these issues) make the global harvesting method impractical. Furthermore, communicating sensor observations using wireless radio is orders of magnitude costlier than local computation [28].

A somewhat better approach involves tasking nodes in the WSAN to perform source-side filtering discarding uninteresting information and only sending packets when interesting observations are recorded. To improve confidence on the event occurrence a node can initiate radio communication to establish whether the event is spatially correlated and consequently produce a notification message for the actuator. The simplest type of filter is realised by thresholds that check whether the sensed observations are above or below a value. Although the simplicity is appealing this technique suffers from severe disadvantages. First, it is sensitive to outliers caused by faulty observations due to inexpensive sensors. Second, it is not capable of handling unknown thresholds such as a case where a user cannot provide a predicate value that distinguishes normal observations from events. Third, it does not scale well as the number of observations and thresholds increase. Last, it is not capable of handling magnitude differences in readings — for instance, two nodes with different temperature sensors will produce readings of different scale requiring two different thresholds.

This chapter discusses methods for event detection in WSANs based on online data mining. First, we depart from the traditional notion of the instantaneous event that can be described by a single occurrence, whether this is a database transaction or a single data point. We introduce the term *pattern* as a finite list of potentially non-unique time-dependent objects. The term sequence as defined by [42] or time series (cf. [14]) refer to the same structure, however the distinguishing factor is that in our case the pattern represents an ordered list of data items that together reveal interesting or unusual activity in the monitored process.

2   Objectives, Motivation and Contributions

Wireless Sensor Actuator Networks (WSANs) have some unique characteristics that distinguish them from other computing devices: first, they are usually powered by commodity batteries or by harvesting energy from the environment. [39], [25]. In both cases, energy is not an abundant resource and its consumption needs to be tightly controlled in a scheduled manner. Related to this restriction is the high energy cost of radio communication which is said to amount to the equivalent of 1000 CPU instructions for sending a bit over the radio [34]. Performing local computations to determine whether a pattern is statistically important or *interesting* is therefore desirable as long as the cost of computation is lower than the cost of transmitting data over the radio. Second, sensor data tends to be noisy due to inexpensive hardware and this brings to surface another requirement: any method for online pattern detection should be tolerant to noise, outliers and missing values while maintaining acceptable *accuracy* and *precision*. Third, WSANs tend to be designed for unattended operation so pattern detection techniques should require little or no human intervention.

In a nutshell, we are examining the problem of efficiently detecting patterns in data that do not conform to a well-defined notion of normal behaviour. The terms *anomaly detection* [7] and *novelty detection* [35] generally encapsulate the objective although the latter extends detection to cover previously unobserved patterns in data. We focus on *online* pattern detection which generally means that detection should take place as close to real time as possible. Since there exist severe resource constraints in a WSAN, pattern detection should investigate segments of sensor data instead of examining data on a global scale which is computationally prohibitive. This is achieved by *windowing*: the application of a sliding window to the streaming sensor data such that a typically smaller data segment is extracted.

The motivating factor for our work is that patterns are ubiquitous across a large number of WSAN applications — a selection of which is reviewed in Section 3 — yet there does not exist a standardised method for their detection. We accept that it is hard to devise an approach that is generalisable across a number of applications since the notion of what constitutes an interesting pattern is subjective and tends to vary according to data characteristics. Despite that, we maintain that there is a significant advantage to be gained by applying pattern detection techniques to WSAN applications. Specifically, we consider methods that borrow concepts from data mining, machine learning and statistics in order to efficiently detect interesting patterns in sensed data. An efficient pattern detection application adds value to a WSAN by contributing to prolonged lifetime of the wireless network and aiding users in identifying which data is important.

Our contribution within the domain of WSAN pattern detection, is a computationally efficient family of methods for pattern detection. These methods cater for pattern detection in both temporal and spatial data, and they require minimal configuration effort. The focus of this chapter is pattern detection in WSANs from a general perspective and in the next section we provide an extensive review of related work that tackles the problem in manners similar to our work. We defer the review of our work [53], [54] to Section 4 where we provide a high-level description of our algorithms. Finally, Section 5 gives our conclusions and identifies open areas of investigation.

Pattern Detection. The term *Pattern Detection* encompasses *Anomaly Detection*, the detection of patterns in data that do not conform to a well-defined notion of normal behaviour, *Novelty Detection*, the detection of previously unforeseen patterns and *Motif Detection*, the detection of patterns in data that recur more often than expected.

3 Review of Work in Pattern Detection in WSAN Data

Before we proceed with the review of related work, some preliminary information related to pattern detection is provided. WSAN data usually has a high degree of spatio-temporal correlation: data from a single node is a linearly ordered sequence where observations are temporally dependent and related in magnitude. Often patterns on the temporal domain differ significantly to their neighbours and this alone may be sufficient for declaring a pattern significantly interesting. On the spatial domain, patterns are usually *collective*: collections of spatially related observations revealing interesting activity in the monitored phenomenon. Sometimes contextual information plays an important role in the task of pattern detection. Consider the example of remote vital signs monitoring of patients at their homes. A pulse oximeter sensor may be used to monitor the heart rate and oxygen saturation level of the patient. A sudden increase in heart rate accompanied by a drop in arterial oxygen saturation may reveal an abnormal situation warranting concern however if the patient is on an exercise bike or treadmill (context) the pattern may be discarded as a false positive. Another example is data centre monitoring that typically involves monitoring environmental conditions such as temperature and humidity. A sustained increase in temperature that lasts for several minutes may indicate a faulty air conditioning unit but if it occurs within a pre-determined maintenance window it may be flagged as a false positive.

The output of pattern detection is usually a *score* or a *label*. The former can be the output of a distance function such as the Euclidean or the Mahalanobis distance that compares a test pattern to either a reference pattern known to be normal or a user-supplied pattern. Although tasking the user to describe patterns of interest may be desirable in some cases, it leaves the system vulnerable to situations where novel patterns are not detected. Furthermore, it burdens users with description of interesting patterns, a task commonly known in expert systems as the *knowledge acquisition bottleneck* [48]. Conversely, if there is confidence that all normal classes are known a priori then emergent patterns can be checked against the normal classes. In both cases, score or distance based detection involves thresholding to determine whether patterns are normal or abnormal. A simplified scenario is to task nodes to discard patterns with distance below a threshold compared to a collection of reference patterns. Assuming the comparison and distance calculation are cheaper than radio communication, then such a technique promotes network longevity.

Labels may be used in a similar manner to determine whether a pattern is normal, interesting or novel. The number of classes employed is user and application dependent. Clustering approaches may be used to cluster patterns with a degree of membership to each class. Such methods are usually employed for outlier detection [21], [4], [45], [52] a function that is conceptually different to pattern detection. We stress that outlier detection is primarily concerned with detecting single observations that deviate from other observation [18] while pattern detection is concerned with identifying interesting temporally contiguous collections of observations.

The solution space for pattern detection is large with techniques such as hypothesis testing, hidden Markov Models, clustering, density estimation, probabilistic matching, statistical testing, neural networks, Bayesian networks, rule-based systems, expert systems, Nearest Neighbour (NN) based techniques and string matching. However, in the following section we focus on the subset of these techniques with either proven or potential applications for WSANs. The works reviewed are summarised in Table 1 and Section 4 offers a discussion of our work which is based on string matching for the temporal domain and stochastic estimation for the spatial domain.

3.1 Spacecraft and Telemetry Data

We start the discussion with systems aiming to detect interesting patterns in spacecraft observations. In [6] the authors describe how they mine scientific data on-board a spacecraft in order to react to dynamic pattern of interest as well as to provide data summaries and prioritisation. Three algorithms are presented that were used on board the Mars Odyssey spacecraft. The first is designed to detect patterns in images for the purpose of thermal anomaly discovery. A thermal anomaly is defined as a region where the surface temperature is significantly warmer or colder than expected, given its location on the planet, the season, and local topography. The second algorithm was developed to identify polar cap edges and illustrates the importance of online pattern detection: transmitting image data is a costly process for a spacecraft, so it is almost always desirable to prioritise by transmitting only the images that reveal interesting activity, in this instance images containing polar cap edges. This algorithm discovered the water ice annulus south of the north polar cap on Mars. The third algorithm was developed to identify high opacity atmospheric events. The opacity (or optical depth) is a measure of the amount of light removed by scattering or absorption as it passes through the atmosphere. The collection of algorithms employ techniques ranging from trivial and dynamic thresholding to Support Vector Machines (SVMs) and reduced set SVMs. Overall the authors present a very mature approach and they explicitly take into account the processing cost and memory requirements. The single criticism is that the three algorithms seem tailored to the specific problems described — it would be interesting to extend the discussion to potential changes needed to generalise the pattern detection performance of the algorithms.

In [15] the authors describe a system based on Kernel Feature Space and directional distribution, which constructs a behaviour model from the past normal telemetry data and monitors current system state by checking incoming data with the model. This type of system is "knowledge-free" in that is not dependent on a priori expert knowledge. Most modern spacecraft, satellites and orbital transfer vehicles transmit *telemetry* data which is multi-dimensional time series. Usually telemetry data is analysed by ground experts but this paper recognises a recent trend that seeks to apply data mining and machine learning in order to perform online pattern detection. The suggested method works as follows: the multi-dimensional telemetry data is divided into subsets using sliding windows. For each subset the method computes the principal component vector and learns the directional distribution modelled as the von Mises-Fisher (vMF) distribution around the optimal direction computed from the principal component vector. Then it computes the occurrence probability of the principal component vector in relation to the current telemetry data mapped into the feature space. If this probability is below a threshold the data is flagged as anomalous. This system is evaluated against simulator-obtained data for three distinct scenarios involving an orbital transfer vehicle designed to make a rendezvous manoeuvre with the International Space Station. All three scenarios involve faults in the thruster engine and one of the scenarios indicates a scenario where the fault would be hard to determine even by a human expert, as the remaining spacecraft thrusters compensate for the underperforming unit. Although the theory behind their approach is sound it suffers from two disadvantages: first, the evaluation is somewhat limited as it only covers three scenarios with failures of the same component. Second, the computation cost is not explicitly modelled: although the authors mention the assumption of building the model offline using previously collected normal telemetry data, they do not explicitly show the cost of the detection computation.

The approach described in [32] presents three unsupervised pattern detection algorithms that have been evaluated offline using historical data from space shuttle main engine — containing up to 90 sensors — for the objective of future inclusion in the Ares I and Ares V launch systems. The usefulness of an online pattern detection approach is highlighted by the fact that sometimes it takes up to 20 minutes until human experts see data from a spacecraft near Mars, time during which catastrophic events could be prevented if automated pattern detection and actuation was performed on-board. The first algorithm is called Orca and it is based on a nearest-neighbour approach for anomaly detection. The second algorithm (Inductive Monitoring System — IMS) is based on clustering and uses distance to flag a data segment as interesting. The final algorithm uses one-class Support Vector Machines (SVMs): it first maps training data from the original data space into a much higher-dimensional feature space and then finds a linear model in that feature space that allows normal data to be on one side and to be separate from abnormal training data. A limitation of this work is related to the performance of the algorithms which varied across different data sets. As the authors identify, it would be useful if in the future the outputs from the different algorithms were combined to give a more coherent picture on the degree of novelty/anomaly of a pattern.

A somewhat similar system aimed at satellite reliability monitoring is described in [12]. This application aims to automate satellite fault diagnosis, a process currently performed by human experts analysing telemetry data periodically transmitted during a fly-by. The diagnosis of faults from, sometimes limited, sensor data is performed by an expert system. The authors describe how the expert system was built even with limited knowledge and its ability to perform inexact reasoning to accommodate sparse sensors. The disadvantage of the system is inherited from expert systems and it involves the effort necessary in describing all the fault states.

3.2 Environmental Pattern Detection

Moving on to ecological monitoring, the approach described in [2] presents a distributed algorithm for detecting statistical anomalies as well as estimating erroneous or missing data. In short, the proposed method performs automatic inference and prediction based on statistical distributions of differences in measurements between a given node and its neighbours. It is assumed that the observed phenomena are spatiotemporally coherent, so that the measurements at neighbouring nodes show a degree of temporal and spatial correlations. The method works in the following manner: at each timer tick a node calculates the differences between its own measurements and those of its neighbours. It also computes the differences between recent and older (local) measurements. By determining the distribution of the differences it can then perform a *p-test* on each new set of measurements and determine whether it is anomalous by comparing to a threshold. The drawback of this method is that it involves considerable radio communication to spatially compare local with remote readings.

In [37] the authors propose a pattern detection system based on elliptical anomalies which are defined by the ellipsoid or hyperellipsoid caused by the region of distance around the mean (cf. [37] for a detailed definition) of two or more monitored variables. They claim that their system is capable of detecting elliptical anomalies in a distributed manner with exactly the same accuracy as that of a centralised solution. Elliptical anomalies are represented using a hyperellipsoidal model. Given the set of column vectors representing sensor observations, the aim of the approach is to partition the set into two subsets: one containing normal observations and one containing anomalous observations. In simple terms, one way to find such a partition is using (Mahalonobis) distance from the sample mean. Moreover, three categories of elliptical anomalies are defined: first-order, second-order and higher-order elliptical anomalies. The authors suggest that the algorithm for first and second order elliptical anomaly detection can be fully distributed in the network. One criticism of this approach is that computational cost is not explicitly considered although the authors seem to target relatively low-end nodes.

The approach presented in [43] attacks the abstract problem of pattern detection using a density test for distributional changes. The main idea is that new, potentially multidimensional, data can be tested against a baseline. For this given baseline data set and a set of newly acquired observations, a statistical test is performed that aims to determine if data points in the newly acquired set were sampled from the same underlying distribution that produced the baseline set. The test statistic is distribution-free and largely based on kernel density estimation and Gaussian kernels. The baseline distribution is inferred using a combination of this kernel density estimator with an Expectation Maximisation algorithm. The strength of the approach is that it seems capable of detecting patterns occurring at multiple data dimensions simultaneously. However it suffers from the disadvantage of high resource and computational requirements making it potentially prohibitive for low-end sensor nodes.

An approach that takes a machine-learning viewpoint to the pattern detection problem is described in [27]. More specifically, the authors describe an

Instance-Based Learning (IBL) model that aims to classify new sets of observations according to their relation to a previously acquired reference data. By storing historical examples of normal observations, the normalcy of emerging observations can be assessed. The authors recognise that such an approach inherently suffers from the high cost overhead of storing multiple "normal" instances and addresses the problem with a combination of instance selection and clustering algorithms that reduce the dictionary size of normal data. The approach also includes a noise suppression filter that removes noise while performing feature selection from the data. The output of the algorithm is a binary decision determining the input as normal or abnormal. Although the authors claim that this solution is highly generalisable to multiple domains, it remains to be determined whether it is suitable for severely resource-constrained environments where the storage size is extremely limited and the cost associated with reading and writing to it is relatively high. Furthermore, there are cases where the user requires more than binary classification and this approach does not cater for approximate detection.

The approach described in [1] presents a model-based predictive system that aims to detect and predict patterns as river flood events in developing countries by deploying sensor networks around the basin area of rivers. The simplest model is based on statistical methods such as linear regression using a portion of data known to be normal. The project aims to cover vast geographical regions of approximately 10,000 km^2 and predict a pattern of interest using a distributed model driven by the collected data. The main drawback of this approach is that it assumes a tiered architecture where resource-constrained sensor nodes transmit summaries and statistics of raw data to a set of computation nodes. The latter determine the correctness of the data, feeds it to the model for prediction and may request additional data from sensors to reduce uncertainty. A somewhat similar tiered system is PRESTO [10] which employs ARIMA (Auto Regressive Integrated Moving Average) time series forecasting models and performs anomaly detection by comparing predicted values to sensor observations.

The work described in [51] introduces contour maps to display the distribution of attribute values in the network and employs contour map matching to determine whether a user-supplied pattern matches node-produced observations. The application scenario is event detection in coal mines, monitoring for the occurrence of gas, dust and water leakage as well as high/low oxygen density regions. A limitation of this approach is that it assumes users capable of perfectly describing the pattern of interest as distributions of an attribute over space and variations of this distribution over time incurred by the event.

3.3 Spatial Structure Pattern Detection

A method aimed at pattern detection over trajectory data is described in [5] where the authors propose a distance metric to determine similarity between trajectory subsequences. Trajectory data is usually a sequence of longitude and latitude readings obtained from Global Positioning System (GPS) readings. Detecting patterns in trajectory data has attracted considerable interest recently due to its numerous applications ranging from remote monitoring of elderly patients to military detection of enemy movements. The contribution of the method is an algorithm that builds local clusters for trajectories in a streaming manner. In detail, it focuses on the problem of determining if a given time window has fewer than k neighbours in its left and right sliding windows, and if so it flags it as an interesting pattern. To improve on efficiency, the authors introduce a data structure (Vantage-Point Tree) that facilitates piecewise rescheduling of cluster joins. Overall, this is a promising approach and the experimental results show the efficiency of the method requiring less than 0.5 milliseconds of processing time for newly acquired observations. The disadvantage of the approach is that it has only been evaluated with offline data lacking the robustness gained by field experiments on real WSAN nodes.

The approach described in [41] focuses on detecting interesting patterns across linear paths of data. A linear path refers to a path represented by a line with a single dimensional spatial coordinate marking an observation point, such as mile markers on motorways or locations along gas pipe lines. Potential applications of pattern detection on this domain would be the discovery of unusual traffic patterns such as accident hubs or proactive infrastructure monitoring on gas or water pipes, tunnels, bridges and so on. The proposed approach is called Scan Statistics for Linear Intersecting Paths (SSLIP) and is in fact a family of algorithms that employ statistical methods for online data mining. With respect to detection of interesting patterns, the method relies on the calculation of an *unusualness* metric that indicates the likelihood ratio or degree of unusualness of a given window of observations in comparison with past data. The windows with highest likelihood ratios are flagged as interesting or unusual. The weakness of this approach is that it is employing heavyweight, in terms of processing, techniques such as Monte Carlo simulations to determine the significance of patterns. Furthermore, the authors clearly state that this method should be used in conjunction with domain experts for the identification of interesting patterns — this restricts the autonomous detection that is highly desired in WSANs and inevitably makes the process somewhat more interactive.

Another relatively interactive approach is described in [47] that introduces *rare category detection*. The main contribution of the proposed method is the capability of detecting both statistically significant and interesting patterns. The central premise of category detection involves tasking a user to label patterns with predetermined categories. A pattern that doesn't belong to any category is novel and is classified to a new category. The presented algorithm aims to identify the rare categories in data using hierarchical mean shift, a variation of the mean shift algorithm which is an iterative procedure that shifts each data point to the average of data points in its neighbourhood [8]. The hierarchical variation involves the iterative application of mean shift with increasing bandwidths, such that a hierarchy of clusters at different scales can be created whilst annotating each cluster with an anomaly score. The two main weaknesses of this approach is that it can be computationally expensive and that it requires a small degree of human interaction. Although the user need not provide information about the data such as the number of categories or their prior probabilities, he or she is still required to classify a statistically significant pattern as interesting.

3.4 Data Centre Monitoring and Context-Aware Computing

Moving on to the emerging research area of data centre monitoring and design, the research described in [36] proposes an online temporal data mining solution to optimise the performance of data centre chillers. Sensor nodes deployed in data centres produce time series data that describe environmental conditions that usually vary in time according to the load of individual servers, storage and networking equipment. This type of system exemplifies the automated control that is typical of a WSAN: according to when/what interesting patterns are sensed a decision must be made to turn on/off chillers, select a utilisation range and generally react to cooling demands. The authors focus their efforts on *motif mining*, that is the identification of frequently occurring temporal patterns. First, they obtain a symbolic representation of the time series data using the Symbolic Aggregate Approximation (SAX) [30] algorithm that is also employed by our own method (Section 4). A Run-Length Encoding (RLE) of the symbol sequence is performed in order to note when transitions from one symbol to another occur. Generally speaking, RLE is a technique that reduces the the size of a repeating string of characters by replacing one long string of consecutive characters with a single data value and count [3]. Frequent episode mining is conducted over the sequence of transitions to identify the underlying efficiency profile of the data centre under different environmental circumstances. This allows the formation of dynamic models of data centre chillers and formulation of control strategies.

A somewhat similar approach [33] targets online novelty detection on temporal sequences utilising Support Vector Regression (SVR) to model the time series data. The authors introduce detection with a confidence score indicating the confidence of the algorithm for the degree of novelty. There is particular consideration to the event duration which is rarely known in advance and must be selected with some care to avoid missing events or spurious matches. The significance of the training window is also identified and its relationship to resource requirements is one the weaknesses of the approach. Another weakness is the limited evaluation that does not provide concrete evidence of the efficacy of SVR for novelty detection.

The approach of [17] largely targets context-aware computing and specifically mining for anomalous structures in observations representing human interactions with their environment. The authors propose the use of a *Suffix Tree* data structure to encode the structural information of activities and their event statistics at different temporal resolutions. The method aims to identify interesting patterns that either consist of core structural differences to previously normal behaviour or differences based on the frequency of occurrence of motifs. With regards to anomalous pattern detection, the authors adopt the view that given a set of normal activity sequences *A*, any subsequence of events is classified as normal as long as it occurred in *A*. Naturally, this relies on training data used to construct the dictionary of legitimate behaviour. Sequences are classed as anomalous using a match statistic computed over the suffix tree. The strength of the approach is that suffix tree traversal can be conducted in linear time, however dynamic suffix tree construction and update is not always suitable to platforms with severe resource constraints that lack dynamic memory allocation capabilities.

3.5 Network Monitoring and Intrusion Detection

Moving on to approaches that aim to ensure WSAN stability, the work described in [11] introduces Artificial Immune Systems (AIS) for misbehaviour detection. AIS are inspired from principles that the human immune system uses for the detection of harmful agents such as viruses and infections. Since WSANs typically lack the infrastructure readily available in wired networks, it is somewhat easier for attackers to maliciously modify a WSAN either by dropping packets or compromising the routing topology. In this context, an interesting pattern would be a sequence of unusual actions for one or more nodes. The method aims to facilitate local learning and detection on WSAN nodes using a gene-based approach. Each node maintains a local set of detectors that is produced by negative selection from a larger set of randomly generated detectors tested on a set of self strings. These detectors are then used to test new strings that represent local network behaviour and detect non-self strings. The approach has been evaluated using MAC protocol messages and has shown encouraging results. Although such an approach is valuable for the detection of local patterns that indicate misbehaviour over some layer of the OSI (Open Systems Interconnection) reference model stack, it is not clear how it can be generalised to apply to sensor-acquired observations.

A somewhat related approach [31], targets intrusion detection in WSANs from samples of routing traffic. First, feature selection of traffic and non-traffic related data is performed in order to learn the distribution of values affecting routing conditions and traffic flows. Second, anomaly detection is performed locally by taking a window of data and examining it against previously collected normal data. This window contains samples that are mapped to points in a feature space and subsequently analysed together with their surrounding region in the feature space. If a point lies in a sparse region of space is classified as anomalous, using a fixed-width clustering algorithm. The method has the capability of detecting previously unseen (novel) patterns while respecting resource constraints by performing detection locally but at the same time shows a weakness in detecting slow attacks happening gradually over a long time scale. Finally, evaluation is somewhat limited and performed exclusively via simulations that do not comprehensively model a large number of attacks.

The approach described in [19], proposes a Principal Component Analysis (PCA) method for detecting anomalies with complex thresholds in a distributed manner. Nodes send their readings to a coordinator node that is responsible for firing a trigger based on the aggregate behaviour of a subset of nodes. The individual nodes perform filtering such that they send readings only when measurements deviate significantly from the last transmitted data. With respect to detection, they propose two window

Table 1 Comparison of pattern detection techniques for sensor data

Approach	Basis	Application	Strength	Weakness
[6]	Dynamic Thresholding and SVMs	Spacecraft image data	Respects constraints	Algorithm tailor-made to problem
[15]	Kernel functions	Spacecraft telemetry data	No prior knowledge required	Comp. costs not explicitly modelled
[32]	NN/Clustering /SVMs	Spacecraft engine data	Pattern mining in data from up to 90 sensors	Performance varies across data sets
[12]	Expert System	Satellite telemetry data	Inexact reasoning	Knowledge acquisition bottleneck common in Expert Systems
[37]	Elliptical Anomalies	Abstract/Environmental data	Distributed solution with same accuracy as centralised	Comp. cost not explicitly modelled
[2]	Statistical p-value tests	Ecological anomaly detection	Automatic Inference & Prediction	Radio communication cost
[43]	Kernel density estimators	Abstract/Multidimensional data	Detects patterns occurring at multiple dimensions simultaneously	High computational requirements
[27]	ML/Instance-based classification	Abstract/Temporal sequences	Noise suppression filter	Binary classification/No approximate detection
[5]	NN/Clustering	Trajectory pattern detection	Efficient processing time for classifying new observations	No evaluation of online operation
[41]	Scan Statistics for Linear Intersecting Paths	Unusual traffic pattern discovery	Unusualness metric	High computational requirements
[47]	Hierarchical Mean Shift	Abstract/Rare category detection	Distinguishes statistically significant and interesting patterns	High computational requirements
[36]	Symbolic conversion/Run-length encoding	Data centre monitoring	Dynamic modelling of data centre chillers	Bias towards motifs rather than novelties
[33]	Support Vector Regression	Abstract/Temporal sequences	Confidence score	High computational costs related to large training windows
[17]	Subsequence matching using Suffix Trees	Context-aware computing	Linear time matching	Not suitable for dynamic tree updates
[11]	Artificial Immune Systems	Misbehaviour detection on network data	Efficiency of local detectors	Not generalisable to sensor-acquired data
[31]	Fixed-width clustering	Network Intrusion Detection	Detects previously unseen patterns	Misses slow-occurring patterns
[19]	Principal Component Analysis	Distributed network pattern detection	Source-side filtering	Relies on coordinator node (SPOF)
[1]	Model-based/Statistical regression	River flood pattern events	Distributed data-driven model	Assumes presence of a resource-rich tier
[51]	Contour-map matching	Coal-mine monitoring	SQL extensions allows users to specify events as pattern	Relies on prior-distribution knowledge by user

triggers that are persistent threshold violations over a fixed or varying window of time series data. Although the approach is aimed at detecting unusual network traffic patterns, it could apply to certain WSAN applications. The main criticism is that it creates single points of failure by assigning the coordinator role to nodes.

4 Symbolic and Stochastic Pattern Detection

Having covered sufficient background and related work, we now focus on our approach to pattern detection for extremely resource-constrained devices. Before we proceed with the details of our approach, we enumerate the requirements for a pattern detection solution in resource-constrained Wireless Sensor Actuator Networks:

- It must be capable of detecting patterns both on the temporal and spatial domain, since sensornets usually monitor phenomena on a spatio-temporal scale.
- It must be capable of detecting both previously unseen patterns and user-submitted patterns with exact and approximate matching semantics.
- It must tolerate outliers, noise, missing values as well as scale differences in the sensor-acquired data.
- It must explicitly take into account the resource-constraints of the execution environment. Specifically, the solution targets low-end nodes such as the TMote Sky [40] and the battery-free Intel WISP [38].
- It must fit well with the existing communication paradigms without requiring any modifications to lower layer protocols.
- It must scale as the number of user-submitted patterns increase as well as the network size grows.

Over the following sections we provide an initial discussion of how the above requirements are addressed. For experimental evidence and in-depth coverage the interested reader is referred to our previous work [55], [53], [56], [54].

4.1 Temporal Pattern Detection Using a Symbolic Representation

First, we convert the pattern detection problem to a pattern matching problem. A mature symbolic representation algorithm — *SAX* [30] — used for numerous data mining tasks is employed to convert sequences of numeric sensor observations to character strings. SAX is a linear time algorithm that first coverts a time series to an intermediate representation using Piecewise Aggregate Approximation (PAA). The resulting string is obtained by converting the PAA representation using a table lookup. Due to space restrictions we will not discuss SAX here in further detail, but we refer to interested reader to the published literature on SAX (cf. [30, 24, 23]. The reasons we convert numeric data to symbols are threefold:

1. A symbolic representation opens up access to a wide variety of mature string matching algorithms,
2. A symbolic representation achieves data reduction and thus requires less space achieving radio communication and storage savings, and

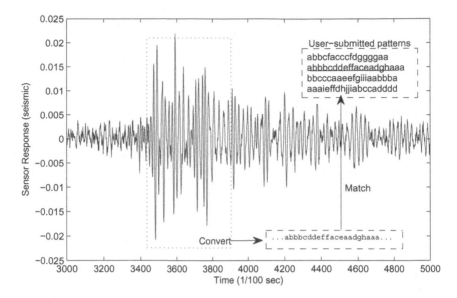

Fig. 1 Example of converting a time series segment to symbols and matching against a database of user-submitted patterns. The user typically enters a pattern as a numeric sequence. This sequence is converted to a string either by an application that acts as the interface between the WSAN and the user, or by the WSAN node itself.

3. A string distance function is defined that lower-bounds the Euclidean distance allowing to perform matching entirely on the symbolic representation without loss of information.

The symbolic conversion algorithm is capable of handling a degree of missing values, noisy data and outliers in the numeric input data. However, no inference of missing values is performed and a high number of missing values would inadvertently affect the matching performance of the algorithm. With regards to detection we offer three methods:

1. *Approximate* or *Exact* Pattern Detection where the user knows in advance the pattern of interest and wishes to be informed when the sensor-acquired data converts to patterns that match the reference pattern either approximately or exactly (example in Figure 1)
2. *Non-parametric* Pattern Detection, where the user need not supply any information in advance but instead the algorithm trains on a window of normal sensor data by constantly observing distances in temporally adjacent strings corresponding to temporally neighbouring time series sequences
3. *Probabilistic* Pattern Detection where a Markov Model is built by monitoring individual character transitions. Strings with improbable character transitions are labelled as unusual.

Further to the above, we cater for Dynamic Sampling Frequency Management (DSFM). Similar to Non-parametric Pattern Detection, this algorithm involves a a training phase to learn the sensed process dynamics and use it to make autonomous local decisions to dynamically increase or decrease the frequency in which the observations are acquired from the sensors. Dynamically adjusting the frequency enables network nodes to conserve energy in periods of relative inactivity for times when interesting patterns occur. We recognise that certain applications can have specific sampling frequency requirements depending on the periodicity of the signal. Having these requirements specified as a sampling interval rather than an absolute value, can help nodes select an appropriate frequency via the DSFM capability, relaxing the pre-deployment need for complex signal processing.

The scalability requirement is met by introducing a *Suffix Array* structure for storage and fast searching of user-submitted patterns. In terms of fitting with standard methods, our approach uses a standard Publish/Subscribe interface that employs state-of-the-art WSAN communication protocols without requiring any modifications or incurring any communication overhead. The computational efficiency of the algorithm is achieved by aggressive optimisations (cf. [56] for a detailed discussion) using integer arithmetic and verified by measuring the execution time on nodes from operational WSAN systems. The additional cost of converting numeric sequences to strings is almost negligible: typically 11 milliseconds of CPU time are required for the conversion of a numeric sequence for a window of 40 data points to a string and this time includes a string distance comparison to a user-submitted pattern. Naturally, larger windows are possible, for instance to capture long-duration pattern events, and can be easily introduced either at pre-deployment or injected dynamically at runtime.

In the case of user-submitted patterns, one point of interest is the relation between the length of the user pattern and the window size employed by the symbolic conversion algorithm. In this context, the length of the user pattern relates to the length of the resulting string, in other words the output of the symbolic conversion. The user submits a pattern in a numeric form and this converts to a string either by the WSAN node itself or by an application that acts as an interface between the WSAN and the user. There exist two possibilities: in Approximate Pattern Detection the length of the user-submitted pattern must be equal to the length of the string produced by a WSAN node. This is due to the string distance function employed for comparison that only accepts strings of equal length. However this can be adapted by using alternative distance functions that can accept strings of different length — the Sequence Alignment algorithms (cf. [16]) are such examples of comparison functions that accept strings of different length. The second option relates to Exact Pattern Detection where the Suffix Array structure is employed. In this case the window size is affected in the following manner: if a user submits a pattern smaller than the resulting string produced by the WSAN node, matching is unaffected since the Suffix Array stores all the suffixes of the user submitted pattern. Conversely, if a user submits a pattern larger than the resulting string produced by the WSAN node, exact matching is affected since strings of different length can never match exactly. To resolve this, the WSAN node adjusts the window size such that the resulting string length is equal to the largest user-submitted string.

Further to the performance evaluation on operational WSAN nodes, in order to refine the granularity of measurements we have opted to simulate the operation of a WSAN by replaying data sets collected in-situ and emulating the operations of a node in software. This approach was used with three data sets representing distinct case studies: first, normalised seismometer and acoustic data from a volcano monitoring application ([50]) was used to quantitatively evaluate the performance against a large number of organic events of varying durations. Indicative results are shown in Table 2 where the double compression refers to running two instances of the symbolic conversion algorithm with different compression settings for accuracy. In fact, the lightweight nature of the approach allows users to execute it simultaneously with different configuration values without penalising performance. This is a significant benefit in cases where the pattern event duration or characteristics are unknown. Second, a case study using environmental data from an indoor network [20] was used to evaluate the performance against data imperfections such as outliers and missing values. We found the accuracy of the algorithm unaffected by imperfections and at the same time real and synthetic patterns were detected on temperature, humidity and light data. Third, ECG and accelerometer data from the UCR data mining archive [46] was used to evaluate the applicability of the algorithm to potential pervasive healthcare and context-aware deployments. Again, we found the algorithm responded well in detecting unseen patterns and pinpointing the change from normal to anomalous. Due to space limitations we cannot discuss the experimental result in great detail however we invite the interested reader to review our results ([53]).

Finally, we have recently adapted the algorithm for the battery-free Intel WISP platform that harvests electromagnetic energy from RFID readers. We have modified the original WISP design by adding a supercapacitor to store harvested energy so that a node can continue operating when it moves outside the range of an RFID reader. We intend to use the pattern detection algorithm for the purpose of activity recognition and more specifically teaching younger children about movements and elementary mechanics (cf. [55] for more details). This verifies the versatility of the pattern detection algorithms for different tasks and their suitability for the extremely resource-constrained platforms.

Table 2 Summary of quantitative detection accuracy results with one and two compression settings.

Compression Setting	Detection Accuracy	
Single compression (4 : 1)	Detected: 733 out of 947	**77.4%**
Double compression (2 : 1 and 4 : 1)	Detected: 878 out of 947	**92.7%**

4.2 Spatial Pattern Detection and Source Location Estimation

For the detection of patterns on the spatial domain, we focus on a specific class of events namely dispersion of pollutants from a single static point source. We assume a WSAN is deployed in the area of the dispersion plume where nodes individually detect the presence of the pollutant in the atmosphere using the algorithm of the previous section or some other method. Since this pattern has a strong spatial element, local detection is not sufficient in itself. The aim is to initiate a "walk" of the network such that a coarse estimate of the source's location is iteratively computed.

The algorithm is similar to local gossip approaches (cf. [29] for the *Trickle* algorithm, an example of a gossip protocol for WSANs) and works by instantiating a Kalman filter that iteratively predicts the state of the dispersion process at neighbouring nodes. At the beginning of the process, the originating node makes an estimate of the observations at its one-hop neighbours (line 2, Algorithm 1). Since no other information is available this estimate is a linear transformation of the local reading. The neighbour that minimises the error is selected as the next hop and receives the necessary Kalman filter parameters and values to continue the process (line 8, Algorithm 1). A geometric computation is employed to take into account the neighbourhood consensus before making the routing decision in order to reduce message cost at each hop (not shown in Algorithm 1 for simplicity). This works in the following manner: after a small (i.e. ≤ 10) number of hops, a convex hull is evaluated for the coordinates of the nodes that participate in the estimation. Provided that the estimation process begins at nearby locations, the convex hull can be computed cheaply without a significant communication overhead. The mean direction of movement is given by calculating the centroid of the convex hull. At this stage only a coarse quadrant direction is necessary: the Cartesian coordinates quadrant in which the majority of nodes participating in the estimation process believe the source is located. This geometric computation adds robustness to the algorithm such that erroneous local routing decisions can be overridden by the majority consensus. Experimental evaluation has shown estimation accuracy of up to 97.33%, where an estimate is considered accurate if it is within 6 meters from the actual position of the source.

The iterative estimation process stops when the estimation error becomes unacceptably high (lines 13-15, Algorithm 1) which indicates that either the node has approached a region close to the source or that it has entered a region where observations differ greatly from the estimate of the process. The process halts and the coordinates and intensity of the last node in the path become the final estimate.

This spatial detection not only detects the pattern but provides useful metadata with respect to the pattern location and intensity. Such estimation tasks are common in applications such as [9] concerned with the *Inverse* problem: given some sensor observations, the goal is to estimate the source location. We have evaluated the spatial detection algorithm over grid and random distributions of different densities and we have found it outperforms a maximum selection algorithm (that selects neighbours with the higher reading) while it is competitive with other heavy-weight

---

Algorithm 1. Spatial Pattern Detection (SED) Location Estimation Algorithm

---

 1: **variables** Estimate Error Covariance P, Measurement Noise Variance R, Process Variance Q, State Transition Matrix A, Measurement Matrix H, Initial Estimate $\hat{x}_k{}^-$, maxhopcount=1, netpath[], counter $c = 0$;
 2: Project state estimate $\hat{x}_k{}^-$ ahead (cf. Equation 4.9 [49])
 3: Project error covariance $P_k{}^-$ ahead (cf. Equation 4.10 [49])
 4: Task *unvisited* neighbours within maxhopcount to report measurement.
 5: **for** (each of replies received) **do**
 6:      calculate innovations $(z_k{}^{(i)} - H\hat{x}_k{}^-)$
 7: **end for**
 8: Select as next hop the node that minimises the innovation.
 9: Compute the Kalman gain (cf. Equation 4.11 [49])
10: Correct (Update) estimate with measurement $z_k{}^{(i)}$ (cf. Equation 4.12 [49])
11: Correct (Update) the error covariance P_k (cf. Equation 4.13 [49])
12: Compute relative error.
13: **if** abs(relative error) $>=$ multiple$\cdot(\mathbb{E}[Rel\ Error])$ **then**
14:      **exit**
15: **else**
16:      Add local address to netpath[c] and increment c.
17:      Send algorithm parameters to selected node (line 8) and task it to start at Line 1.
18: **end if**

---

location estimation approaches. For further information and experimental results the interested reader is referred to [54].

5 Conclusions

In this chapter we have outlined the need for efficient pattern detection in Wireless Sensor Actuator Networks (WSANs) that lack core infrastructure services such as reliable communication and time synchronisation. The majority of middleware approaches that attempt to deal with the pattern detection problem from a composite event calculus perspective are not suitable for severely resource-constrained execution environment. Instead, pattern detection techniques that adapt statistical, machine-learning and data mining approaches are much more suitable.

We presented a large collection of pattern detection approaches from different application domains that address the problem using a multitude of techniques. A universal solution to the problem capable of detecting patterns in different types of data is extremely difficult since application requirements vary. To partially address this problem, we described in Section 4 a data-mining inspired technique that employs string matching and is capable of detecting patterns in sensor data of different modalities across both temporal and spatial domains. This method requires little or no configuration and therefore fits well the long-term vision that anticipates WSANs comprised of millions of inexpensive nodes. Furthermore, it leverages the development of existing state-of-the-art communication methods using standard interfaces such as Publish/Subscribe for the notification of interesting patterns and without requiring any modifications to lower layer protocols.

The discussion of Pattern Detection for WSANs leaves a few directions open for further exploration. First, the impact of local coordination in relation to Non-Parametric and Probabilistic Pattern Detection has to be investigated further. Local coordination refers to geographical adjacent WSAN nodes exchanging information that facilitates the training phase of the algorithms. Second, we aim to investigate a direction based on real-time classification of WSAN data. There is preliminary work in this area utilising string algorithms [44] and numerous applications exist ranging from assisted diagnosis [13] to augmenting learning processes for children [55]. The final research direction involves the investigation of mixed methods where multiple detection algorithms run in parallel within a WSAN in order to improve detection accuracy. Addressing the above directions through evaluation on operational WSANs will extend the work on Pattern Detection and benefit users of reactive applications across a number of application domains.

References

1. Basha, E.A., Ravela, S., Rus, D.: Model-based monitoring for early warning flood detection. In: SenSys 2008: Proceedings of the 6th ACM conference on Embedded network sensor systems, pp. 295–308. ACM, New York (2008)
2. Bettencourt, L.M.A., Hagberg, A.A., Larkey, L.B.: Separating the wheat from the chaff: practical anomaly detection schemes in ecological applications of distributed sensor networks. In: Aspnes, J., Scheideler, C., Arora, A., Madden, S. (eds.) DCOSS 2007. LNCS, vol. 4549, pp. 223–239. Springer, Heidelberg (2007)
3. Bose, R.: Information theory, coding and cryptography. Tata McGraw-Hill, New York (2002)
4. Branch, J., Szymanski, B., Giannella, C., Wolff, R., Kargupta, H.: In-network outlier detection in wireless sensor networks. In: 26th IEEE International Conference on Distributed Computing Systems, ICDCS 2006, pp. 51–59 (2006)
5. Bu, Y., Chen, L., Fu, A.W.-C., Liu, D.: Efficient anomaly monitoring over moving object trajectory streams. In: KDD 2009: Proceedings of the 15th ACM SIGKDD international conference on Knowledge discovery and data mining, pp. 159–168 (2009)
6. Castano, R., Wagstaff, K.L., Chien, S., Stough, T.M., Tang, B.: On-board analysis of uncalibrated data for a spacecraft at mars. In: KDD 2007: Proceedings of the 13th ACM SIGKDD international conference on Knowledge discovery and data mining, pp. 922–930 (2007)
7. Chandola, V., Banerjee, A., Kumar, V.: Anomaly Detection: A Survey. ACM Computing Surveys (2009)
8. Cheng, Y.: Mean Shift, Mode Seeking, and Clustering. IEEE Trans. Pattern Anal. Mach. Intell. 17(8), 790–799 (1995)
9. Chin, J.-C., Yau, D.K.Y., Rao, N.S.V., Yang, Y., Ma, C.Y.T., Shankar, M.: Accurate localization of low-level radioactive source under noise and measurement errors. In: SenSys 2008: Proceedings of the 6th ACM Conference on Embedded Network Sensor Systems, pp. 183–196. ACM, New York (2008)
10. Desnoyers, P., Ganesan, D., Li, H., Li, M., Shenoy, P.: PRESTO: A predictive storage architecture for sensor networks. In: Tenth Workshop on Hot Topics in Operating Systems, HotOS X (2005)

11. Drozda, M., Schaust, S., Szczerbicka, H.: Is AIS based misbehavior detection suitable for wireless sensor networks. In: Proc. IEEE Wireless Communications and Networking Conference (WCNC), Citeseer (2007)

12. Durkin, J., Tallo, D., Petrik, E.J.: FIDEX: An expert system for satellite diagnostics. In: In its Space Communications Technology Conference: Onboard Processing and Switching, pp. 143–152 (1991) (see N92-14202 05-32)

13. Dutta, R., Dutta, R.: Maximum Probability Rule based classification of MRSA infections in hospital environment: Using electronic nose. Sensors and Actuators B: Chemical 120(1), 156–165 (2006)

14. Faloutsos, C., Ranganathan, M., Manolopoulos, Y.: Fast subsequence matching in time-series databases. SIGMOD Rec. 23(2), 419–429 (1994)

15. Fujimaki, R., Yairi, T., Machida, K.: An approach to spacecraft anomaly detection problem using kernel feature space. In: KDD 2005: Proceedings of the eleventh ACM SIGKDD International Conference on Knowledge Discovery in Data Mining, pp. 401–410 (2005)

16. Gusfield, D.: Algorithms on Strings, Trees and Sequences: Computer Science and Computational Biology. Cambridge University Press, Cambridge (1997)

17. Hamid, R., Maddi, S., Bobick, A., Essa, I.: Unsupervised analysis of activity sequences using event-motifs. In: VSSN 2006: Proceedings of the 4th ACM International Workshop on Video Surveillance and Sensor Networks, pp. 71–78 (2006)

18. Hawkins, D.M.: Identification of outliers. Monographs on applied probability and statistics. Chapman and Hall, Boca Raton (1980)

19. Huang, L., Garofalakis, M., Hellerstein, J., Joseph, A., Taft, N.: Toward sophisticated detection with distributed triggers. In: MineNet 2006: Proceedings of the 2006 SIGCOMM workshop on Mining network data, pp. 311–316. ACM, New York (2006)

20. Intel. Lab Data, Berkeley (2004),
    `http://db.csail.mit.edu/labdata/labdata.html`

21. Janakiram, D., Reddy, V.A., Kumar, A.: Outlier detection in wireless sensor networks using bayesian belief networks. In: First International Conference on Communication System Software and Middleware, Comsware 2006, pp. 1–6 (2006)

22. Karpiński, M., Cahill, V.: Stream-based macro-programming of wireless sensor, actuator network applications with SOSNA. In: DMSN 2008: Proceedings of the 5th workshop on Data management for sensor networks, pp. 49–55. ACM, New York (2008)

23. Keogh, E., Lin, J., Fu, A.: HOT SAX: Efficiently Finding the Most Unusual Time Series Subsequence. In: IEEE International Conference on Data Mining, pp. 226–233 (2005)

24. Keogh, E., Lonardi, S., Ratanamahatana, C.A.: Towards parameter-free data mining. In: KDD 2004: Proceedings of the tenth ACM SIGKDD international conference on Knowledge discovery and data mining, pp. 206–215. ACM, New York (2004)

25. Kompis, C., Aliwell, S.: Energy Harvesting Technologies to Enable Wireless and Remote Sensing — Sensors & Instrumentation KTN Action Group Report (June 2008),
    `http://server.quid5.net/ koumpis/pubs/pdf/ energyharvesting08.pdf`

26. Krishnamachari, B.: Networking Wireless Sensors. Cambridge University Press, Cambridge (2005)

27. Lane, T., Brodley, C.E.: Temporal sequence learning and data reduction for anomaly detection. ACM Trans. Inf. Syst. Secur. 2(3), 295–331 (1999)

28. Levis, P., Culler, D.: Maté: A Tiny Virtual Machine for Sensor Networks. In: ASPLOS-X: Proceedings of the 10th International Conference on Architectural Support for Programming Languages and Operating Systems, New York, NY, USA, pp. 85–95 (2002)

29. Levis, P., Patel, N., Culler, D., Shenker, S.: Trickle: a self-regulating algorithm for code propagation and maintenance in wireless sensor networks. In: NSDI 2004: Proceedings of the 1st Conference on Symposium on Networked Systems Design and Implementation, vol. 2, USENIX Association, Berkeley (2004)

30. Lin, J., Keogh, E., Lonardi, S., Chiu, B.: A symbolic representation of time series, with implications for streaming algorithms. In: DMKD 2003: Proceedings of the 8th ACM SIGMOD workshop on Research issues in data mining and knowledge discovery, pp. 2–11. ACM, New York (2003)

31. Loo, C.E., Ng, M.Y., Leckie, C., Palaniswami, M.: Intrusion detection for routing attacks in sensor networks. International Journal of Distributed Sensor Networks 2(4), 313–332 (2006)

32. Oza, N., Schwabacher, M., Matthews, B.: Unsupervised Anomaly Detection for Liquid-Fueled Rocket Propulsion Health Monitoring. Journal of Aerospace Computing, Information, and Communication 6(7), 464–482 (2007)

33. Ma, J., Perkins, S.: Online novelty detection on temporal sequences. In: KDD 2003: Proceedings of the ninth ACM SIGKDD international conference on Knowledge discovery and data mining, pp. 613–618. ACM, New York (2003)

34. Madden, S., Franklin, M.J., Hellerstein, J.M., Hong, W.: The design of an acquisitional query processor for sensor networks. In: SIGMOD 2003: Proceedings of the 2003 ACM SIGMOD international conference on Management of data, pp. 491–502. ACM, New York (2003)

35. Markou, M., Singh, S.: Novelty detection: a review–part 1: statistical approaches. Signal Processing 83(12), 2481–2497 (2003)

36. Patnaik, D., Marwah, M., Sharma, R., Ramakrishnan, N.: Sustainable operation and management of data center chillers using temporal data mining. In: KDD 2009: Proceedings of the 15th ACM SIGKDD international conference on Knowledge discovery and data mining, pp. 1305–1314 (2009)

37. Rajasegarar, S., Bezdek, J.C., Leckie, C., Palaniswami, M.: Elliptical anomalies in wireless sensor networks. ACM Trans. Sen. Netw. 6(1), 1–28 (2009)

38. Intel Research. WISP: Wireless Identification and Sensing Platform (2008), http://seattle.intel-research.net/wisp/

39. Roundy, S., Wright, P.-K., Rabaey, J.: Energy Scavenging for Wireless Sensor Networks: with Special Focus on Vibrations, 1st edn. Springer, Heidelberg (2003)

40. MoteIV (later renamed to Sentilla). TMote Sky Datasheets and Downloads (2008), http://www.sentilla.com/pdf/eol/tmote-sky-datasheet.pdf

41. Shi, L., Janeja, V.P.: Anomalous window discovery through scan statistics for linear intersecting paths (SSLIP). In: KDD 2009: Proceedings of the 15th ACM SIGKDD international conference on Knowledge discovery and data mining, pp. 767–776 (2009)

42. Sipser, M.: Introduction to the Theory of Computation. PWS Pub Co, Boston (1996)

43. Song, X., Wu, M., Jermaine, C., Ranka, S.: Statistical change detection for multi-dimensional data. In: KDD 2007: Proceedings of the 13th ACM SIGKDD international conference on Knowledge discovery and data mining, pp. 667–676. ACM, New York (2007)

44. Stiefmeier, T., Roggen, D., Tröster, G.: Gestures are strings: efficient online gesture spotting and classification using string matching. In: BodyNets 2007: Proceedings of the ICST 2nd international conference on Body area networks, pp. 1–8 (2007)

45. Subramaniam, S., Palpanas, T., Papadopoulos, D., Kalogeraki, V., Gunopulos, D.: Online Outlier Detection in Sensor Data Using Non-Parametric Models. In: Dayal, U., Whang, K.-Y., Lomet, D.B., Alonso, G., Lohman, G.M., Kersten, M.L., Cha, S.K., Kim, Y.-K. (eds.) VLDB, pp. 187–198. ACM, New York (2006)

46. Riverside University of California. The UCR Time Series Data Mining Archive (2008),
    http://www.cs.ucr.edu/~eamonn/TSDMA
47. Vatturi, P., Wong, W.-K.: Category detection using hierarchical mean shift. In: KDD
    2009: Proceedings of the 15th ACM SIGKDD international conference on Knowledge
    discovery and data mining, pp. 847–856 (2009)
48. Wagner, W.P.: Issues in knowledge acquisition. In: SIGBDP 1990: Proceedings of the
    1990 ACM SIGBDP conference on Trends and directions in expert systems, pp. 247–
    261 (1990)
49. Welch, G., Bishop, G.: An Introduction to the Kalman Filter. Technical Report 95-041.
    Chapel Hill, NC, USA (1995)
50. Werner-Allen, G., Dawson-Haggerty, S., Welsh, M.: Lance: optimizing high-resolution
    signal collection in wireless sensor networks. In: SenSys 2008: Proceedings of the 6th
    ACM conference on Embedded network sensor systems, New York, NY, USA, pp. 169–
    182 (2008)
51. Xue, W., Luo, Q., Chen, L., Liu, Y.: Contour map matching for event detection in sensor
    networks. In: SIGMOD 2006: Proceedings of the 2006 ACM SIGMOD international
    conference on Management of data, pp. 145–156. ACM, New York (2006)
52. Zhang, J., Wang, H.: Detecting outlying subspaces for high-dimensional data: the new
    task, algorithms, and performance. Knowledge and Information Systems 10(3), 333–355
    (2006)
53. Zoumboulakis, M., Roussos, G.: Efficient pattern detection in extremely resource-
    constrained devices. In: SECON 2009: Proceedings of the 6th Annual IEEE communica-
    tions society conference on Sensor, Mesh and Ad Hoc Communications and Networks,
    pp. 10–18 (2009)
54. Zoumboulakis, M., Roussos, G.: Estimation of Pollutant-Emitting Point-Sources Using
    Resource-Constrained Sensor Networks. In: Trigoni, N., Markham, A., Nawaz, S. (eds.)
    GSN 2009. LNCS, vol. 5659, pp. 21–30. Springer, Heidelberg (2009)
55. Zoumboulakis, M., Roussos, G.: In-network Pattern Detection on Intel WISPs (Demo
    Abstract). In: Proceedings of Wireless Sensing Showcase (2009)
56. Zoumboulakis, M., Roussos, G.: Integer-Based Optimisations for Resource-Constrained
    Sensor Platforms. In: Hailes, S., Sicari, S., Roussos, G. (eds.) S-CUBE 2009. LNICIST,
    vol. 24, pp. 144–157. Springer, Heidelberg (2010)

Smart Patient Care

Diogo Guerra, Pedro Bizarro, and Dieter Gawlick

Abstract. The creation, management, and use of Electronic Medical Records (EMR) is a central issue for the medical community and is a high priority for many governments around the world. Collecting, storing, and managing EMR is expensive and difficult due to a set of demanding requirements for quality attributes (reliability, availability, security), multiple types of data (real-time data, historical data, medical rules, medical vocabularies), and data operations (raising alarms, pattern detection, or predictions). The traditional approach uses a combination of multiple data management systems such as databases, rule engines, data mining engines, event processing engines and more. Having multiple data management systems leads to "islands of data", missed correlations, and frequent false alarms. However, recent advances in database technology have added functionality to database systems such as temporal support, continuous queries, notifications, rules managers, event processing and data mining. This chapter describes a prototype, SICU, that using those advanced functionalities, implements a complete, single-system EMR engine to monitor patients in emergency care units. SICU was designed as a proof-of-concept EMR system that manages real-time data (vitals and laboratory data), historic data (past clinical information), medical knowledge (in the form of rules) and issues appropriate alarms with the correct level of criticality and personalized by doctor or patient. In addition, using data mining models built from real patient profiles, SICU is able to predict if patients will have a cardiac arrest in the following 24 hours. The prototype has shown a way to significantly enhance evidence based

Diogo Guerra
FeedZai, Portugal
e-mail: diogo.guerra@feedzai.com

Pedro Bizarro
University of Coimbra, Portugal
e-mail: bizarro@dei.uc.pt

Dieter Gawlick
Oracle, California
e-mail: dieter.gawlick@oracle.com

S. Helmer et al.: Reasoning in Event-Based Distributed Systems, SCI 347, pp. 217–237.
springerlink.com   © Springer-Verlag Berlin Heidelberg 2011

medicine and is therefore of great interest to the medical community. The lessons can be applied to other domains such as smart utility grids.

1 Introduction

Collecting, storing, and managing EMR is expensive and difficult because it needs two types of data management requirements. First, as with other sensitive datasets, managing EMR requires reliability–records can never be lost, availability–records must be available at all times, and security–records can never be available to unauthorized people or systems. However, reliable, secure access to EMR records at all times is not enough [7]. Indeed, a complete EMR system should have a second set of requirements aimed to support doctors: active analysis of the many types of patient data, managing and checking medical rules, or predicting current and future patterns. While the first set of requirements is usually supported by a commercial database management system, up until now the second set of requirements was normally supported with the help of custom-code or additional data management systems such as rule engines, data mining engines, or event processing engines.

Besides being more expensive and harder to manage, having multiple data management systems frequently leads to the inability of finding correlations across datasets which in turn leads to false alarms and false negatives. However, recent advances in database research have added functionality to database systems such as temporal support, continuous queries, notifications, rules managers, event processing and data mining.

This chapter describes a prototype, SICU, that using multiple advanced functionalities, implements a complete, single-system EMR engine to monitor patients in emergency care units. Created by the University of Coimbra in cooperation with the University of Utah Health Science Center (UUHSC) and Oracle Corporation, SICU is a single integrated EMR prototype system built inside an Oracle database. SICU was designed as a proof-of-concept EMR system that manages real-time data (vitals and laboratory data), historic data (past clinical information), medical knowledge (in the form of rules) and issues appropriate alarms with the correct level of criticality and personalized by doctor or patient. Using data mining models built with 725 real patient profiles collected over 4 years, SICU aims to predict if patients will have a cardiac arrest in the following 24 hours

The prototype focuses on new ways to represent data (both patient data as well as medical rules), extraction of evidence from available data (using rules, classifications, and models), and on allowing customized interpretations over the same data. These characteristics of the prototype were implemented with an architecture based on the Oracle Database and focused on four of its features:

- *Total Recall (TR)* – a recent feature of Oracle database, responsible for automatically managing the history of all data changes [16];
- *Continuous Query Notification (CQN)* – a feature responsible for detecting and announcing changes in data [2];

- *Business Rules Manager (BRM)* – a new complex event processing (CEP) engine inside the database which is responsible for classification and personalization of alerts [1];
- *Oracle Data Mining (ODM)* – responsible for detecting complex patterns and identifying new classes in the data. It has been a feature of the Oracle Database for many releases [4].

Figure 1 shows the flow of data and the organization of the different modules in the system.

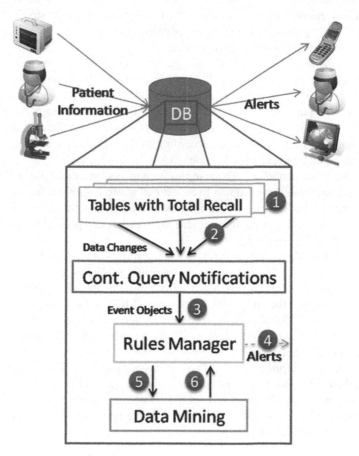

Fig. 1 Architecture of the prototype

1.1 *Contributions*

The major contributions of SICU are:

- *Automatic data history* – Using Total Recall (Section 2.1.1), past clinical data is automatically and transparently managed by the system and is accessed using minimal SQL enhancements. Then, using Continuous Query Notification (Section 2.1.2), the historic data is used to detect significant changes (or lack thereof) between any two or more versions of a value.

- *Canonical event type model* – In spite of the multiple data types (e.g., heart rate, blood pressure, temperature, urine information) and multiple rule types, the implementation reduced all new data to a single type of event (a "new reading") and to a single type of rule greatly simplifying the architecture. (See Section 4.1.)

- *Representation of vocabularies and domain knowledge* – Vocabularies represent the semantics of data specific to the domain. SICU stores the terms and concepts of the domain knowledge using rules that conform to generally accepted medical rules. (See Section 4.2.)

- *Classification and Customization framework* – Although there are general rules that always apply, each doctor has her own interpretation of the importance of events (e.g., is a heart rate of 100 beats per minute too much for some patient?) and the frequency and urgency of warnings. SICU allows the customization of thresholds, alarms, timeouts and the classification of events into multiple classes according to their criticality such as guarded, serious and critical. (See Section 4.3.)

- *Rules Composability* – The complex scenarios seen in real medical systems are sometimes too complex to be defined using a single rule. As such, the prototype builds upon a rule composability property that allows the definition of complex rules to represent events such as "possible cardiac arrest situations". (See Section 4.4.)

- *Predictive Models and Data Mining Integration* – Although there is plenty of medical knowledge that can be captured using simple and composable rules, there are situations where a predictive model is able to capture knowledge that not even the specialists knew about. SICU uses a data mining model that, using 725 real patient profiles collected over 4 years, is able to predict with 67% accuracy if patients will have a cardiac arrest in the following 24 hours. (See Section 4.5.)

- *Declarative Applications* – This prototype illustrates a new style of rule driven applications that provides timely access to critical information and that can handle enormous complexity while remaining highly customizable and extensible and while using a declarative approach. (Described better in another publication [11].)

In addition to the above major contributions, the prototype has shown the importance of database technology to event processing. In fact, by taking advantage of the declarative nature of database technology, and in spite of using so many new features and being implemented by a four-site team (in Utah, California, Massachusetts, and Portugal), the prototype was implemented in just 4 months. In the process, the implementation of the prototype has also shed new light on techniques for evidence discovery.

1.2   Chapter Outline

Although we assume that readers are generally familiar with database technology, Section 2 highlights important functional and operational characteristics of the Oracle database. Then, and in more detail, Section 2 describes some of the more modern features of databases that readers may not be familiar with. Additionally, section 2 describes and defines *classification* and *customization*.

Section 3 describes the overall architecture of the prototype. The approach of this section tries to explain to the reader the architectural decisions made along with the prototype development and how they were executed. The section will highlight the way that the components described in Section 2 were linked and integrated in a unified system.

Section 4 describes in detail the implementation, how data are captured with Total Recall, how events are triggered with Continuous Query Notification, how events are further processed by the Business Rules Manager, and how the Business Rules Manager is also used to customize and compose rules. The last subject is the integration of Oracle Data Mining, the creation of models as well as the on-line scoring of incoming patient data to compute the probability that the patient will have a cardiac arrest.

Section 5 describes alternative implementation approaches. Section 6 gives an overview of the state of the prototype, Section 7 give our conclusions and highlights future work.

2   The Technology

The prototype was built on top of the Oracle database and of the Glassfish [9] application server. This section describes features from the Oracle database that are important to the prototype and are not well known in the community.

2.1   Oracle Database

Databases are known for their operational characteristics, and high performance, scalability, security, and availability, including disaster tolerance. Any of these characteristics are required for the management of EMR (Electronic Medical Records).

In addition, databases have to support the appropriate data types and models for each specific application. Oracle database initially supported only SQL 92 [12], a simple relational model. This support has been dramatically extended. As of now, the Oracle database supports SQL 99 [13], XML [14], and RDF [15]. Furthermore, the database provides support for extensibility; e.g., users can add new data structures as well as operators to manage and access data represented by these data structures. Oracle has used this extendibility support to support text processing, video, spatial, as well as other data types. In the medical environment the extensibility can be used to manage and access domain specific data such as MRIs (magnetic resonance imaging) or to extend the richness of existing data types to support classification and customization (which we discuss further in Section 4.3.)

Keeping up with the evolution on data types and models, the Oracle database now also provides a full array of data management services. That is, the database is not just able to manage and store transactional data, but can also support the (online) maintenance of warehouses, automatically manage and provide access to the history of records (with Total Recall-TR), support fast complex event processing based on registered queries (with Continuous Query Notifications-CQN), support storing, managing and evaluating business rules (with the Business Rules Manager-BRM- module), support the development, testing and scoring of non-hypothesis driven and/or predictive models (with Oracle Data Mining-ODM), and supports the dissemination of information through queuing mechanisms.

The following section will introduce TR, CQN, BRM, and ODM briefly.

2.1.1   Total Recall

Total Recall [16] is a database feature to automatically manage the database history. Whenever a record is changed or deleted the previously existing version will be transparently stored into a history table. Automatic management of database history represents significant savings in application development because programmers can have the benefits of accessing any past versions without worrying with the details on how to store, index, or access those versions.

Total Recall uses the application schema (vocabularies) and extends it with special time fields. Thus, each record has a start time, the time of the creation of the version, and an end time, the time a version has been replaced by a newer version or the record has been deleted. The end time for currently valid records is infinite or undefined.

Users can access previous records by using so called *flashback queries*, that is, standard SQL queries extended by an optional AS OF *timestamp* clause (see Figure 2). The execution of any query decorated by AS OF will access the versions that were valid at the specified timestamp. This allows reviewing the status of a patient as of a specific time in the past. Other sophisticated expressions can evaluate the changes in a field between two points in time. The default for AS OF is the current time, e.g., programs that are unaware or do not use this feature will work as if Total Recall does not exist.

The second language extension for Total Recall is a VERSIONS clause with two time parameters representing a time interval. VERSIONS provides access to time series. A version query asking for a specific record between t1 and t2 will return all versions of this record that existed during this interval (see Figure 2).

Total Recall preserves the performance of the active data by separating the currently active versions from previous versions.

Efficient storage management is especially important for historic data. Total Recall uses compression technologies to minimize storage consumption; e.g., it will use compression algorithms that are optimized for read only. It also takes

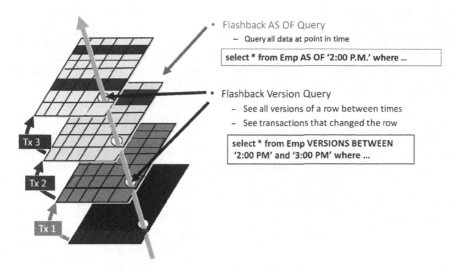

Fig. 2 Oracle Total Recall

advantages of the high compression rates that are common in columnar stores by treating several versions of a field as a column.

2.1.2   CQN - Continuous Query Notification

Many applications have requirements for taking actions when something happens in the database such as when a record is changed, deleted or added. Usually this type of functionality can be achieved with the use of database triggers.

An alternative to triggers is Oracle CQN. Oracle CQN allows an application to register queries with the database for either object change notification or query result change notification. An object referenced by a registered query (e.g., a table) is a registered object.

If a query is registered for object change notification (OCN), the database notifies the application whenever a transaction potentially changes an object that the query references and commits those changes, whether or not the query result changed. On the other hand if a query is registered for query result change notification (QRCN), the database notifies the application only for committed transactions that indeed changed the result of the query.

CQN complements existing methods of event discovery in the database. Unlike triggers that see *dirty data* (i.e., uncommitted data) and may have issues scaling with the number of conditions, CQN sees only committed data and scales better. Log mining is another standard event discovery technology: it is, however, typically optimized for batch processing and the evaluation of single elements and is not considered further here.

2.1.3 Rules Manager

The Oracle Business Rules Manager (BRM) provides a set of tools and techniques to define, manage, and enforce business rules from core business components. The Rules Manager enables better integration of business components such as CRM, ERP, and Call Center and enables automated workflows [1].

BRM also introduces PL/SQL APIs to manage rules in an Oracle database. It also defines an XML and SQL based expression algebra to support complex rule applications. The business Rule Manager leverages the Expression data type and the Expression Filter index [1] to store and evaluate the rule conditions efficiently.

BRM allows application developers to define groups of rules that are related; e.g., all rules related to cardiac arrest. Policies can define which result will have preference if there are conflicting results.

BRM also allows the integration of external data (events) with OCN events; e.g., look at room conditions in accessing a patient's situation, assuming room conditions are external data.

2.1.4 Oracle Data Mining

Oracle Data Mining (ODM) implements a variety of data mining algorithms inside the Oracle database. These implementations are integrated right into the Oracle database kernel and operate natively on data stored in the relational database tables. This integration eliminates the need for extraction or transfer of data between the database and other standalone mining or analytic servers.

The relational database platform is leveraged to manage models and efficiently execute SQL queries on large volumes of data. The system is organized around a few generic operations providing a general unified interface for data mining functions. These operations include functions to create, apply, test, and manipulate data mining models. Models are created and stored as database objects, and their management is done within the database - similar to tables, views, indexes and other database objects.

In data mining, the process of using a model to classify, or "score", existing data is called "scoring". Scoring is traditionally used to derive predictions or descriptions of behavior that is yet to occur.

Traditional analytic workbenches use two engines: a relational database engine and an analytic engine. In those setups, a model built in the analytic engine has to be deployed in a mission-critical relational engine to score new data, or the data must be moved from the relational engine into the analytical engine. ODM simplifies model deployment by offering Oracle SQL functions to score data stored right in the database. This way, the user/application developer can leverage the full power of Oracle SQL (e.g., in terms of parallelizing and partitioning data access for performance).

ODM offers a choice of well known machine learning approaches such as Decision Trees, Naive Bayes, Support vector machines, Generalized Linear Model

(GLM) for predictive mining, and Association rules, K-means and Orthogonal Partitioning Clustering and Non-negative matrix factorization for descriptive mining. Most Oracle Data Mining functions also allow text mining by accepting Text (unstructured data) attributes as input.

2.1.5 Operational Characteristics

Databases are known for their rich support of operational characteristics. These characteristics are available to applications without any additional development, and indeed operational characteristics can significantly reduce the development effort, such as using Total Recall to provide automatic and complete auditing and tracking. This section highlights some of the major operational characteristics that are important to the prototype.

Performance. Databases can store tens of thousands of records per second. The response times for storing and retrieving records and for evaluating a large number of registered queries are well below human awareness.

Scalability. Databases can store very large amounts of data, handle large amount of clients concurrently, and evaluate tens of thousands of registered queries. Compression and other technologies are used to minimize storage requirements.

Availability. Databases are known for not losing records. Recovery, restart, fault tolerance for continuous availability, as well as disaster tolerance are all standard features.

Security. Databases provide many security-related features, such as access control, fine grain security - e.g., a doctor can only see patients that are assigned to him/her, contextual security - e.g., a doctor can see only data when she is in the hospital. Additionally, all data items as well as all journals can be encrypted in a way such that not even IT personnel with unlimited access to the metadata will be able to decode the encrypted data.

Auditing and tracking. Journals provide a full account of any activity in the database; e.g., there is a full record of who did what at what time. Total Recall and a feature called Audit Vault in conjunction with Total Recall provide online access to any historic data as well as to the information of who did an action at what time.

2.2   Other Technologies

During the design of the prototype, we learned that doctors wanted to be informed of situations such as "patient has a critical temperature for more than three minutes". This type of semantic interpretation of observed data, i.e., "critical temperature", poses two challenges: classification and customization. *Classification* consists of assigning a class (e.g., normal, guarded, serious, or critical) to a range of observed values. *Customization* consists of having two or more health care professionals with potentially different classes or potential different ranges of the same class. E.g., for one doctor, a temperature above 40 C may be considered "critical", while for another doctor temperatures up to 41 C should be only "serious". The next two subsections further clarify these two concepts and describe how the prototype incorporated them.

2.2.1   Classification

When people interpret data, they tend to associate individual values or intervals to some classification. For example, doctors frequently use terms such as serious and critical to identify values of concern or for communication between colleagues. Decision tables can be used to define the mapping between values and classes for a given set of attributes. In some cases, such as an EKG (electrocardiogram), there are already classifications. In this case one can map the EKG specific classifications to more generic classifications.

Using classifications simplifies and generalizes the formulation of queries and rules. With classification one can ask for all patients with at least one critical value or for the immediate notification when a patient has at least one critical value as well as a periodic reminder. Note that using classification has the added benefit of reducing the number of rules needed to monitor the systems. That is, a single rule, e.g., "alert when a parameter is critical", can be used to monitor hundreds of different parameters.

Last but not least the model can be used to classify the urgency of the information. If a rule is marked either as serious or as critical doctors will be alerted right away. The difference between critical and serious will tell them how fast they have to react. If a rule is marked guarded, doctors will not be immediately alerted at all, however, an entry will be added to the patient records and they will see the information when they are looking at these records the next time. Obviously, this will reduce unnecessary interruption and most importantly, it reduces alert fatigue.

The prototype has shown that the use of classification fundamentally reduces the complexity of the acquisition of data; classification is an important part of the metadata. Storing these metadata with TR (Total Recall) provides a full history to the classification; e.g., it becomes clear what the classification was at any point in time. The prototype supports only a specific classification: normal, guarded, serious, and critical. In general one would allow users to specify one or more classifications.

In some cases, classification was performed taking the specific situation of patients into account. For example, a heartbeat of 150 can be classified as normal or guarded for a baby but as serious or critical for a 90+ year old patient.

Classification could be extended to time series. For example, classes could be assigned to deteriorating (i.e., monotonically decreasing) attributes or classes could be assigned depending on the speed of deterioration; e.g., a value is slowly/rapidly deteriorating. These classification techniques were not used in the prototype.

2.2.2 Customization

The use of classifications simplifies significantly the interpretation of data and the specification of very generic queries and rules for ad-hoc information as well as for event processing. However, not all institutions and doctors will classify the same information in the same way. The solution to this problem is customization of the classification.

The prototype provides generic classification; e.g., classification that applies if there is no overriding customization. There are four levels of customization: generic, protocol, doctor, doctor/patient

- *Generic* – a classification used for any patient, regardless of protocol or doctor;
- *Protocol* – a classification that has been adjusted to the treatment plan that is used for patients of a specific protocol;
- *Doctor* – a classification that reflects the preferences of a doctor independent of the patient. It reflects the personal view of a doctor in respect to the interpretation of patient data;
- *Doctor/Patient* – a classification that is tailored to the view of a specific patient by a specific doctor.

In all cases the most restrictive classification will be used; e.g., doctor/patient overrides doctor customization, which overrides protocol customization which overrides the generic customization.

Customized classification allows doctors to fine-tune systems to their needs and their view of medicine. This is a very important step in increasing the acceptance of any system as well as reducing alert fatigue.

A consequence of customization is that two doctors looking at the same data may get different results. Obviously, they may get different alerts as well. One of the major issues is the very easy and intuitive customization of values. The prototype provides this by showing the current setting for the majority of vital signs in the form

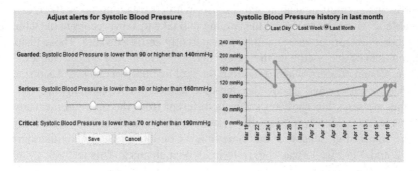

Fig. 3 Sample interface for customization

of sliders and numerical values. Doctors can easily adjust the settings by moving the sliders; the numerical values will be automatically adjusted (see left side of Figure 3). The prototype ensures that the new settings are consistent; e.g., that the values for critical are higher/lower than the settings for serious. Once a doctor saves the settings they will be used as the doctor/patient classification.

The prototype has a fixed customization hierachy (doctor/patient, doctor, protocol, generic). In a product one would allow users to define their own customization structure.

3   Architecture

In this section the high level architecture of the SICU prototype will be explained, including relevant characteristics of each module as well as some implementation decisions. As mentioned, the architecture of the prototype is database centric. This means that all the components of the system are running embedded in the database engine. This also means that operational benefits such as transactionality, logging, high availability, and high performance are implemented by the database engine and the prototype inherits them with no extra code or effort. The architecture of SICU relies only on modules of the Oracle Database. Without ignoring other important modules, and as mentioned previously, there are four main technologies of the Oracle Database that were used in the core of the system: Total Recall, Continuous Query Notification, Business Rules Manager and Oracle Data Mining.

One of the first concerns in this prototype was how to maintain a history of the data without having extra logic in the application (i.e., without using support tables or filtering timestamps). This was a decision with a potential impact on overall performance. Total Recall was chosen because it can maintain history transparently and with very good performance. Enabling this option (See step 1 in Fig. 1) will cause all operations (inserts, deletes or updates) to be recorded transparently with a timestamp enabling change tracking in time.

In a classic database, if there is a need to store current and historical values of a sensor the normal approach is to add application logic and write time-stamped records. In this traditional approach, a sensor with 100 readings takes 100 records in the table(s). However, using TR the reading will take just one record in the table and the other 99 historical values will be transparently kept elsewhere. Those historical versions can be accessed with the *AS OF* and *VERSIONS BETWEEN* clauses.

The second concern when designing the architecture was how to keep track of changes in sensors and ensure that the data treated was consistent (i.e., the sensor that measures the cardiac rhythm is different from the one that measures the respiratory rate. However when the system needs to analyze new data, it should see the status of all sensors at the same time). Continuous Query Notification (See steps 2 and 3 in Fig. 1) was chosen to track data changes. This technology allows the database engine to notify clients about new, changed or deleted data at commit time. Note that CQN has some similarities with triggers. However, while triggers fire when SQL statements are executed, CQN only notifies once data is committed and consistent.

For every new new reading, the incoming data needs to be processed, classified and matched with some intelligence loaded into the system to produce valuable information to the users. Rules Manager (See step 4 in Fig. 1) is a rules engine running in the Oracle database and allows matching data against rules.

Because Rules Manager (RM) is running embedded in the database, it has access to all the information stored in it. This allows RM to take decisions considering both the new information as well as the history of that patient and/or of groups of patients that share physiological or clinical characteristics. Some of the rules designed for the prototype notify Oracle Data Mining to re-score data and detect critical situations.

When Rules Manager triggers Oracle Data Mining (See step 5 in Fig. 1), a query gathers the information needed from the sensors specified by the rules and then ODM scores the data for a probability of occurrence of a dangerous future event. The algorithm also provides information for the users to understand how that probability was calculated and the weight of each parameter. At the end of scoring, the result is sent back as a new event to RM (See step 6 in Fig. 1) where it will be analyzed.

Finally, the information is delivered to the end user through a Java EE application running on an Application Server. (See step 4 in Fig. 1.)

In short, sensors continuously send new readings to the database which are then stored with the help of Total Recall. Continuous Query Notification is always monitoring incoming data and notifies Rules Manager only of the changed values considered relevant. Rules Manager is where all the knowledge (represented as rules) resides and with those rules the incoming data is correlated (with a context) and classified, and alerts are sent to users. Some Rules Manager rules trigger the predictive Data Mining models, which score incoming data and detect possible complex scenarios such as cardiac arrests.

4 Implementation

The development phase of the prototype took about 4 months of a developer with the support of an Oracle technical team. This section describes the implementation details of the prototype.

4.1 Event Triggering

Rules Manager bases its processing on event objects, therefore it was decided for more flexibility the system would use CQN to generate those events and send them to be consumed by RM calling its *add_event* function.

The sources of information are very heterogeneous (bed side monitors, imaging, text-based doctors reports, lab results, etc) and the system can be monitoring either numeric simple values or text based reports. This leads to a problem when defining rules for extracting the desired information from the raw data because it would lead to an explosion of event data types and rule data types. (In CQN rules have data types, which must match the types of the events that may trigger the rule evaluation.) To simplify the design of the system, all incoming pieces of data from sensors were considered to be of the same datatype: a measurement (the value of the reading). All rules where considered to be rules of type array of measurements.

The two main advantages of this decision are: i) all rules are defined based on only one type, and ii) the events are smaller (if a row in the database is updated but only a column is changed then only that value generates an event).

The Continuous Query Notification is the module responsible for comparing the current and new values of the information received, generating the events and sending them to RM at commit time. This processing is defined in a callback function generated dynamically for each table that CQN is monitoring.

4.2   Rules Evaluation

Rules Manager is the main processing component of the prototype. This component provides a flexible environment to evaluate rules with several options for deciding the ordering of rule execution (when multiple rules can fire), event consumption policies or duration of events.

Ordering of rule execution is determined as follows. When an event arrives, all rules that can be triggered by the event are identified as candidate rules. Then, a conflict resolution module is executed to determine which candidate rules should be triggered and in which order they should fire. In the prototype, rules are grouped by a priority group number and are also classified by criticality level. Starting with the highest priority group with candidate rules, the most critical candidate rule is identified and fired. The process is then repeated for all other groups, ordered by priority, where for each group, only the most critical candidate rule fires. The priority of rules can be changed through a user interface.

Another configuration option implemented in the prototype was the concept of duration of alerts: when an alert is issued it has a duration period and within that period another instance of the same alert cannot be issued. This feature reduces alert fatigue. To achieve this functionality each time a rule is candidate for triggering it is also compared with the last time an instance of that rule/alert was triggered to see if the duration was expired or not.

After the full evaluation and conflict resolution the events are kept in the pool of events because they can be used by other rules in conjunction with other events. For example, the event "Temperature is above 41C" should be not consumed by the rule "Trigger serious alert if temperature is above 41C" because there might be another rule, e.g., "Trigger critical alert if bpm is above 100 and temperature is above

41C", that might need the event. However, events could not be left unconsumed in the system forever. Thus, a time-to-live or event duration was designed into the system. The final event duration was decided by the medical personnel as 24 hours. After these 24 hours in a intensive care unit, any values are considered obsolete and should be considered part of the patient history. Although event duration was set to 24 hours, RM supports other ways to specify event duration such as by call, session or transaction.

4.3 Rules Customization

Medical staff were particularly interested in a system with a high degree of customization of rules. In fact, due to the lack of personalization, doctors cannot customize the alerting modules of most other current EMR systems or can only make simple, threshold type customizations. That lack of personalization in current EMR systems in turn leads to a high number of false alerts. The customization desired by doctors led to the implementation of the following options that determine when rules fire or not:

- If the rule belongs to a group of already defined rules or if it is in a new group;
- The priority inside the group;
- If that rule will apply to all patients, only for that patient, only for the patients of the doctor that is defining the rule, or only for a specific patient under that specific doctor supervision;
- The time to live of the alerts related with that rule;
- The description of the rule;
- The definition of the rule parameters to evaluate.

All of these options are defined as metadata in the rule definition table and used by the *RM* evaluation process or in the conflict resolution logic to achieve the defined functionality. With this level of customization the system can be fully personalized by the medical staff and improve the patient care due to the optimization of doctors' and nurses' time.

4.4 Rules Composability

Rules Manager provides extensive support for the design of complex scenarios based on simple rules and patterns. The rules are written in XML and allow relating events and performing aggregations over windows and also specifying situations of non-events.

However by composing rules it is possible to define more complex scenarios than the ones the language allows in a simple rule. Below is a sample of a scenario in medical technical language to identify possible cardiac arrest situations. The complete rule takes more than a page to be described in technical English.

II	MAP, SBP or DPB in serious/critical range for >30min
II	MAP, SBP or DBP change from serious to critical
III	Asystole (AYST)
I	Supraventricular tachycardia (SVT)
I	Arterial Fibrillation (AFIB)
I	Arterial Flutter (FLUT)
III	Ventricular Fibrillation (VFIB)
III	GCS <= 8
II	Critical low or high temperature
I	Trauma of Cardiothoracic patient
...	... (continues) ...

To represent the rule above, the prototype uses what we call complex rules. Complex rules are rules composed of other rules. In the scenario above, the corresponding complex rule can be implemented in RM with about 50 rules. The evaluation of such complex rules happens in the following way. The roman numbers on the left column above represent different rule categories with different weights. In the above example, the complex alert should occur when one rule of category III or three rules of category II or five rules of category I fire.

4.5   Data Mining Integration

Another feature of the system is the integration of data mining models that score the status of the patients for the probability of risk situations such as Cardiac Arrest

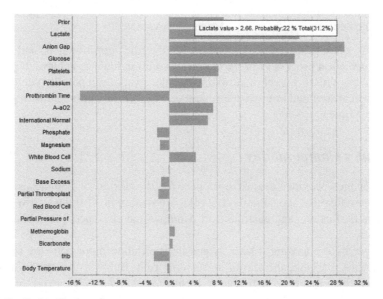

Fig. 4 Prediction Explanation

or Respiratory Failure. Since scoring the status of patients on every data change is too expensive, the process is done in two steps. First, an inexpensive rule filters the majority of non-relevant data changes. However, in some cases, the rule will trigger the execution of the more expensive scoring using the data mining models.

Scoring returns the probability of those risk situations and returns also the factors that contributed for that prediction. In Figure 4 is an example of a prediction of a Cardiac Arrest with 96% probability. The physician can look at the chart and see that the negative fractions (to the left) are values that are normal, or that do not contribute to the condition of Cardiac Arrest and the positive ones are the ones that contribute to the cardiac arrest.

5   Alternatives

The classical approach for applying event processing in an ICU (Intensive Care Unit) is to focus on vitals; e.g., blood pressure, heart beat, EKG (electrocardiogram). These vitals are shown on monitors in the patients' room and supervised in a central location. Simple limits are set for individual attributes and whenever there is an out of range value an audio alarm – an alert – is triggered. However, frequently there are too many false and simple alarms. This leads to alert fatigue and consequently even true alerts are mostly ignored.

ICUs are a favorite use case of the Event Processing community [10]. The typical approach is to capture data from the monitors and import them as streams into a CEP environment. The measurements are considered as being events. Rules and (in advanced environments) models are used to evaluate or score the incoming data to create (complex) events, which will be brought to the attention of the medical personal. In some cases, information will be additionally stored in a data base. Because of its many flaws, we dismissed this approach for our prototype. Here are some of those flaws:

- The medical personnel has to work with two inconsistent data models, one using measurements as events, one using measurements as (historical) data.
- Alerts may be created before data have been committed; e.g., auditing and tracking may not work in the presence of failures.
- Alerts are often only meaningful when new information is considered in the context of other patient data - doing so would create significant performance issues since it requires pull technology on the patient data which does not work well with the push-based, stream oriented nature of most CEP tools.
- Predictive models which are limited to vitals have in most cases marginal value; e.g., the predictive model for a 24 hour prediction of cardiac arrest showed no correlation to vitals.
- There is no classification and customization in CEP tools; the fundamental reason for alert fatigue has not been addressed.

- This approach has only very marginal value for the ICU environment while the approach taken by our prototype can be applied to any patient care.

6 State of Work

The current implementation of the application is running only for demonstration purposes, although there is an interface for the users to input and simulate data as it would occur in a real scenario and then see how the system reacts.

The main features of the implemented application are: monitor the patient vitals, labs and information; easily define custom ranges for alerts for any combination of patient, patient-doctor, or doctor; query patient information history; reduce alert fatigue and support predictions using data mining models.

Currently the prototype has about 200 simple rules and 190 classification rules deployed. For the complex rules definition there are 160 additional deployed partial rules to trigger possible Cardiopulmonary Arrest and Respiratory Failure situations. All the rules were defined by doctors, therefore are triggering real danger situations and not only test situations.

As for the data mining integration, the prototype used a model developed by Dr. Pablo Tamayo from the Whitehead Institute/MIT Center for Genome Research, in Cambridge, Massachusetts, EUA. This model was built with a Bayesian Network using a training set of 725 patients treated at the Surgical Intensive Care Unit of the *University of Utah Health Sciences Center* in the 2004-2008 period. The model was then tested using a test set collected from 482 different patients. The results shown to predict Cardiac Arrests within 24 hours with a certainty of 67%. This represent an advance over the state-of-the-art as previsions prediction systems could only suggest a cardiac arrest with a few hours in advance [20].

7 Conclusions and Future Work

This prototype illustrates a new style of applications that provides critical information in time. These applications can handle enormous complexity, are highly customizable and extensible, and are fast to implement. In the specific implemented prototype the system first creates a knowledge base, based on vocabularies, rules (registered queries), (predictive) models, classifications, customizations, and visualization. Then, the system can improve the knowledge base by allowing users to customize or fine-tune rules and thresholds. Next, data is stored in a temporal database such that both current and past values are accessible and can be used in rules. Finally, data changes are detected and rules fire and apply predictive models as needed.

Rick Hayes-Roth coined the term VIRT (Valuable Information at the Right Time) [19]. The term VIRT seems to express the reason for and the value of event processing. The prototype shows a modern implementation of the VIRT principle.

The development of the prototype took only four months of development that focused on putting all the pieces together in one single system and delivering the functionality to the end user. This short development period shows that there are now sufficiently advanced custom-off-the-shelf components that can be put together to build sophisticated VIRT applications.

Classification is a key issue of the developed solution. Due to the prominence of classification, some more work on how to improve the ways to classify the information and how to discover new patterns to make this classification should be done. Doctors need to classify the information easily and above all the system should help and guide them to correctly classify data.

The integration of data mining models with other data acquisition and data processing systems is not trivial to achieve but has a great potential. This prototype showed that that integration is possible and can produce potential life-saving warnings.

The medical part of the team that collaborated with us showed great interest in the prototype volunteering with many extra work hours to provide feedback, improve the prototype, and demonstrate the system. In fact, Dr. Edward Kimball, from the University of Utah Health Science Center, was quoted saying "This [prototype] has the potential to revolutionize medical IT and significantly improve clinical care of a broad range of patients". Still, the system's success rests on the acceptance by the medical community in general. This means that the system needs to be easy to use and be a help to them and not something that is stopping their work. This said, the user interface to define rules and customize all the system is a part of the system that needs improvements to give this flexibility to the doctors.

One of the problems when dealing with history is the testing and load of previous history. A solution was found to move forward in the development but the system needs to have support for simulation (run in fast forward mode) and also to support load of historical data in bulk mode for loading previous data of patients when deploying the system.

Acknowledgments

The authors would like to acknowledge the great contribution of Ute Gawlick MD/PhD at the UUHSC (University of Utah Health Services Center). Ute made sure that the prototype reflected the thinking in the medical community; Ute guided the team through many challenging discussions and was instrumental in developing the ideas related to classification and customization, the visualizations, and the predictive models. We also like to thank Dr. Sean Mulvihill, Dr. Edward Kimball, and Dr. Jeffrey Lin from the UUHC for insisting on solving difficult questions and for providing suggestions solving them. Finally the authors like to thank members of the Oracle staff whos technical support was very crucial to the success of this prototype: Venkatesh Radhakrishnan, Pablo Tamayo, Srinivas Vemuri, and Aravind Yalamanch (Aravind is now working for amazon.com). Pablo developed the non hypothesis driven model for cardiac arrest, and designed the online scoring model.

References

1. Oracle® Database Rules Manager and Expression Filter Developer's Guide 11g Release 1 (11.1), http://download.oracle.com/docs/cd/ B28359_01/appdev.111/b31088/toc.htm (cited May 2009)
2. Oracle® Database PL SQL Packages and Types Reference 11g Release 1 (11.1), http://download.oracle.com/docs/cd/B28359_01/appdev.111/ b28419/d_cqnotif.htm (Cited May 2009)
3. Oracle® Database Advanced Application Developer's Guide 11g Release 1 (11.1), http://download.oracle.com/docs/cd/B28359_01/appdev.111/ b28424/adfns_flashback.htm (Cited May 2009)
4. Oracle® Database PL/SQL Packages and Types Reference 11g Release 1 (11.1), http://download.oracle.com/docs/cd/B28359_01/appdev.111/ b28419/d_datmin.htm (Cited May 2009)
5. Bizarro, P., Gawlick, D., Paulus, T., Reed, H., Cooper., M.: Event Processing Use Cases Tutorial. In: The Proceedings of the Third ACM International Conference on Distributed Event-Based Systems, DEBS 2009, Nashville, Tennessee, USA, (July 6-9, 2009)
6. Guerra, D., Gawlick, U., Bizarro, P.: An integrated data management approach to manage health care data. In: The Proceedings of the Third ACM International Conference on Distributed Event-Based Systems, DEBS 2009, Nashville, Tennessee, USA (July 6-9, 2009)
7. DesRoches, C.M., et al.: Electronic Health Records in Ambulatory Care - A National Survey of Physicians. The New England Journal of Medicine 359, 50–60, http://content.nejm.org/cgi/content/full/NEJMsa0802005 (cited July 2010)
8. Steinbrook, R.: Health Care and the American Recovery and Reinvestment Act. The New England Journal of Medicine 360, 1057–1060, http://content.nejm.org/cgi/content/full/NEJMp0900665 (cited July 2010)
9. Glassfish Application Server, https://glassfish.dev.java.net/ (cited June 2010)
10. Internet Evolution. Exploring IBM Research Labs (Part 1) - Health care analytics (January 23, 2010), http://www.internetevolution.com/ document.asp?doc_id=187007&f_src=ieupdate (cited June 2010)
11. Gawlick, D.: Healthcare beyond record keeping. In: 13th International Workshop on High Performance Transaction Systems. HPTS 2009, Pacific Grove, CA, USA (October 25-28, 2009), http://www.hpts.ws/session11/gawlick.pdf (cited July 2010)
12. Wikipedia: SQL-92, http://en.wikipedia.org/wiki/SQL92 (cited April 2010)
13. Wikipedia: SQL:1999, http://en.wikipedia.org/wiki/SQL:1999 (cited April 2010)
14. Wikipedia: XML, http://en.wikipedia.org/wiki/Xml (cited April 2010)
15. Wikipedia: Resource Description Framework, http://en.wikipedia.org/wiki/Resource_Description_Framework (cited April 2010)
16. Oracle Total Recall, http://www.oracle.com/us/products/database/ options/total-recall/index.htm (cited April 2010)

17. Hayes-Roth, F.: Valued-Information at the Right Time: Why less volume is more value in hastily formed networks. NPS Cebrowski Institute (2006), http://faculty.nps.edu/fahayesr/virt.html (cited April 2010)
18. Etzion, O., Niblett, P.: Event Processing in Action. Manning Publishing Company (July 2010)
19. Wikipedia: Rick Hayes-Roth, http://en.wikipedia.org/wiki/Rick_Hayes-Roth (cited April 2010)
20. Gawlick, U.: A Novel Approach to ICU Surveillance and Prediction of Disease. Presentation given on June 19, 2009 at the University of Utah Health Sciences Center (2009)

The Principle of Immanence
in Event-Based Distributed Systems

Pascal Dugenie and Stefano A. Cerri

Abstract. This chapter focuses on the principle of *immanence* for autonomic event-based distributed systems such as collaborative environments on the GRID. On the one hand, GRID provides a sound infrastructure for coordinating distributed computing resources and Virtual Organisations (VO). On the other hand, *immanence* is a principle that emerges from the internal behaviour of complex systems such as social organisations. Although several existing VO models specify how to manage resources, security policies and communities of users, none of them has considered mechanisms that reflect the internal activity to constantly improve the overall system organisation. The AGORA model, proposed in 2004, has been integrated in an experimental collaborative environment platform. After several years of experimentation with communities of scientists from various domains, the AGORA architecture has been enhanced with a novel design approach for VO management. The model is a dynamic system in which the result of interactions are fed back into the system structure. The basic idea is to specify a set of mechanisms to catalyse the collective intelligence of active communities in order to enable self-organisation of the collaborative environments.

1 Introduction

1.1 The Principle of Immanence

Usually, *immanence* refers to philosophical and metaphysical theories, often related to religious doctrines. The general idea behind the notion of *immanence* is that the cause of the development of an object occurs inside this object. This approach presents much interest in complex systems theory to explain how the flow of events

Pascal Dugenie · Stefano A. Cerri
CNRS, Centre National de Recherche Scientifique
LIRMM, Laboratoire d'Informatique de Robotique et de Microelectronique de Montpellier
161 rue Ada, 34392 Montpellier Cedex 5 France
e-mail: {dugenie,cerri}@lirmm.fr

S. Helmer et al.: Reasoning in Event-Based Distributed Systems, SCI 347, pp. 239–256.
springerlink.com

inside a system and its activity may engender the emergence and an organisation that is continuously in self-adapting. A living body is a typical example of such an immanent system. The paradigm of complex system theory originally appeared in the domain of biology [31] and has been adapted later in the domain of social theory [23]. The term autopoiesis, coined in cognitive biology by *Maturana* and *Varela* [24], was also used to express that a complex system maintains its distinctive identity by constantly considering in its communication what is meaningful and what is not.

According to the sociologist *Niklas Luhmann*, social systems are autopoietically closed in the sense that they use and rely on resources from their environment. The principle of immanence in a social organisation emerges from this process of selection of elements in the system filtered from an over-complex environment. *Niklas Luhmann* considers mutual confidence between actors as the main factor to reduce the complexity of the system [23]. In summary, this literature suggests the existence of a strong interdependence between the *organisation* and the *activity* of a complex system: *the organisation of a social system is immanent to the activity within that system*. A circular causality[1] exists between the organisation and the activity of a system: the organisation enables the generation of activity, meanwhile the activity constantly seeks to improve the organisation. The effect of immanence is the living link between the organisation (*i.e* the static part) and the activity (*i.e:* the dynamic part) of the system. In contrast, a system whose behaviour would be completely determined from initial conditions with no feedback effect of its activity on its own structure is not an immanent sytem and has no chance to be self-adaptive in case of changes of conditions of its environment.

1.2   Immanence on the Web

Talking about *immanence* in the domain of informatics tended to appear quite utopic only a few years ago. Since then, *Pierre Levy* has introduced the notion of immanence by depicting the Web as a common infrastructure (*i.e* the material part) that is immanent to a global collective intelligence (*i.e* the immaterial part) [22]. This dichotomy has been exacerbated with the recent emergence of social networks in the Web 2.0. This is one illustration of the many new forms of social interactions in collaborative environments. A social network on the Web continuously behaves so as to adapt its own structure throughout its internal activity. The structure is composed of a complex network of communication channels while the activity consists of the numerous interactions passing through these communication channels.

Nowadays, several factors encourage much interest for modelling collective behaviour: the rise of the holistic modelling approach inherited from research in MAS[2], or the maturity of distributed computing technologies, especially the GRID. Agents may share their knowledge and expertise, while the GRID provides computing resources and various modalities of communication. Immanence occurs as

---

[1] A circular causality means that the effects cannot be separated from their causes.

[2] MAS: Multi-Agent Systems.

a side effect in most collaborative situations. In the case of a collaborative environment operating in a GRID infrastructure, actors may act alternatively as system designers as well as users. A computing element of GRID may exchange services with the actors in the form of a given result to a given request. Similarly an actor may provide a service in the form of an answer to a question based on a particular competence. The GRID aims to organize all these kinds of heterogeneous services. The GRID provides mechanisms for instantiating services within its infrastructure, with a proactive behaviour. Thus the activity of the system, represented by the aggregation of all interactions of the GRID services, generates a logical form of organisation. One major distinction between the GRID over the Web is the possibility for the GRID to deploy stateful resources necessary to operate autonomous services.

In this approach, interactions between services are contextualised within a generic collaborative process composed of both humans and artificial processes. The notion of agent is extended in the literature to cover both types of service producers (i.e humans and artificial processes) [2, 17]. Agents interact within a collaborative environment by providing or using services. One essential condition for a collaborative environment to become immanent is that any agent of the system may always play an active role in the elaboration of the organisation [8, 5, 6, 7]. For instance, both system designers and expert-users may feed back their experience into the cycle of development and validation of a complex application. They interact by providing services to each other via a common collaboration kernel. They may develop their point of view in the context of a collaboration process and their role may evolve if necessary. Without a flexible self-adaptiveness, such a system would not operate efficiently.

A collaborative environment on a WEB or a GRID infrastructure is a typical case of event-based distributed systems composed of social links, where events are the interactions between agents.

In order to interconnect concepts related to event-based distributed systems and the behaviour of a collaborative environment, this chapter briefly reviews in Section 2 the state-of-the-art of virtual collaborative environments and the deployment aspects of collaborative services on the GRID. Through a description of the AGORA model, the Section 3 then presents four interaction mechanisms which contribute to specify the principle of immanence.

2   Background

The scope of this chapter is in the intersection between three distinct research domains:

- the domain of CSCW[3] that explores concepts related to VCE[4],
- the domain of GRID that specifies a kind of SOA[5] and management models for VO (Virtual Organisations),

---

[3] CSCW: Computer Suported Collaborative Work.

[4] VCE: Virtual Collaborative Environment.

[5] SOA: Service-Oriented Architecture.

- the domain of MAS that analyses collective behaviour in distributed computing systems.

2.1 Virtual Collaborative Environments on GRID

Virtual Collaborative Environments (VCE) have emerged along with event-based distributed computing systems. A VCE aggregates resources, services and interfaces to provide a dedicated environment for collaboration. These interfaces may support several modalities of communication such as audio-video, shared visualisation, instant messagging, notification and shared file repositories. GRID is a pervasive technology allowing seamless access to distributed computing resources. A VCE on the GRID is an ubiquitous system[6] composed of stateless terminal elements where all computing resources are delivered by the infrastructure [12, 29, 20]. GRID has the capability to maintain the state of the communications independently from the type or the location of the terminal elements.

Access Grid[7] (AG) is the largest deployed GRID VCE solution world-wide. AG operates on *Globus* [11], the most popular GRID middleware. The topology of the AG infrastructure consists of two kinds of nodes: the venue clients and the venue servers [25]. AG venue clients can meet in a venue server to set up a meeting. AG uses the H.263 protocol [16] for audio and video encoding and a multicast method to distribute the communication flow between sites. The display from multiple H.263 cameras in every site gives a strong feeling of presence from every other site. The modular characteristic of AG allows the addition of new features such as application sharing (shared desktop, presentation, etc.) and data sharing. AG focusses on the principles of *awareness* and *ubiquity*. However, AG does not include a powerful means for VO management. VO are managed in an *ad hoc* manner at the venue server side. This requires much technical administrative work from computer experts in this domain.

2.2 VO Management Models

In its original definition, a VO is a community of users and a collection of virtual resources that form a coherent entity with its own policies of management and security [15]. A rudimentary VO management system has been originally built into *Globus* but has little potential for scalability. In order to resolve these limitations several VO management models have been proposed within the GRID community.

CAS (Community Authorization Service) has been specifically designed to facilitate the management of large VOs [27]. The functionalities for VO membership and rights management are centralised in a LDAP[8] directory. Since the structure of the VO is strongly hierarchichal, it is difficult to reorganise the initial tree once the services are deployed.

---

[6] an ubiquitous system enables the access to the VCE from anywhere at anytime.

[7] Access Grid: www.accessgrid.org

[8] Lightweight Directory Access Protocol.

VOMS (VO Membership Service) is deployed in more recent GRID infrastructures such as EGEE [28]. It resolves some of the problems of CAS, such as membership management by providing a more flexible relational database instead of a flat tree structure. For instance, a database has the possibility to specify complex, evolving relations between concepts such as users, groups and rights. The user management in a tree such as LDAP is not easily evolving because the concepts are embedded into the structure of the tree. However, VOMS still presents some conceptual limitations such as an inheritance link between a parent VO and its children. The subdivision of VO into groups often creates confusion in the management of rights and does not enable a complete independence between the groups and the VO. For instance, the lifetime of a group is determined by the lifetime of the parent VO.

2.3 GRID *and MAS Convergence*

The convergence of GRID and MAS research activities has been claimed in 2004 [13] as a major research objective that sustained in the early years 2000 [3, 4]. Both domains share an approach based on SOA which enables to abstract the underlying technology behind a common concept: the *service*. Since then, a sustained research activity has attempted to formalise the mapping of GRID and MAS concepts [19, 18]. This work opens new perspectives to specify immanent systems.

As established in the OGSA (Open GRID Service Architecture) framework for resource and security management, GRID is a kind of SOA [14]. MAS focusses on complex organisation models far more advanced than the GRID ones. The well-known MAS conceptual model Agent-Group-Role (AGR) emphasises the characteristic of flexibility required for a self-organised organisation model [10]. Further to the suggestion for convergence between GRID and MAS concepts their integration has required an extensive effort with AGIL (Agent-GRID Integration Language) [19, 18] to formalise the organisation and the interactions of distributed resources and agents and propose the service as a pivot concept. AGIL has been originally proposed in 2006 as an ontology [9], then has been developed more recently as a language. The abstraction of GRID and MAS concepts allows the development of complex architectures. The AGORA model is an example of such architecture represented with AGIL concepts.

2.4 *Discussion*

Existing VCE solutions such as AG are usually studied in the CSCW research domain which focuses on the *collaboration processes* aspects. The aspect of *VO management* such as CAS or VOMS is studied within GRID research communities. In order to efficiently identify the impact of the principle of immanence, many issues of these two complementary aspects would be better understood if they were combined rather than studied in separate domains.

Moreover, a SOA is clearly a convenient approach for self-adaptive systems. Models based on classical client-server architecture often present a lack of flexibility resulting in restrictions in their evolution. Technological choices are adopted more or less arbitrarily by the designer of the architecture to respond to an initial need. However as soon as the need changes, the architecture must follow a different pattern. For example, identity management is clearly difficult to specify at the beginning of the life of a system. Many choices often result from technical limitations. Some architectures adopt rights-centered management of groups where the members of a given group benefit from a set of rights over a service (*e.g* a group that would include all moderators of a service). Typically, this approach as used in AG does not include a powerful means for VO management.

3 About AGORA

AGORA is a VCE architecture model which exhibits all together the principles of *immanence*, *ubiquity* and *awareness*, in order to resolve the limitations described in the previous section. The development began in 2004 in the context of the european project ELeGI[9]. Initially, the challenge consisted of specifying a generic VCE on GRID that enables various kinds of communities of scientists to freely collaborate while minimizing the intervention of software specialists. Later the rationale of this project focused on the question of self-organisation of the communities as this aspect presented a growing interest in the domain of workflow management and concurrent access to distributed resources.

Besides *immanence*, *ubiquity* and *awareness* are the two other key principles identified that outlined the future AGORA platform. *Ubiquity* in computing science means that the location of resources is independent of the conditions of temporal and spatial use. In other words, *ubiquity* enables access to the VCE from anywhere at anytime. This implies that the resource used by the VCE has to be provided exclusively by the infrastructure and not from the terminal equipment side. The resource includes capacities of memory, processing and transmission as well as access to a large and heterogeneous instrumentation. Although this principle has been envisaged several years ago [32], the concrete deployment of operational solutions has been feasible only recently by means of pervasive technologies such as GRID. The principle of *awareness* indicates that members of a community seem to be more present to each other. The usual way to enhance the presence on the Web is to choose various modalities of communication with synchronous services.

The experimental VCE platform called AGORA UCS (Ubiquitous Collaborative Space) is the unique solution that proposes a fully integrated range of asynchronous and synchronous services. AGORA UCS includes a pool of resources accessible asynchronously and a range of services to operate on these resources concurrently and synchronously. Each member of a community may access to these resources via a viewer desktop coupled with a shared service for audio visual communication with

---

[9] ELeGI (European Learning Grid Infrastructure)[26, 30], European Framework Program n6 on IST (Information Services and Technologies), 2004-2007.

other members. AGORA UCS is also a proof-of-concept for an architecture based on user-centric requirements that have been identified by specialists of cognitive science[1].

Extensive experiments of the AGORA UCS prototype have been performed with more than eighty users accross the world [5] and about twenty communities in various scientific domains (organic chemistry, earth studies, optical physics, microelectronics). For the last two years, experiments have been extended to social and human sciences in order to improve the mechanisms underlying the immanence principle.

The AGORA architecture is composed of:

- a conceptual model including five concepts linked by four relations,
- a set of six persistent core services to manage the collaborative environment,
- four protocol mechanisms aiming for ensuring self-organisation of the communities.

3.1 AGORA *Conceptual Model*

3.1.1 Overview

The AGORA conceptual model presented on Figure 1 consists of a set of five concepts and four relations. This model includes a generator of events that are originated by humans or artificial processes. The organisation of agents in groups specifies how the flow of actions and the access to the resources are scheduled.

3.1.2 Definitions of AGORA Concepts

1. ⌈**Agent**⌋ constitutes the active component of the system since every action performed by an agent has an impact on the state of the system. In a collaborative

Fig. 1 The AGORA conceptual model

context and from the system point of view, all these actions may be performed artificially or by humans. For this reason, the meaning of agent in AGORA is clearly different than user or member. The user is one of the two kinds of agents: the human user and the artificial process. An agent holds its own identity outside the context of a community. An agent belonging to a community becomes a member of that community. A member specifies the relation (the membership relation) between an agent and a community.

2. Group or Community is a subset of agents. An agent may be a member of several communities. Once it is member, agents may undertake an activity in the context of the community.

3. Organisation is a composition of one given group and one unique set of resources. This concept is analogous to **VO** in GRID terminology (OGSA standardisation).

4. Resource is a set of means to carry out tasks, having the capacity of memory, processing, broadcasting, and access to a variety of instrumentation. This concept has been formalised as a **service container** by the GRID (OGSA).

5. Activity describes the flow of events and interactions within a community. Activities are always carried out by agents in the context of a community. This involves the notion of authorisations since any activity requires resource for completing a series of operations.

3.1.3 Definition of AGORA relations

1. Agent - Group - Activity is a ternary relation that links an agent to a community and to a number of activities according to the context of that community.

2. Activity - Resource indicates that any activity is attached to some resource. An activity always needs various kinds of resources like processing power, storage facilities or access to peripherials (network or specific instruments).

3. Group - Organization indicates that a community is attached to a proper organization. The bijective nature of this relation expresses the fact that an organization is unique for a given community.

4. Resource - Organization indicates that a set of resources is attached to a unique organization. This relation is symetrically bijective: a resource constitutes one of the two parts of an organization, the second one being the community.

Usually, existing VO management solutions specify two binary relations for organising agents into communities. A first binary relation corresponds to the agent membership in the community (e.g the agent A is a member of the community C). A second binary relation corresponds to the level of authorisations (sometimes called the role) of this agent (e.g A is *moderator* or A is *user*, etc.). However, in many situations it becomes complex to manage relations of pairs that mix three concepts, in particular when agents play different roles in several communities. Also, a role does not have a universal significance. A *moderator* may have a level of authorisation in one community that is different than in another. The ternary relation proposed in the

AGORA conceptual model allows these conceptual limitations to be overcome by allocating unique identifiers to each set of triplets agent-activity-community. Also, the concept of activity is more appropriate than role to express a series of actions that requires a certain level of authorisation in a given community.

3.2 Persistent Core Services

AGORA includes six PCS (Persistent Core Services) instantiated in the service container of every community. They are necessary for bootstaping and maintaining a collaborative environment. Figure 2 is a representation of a service container based on the AGIL formalism.

1. [A] uthorisations: Members of a community may have a different level of permission on services. This service is in charge of assigning rights to members including the permissions over the PCS.
2. [M] embers: A community is composed of members. This PCS manages the description of members, adding or removing members of a community.
3. [C] ommunity: A number of properties (identifier, description, etc.) are necessary to describe a community at a metalevel. Also, the creation of a new community must be done in the context of another community. This PCS is in charge of two kinds of operations: intra community operations (modifying the properties of the current community) and extra community operations (instantiating a new community). There is a community instantiated at the initialisation of the system and only dedicated to bootstrapping the first community.
4. [H] istory: All data belonging to a community must be stored, maintained and indexed. This PCS is in charge of keeping track of changes, logs of events and also of recording collaboration sessions. This PCS also aggregates the events to provide an evaluation of the frequency of access to the resources, that is used for the mechanism of *implication* (see Section 3.3).
5. [E] nvironment: A community may personalise its environment. This environment operates in a service container. This PCS is in charge of adding or removing services (excluding the PCS).
6. [N] otifications: Communication between members of a community and services is performed via notifications. This service handles the flow of notifications of every kind of communication and manages the states of the exchanged messages. These notifications are particularly useful for triggering events produced during the different decision processes. For instance, all the interaction mechanisms described in Section 3.3 require events triggered by community members in response to an initial request.

3.3 Four Interaction Mechanisms

AGORA aims to establish and maintain a high level of confidence inside the community by adopting protocols based on four basic interaction mechanisms called *cooptation, implication, delegation* and *habilitation*. The choice of these mechanisms

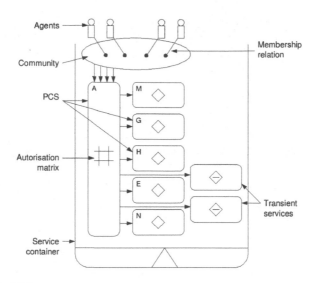

Fig. 2 The six PCS

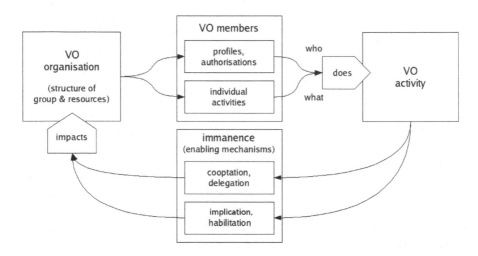

Fig. 3 AGORA's top level VO management model

has been adopted further to extensive analyses of various case studies[10]. Thus, they contribute all together to the principle of immanence by resolving most constrains for self-organizing agents and resources within communities. On the left part of Figure 3 the VO management system determines the overall system organisation. On the middle part, mechanisms for VO members management and, on the right

[10] The contributions to these cases studies by Erion Bregu, Houda Mourad, Amel Ghezzaz, Lea Guizol and Ahmed Bendriouech, is greatly acknowledged.

part, the resulting activity are directly dependent on the initial organisation. At this stage, there is no possibility to re-introduce activities back into the system organisation. The bottom part of the figure shows the mechanisms for enabling the immanence principle. For example, an agent *a* which is not yet a member of a community *C* may initially be coopted by an agent *b* (already a member of *C*). If *a* accepts the invitation from *b*, *b* may decide to delegate a set of rights on services for *a*. Once *a* becomes member of *C*, other members may decide to extend the habilitation of *a* over more services.

3.3.1 Cooptation

The mechanism of *cooptation* consists of appointment of new members in a community by members who already belong to it. The choice of *cooptation* as a method for members management (the $\boxed{M}$ service) is explainable through a well-known example of a collaborative editing environment such as a wiki. Like in many kinds of collaborative environments, there are usually two methods to introduce new members into a wiki community. The first one is to allow freedom to anyone to become a member and allocate default rights to them for editing pages of the community. The second one is to declare a restricted number of moderators who have the capability to validate the requests from external users to become members. Editing rights can be further granted or refused by the moderators. However, none of these two methods is sufficiently flexible to allow a fair workflow process of contributions since the choice of the moderators is established *de facto* and seldom revised. One of the challenges for the $\boxed{M}$ service enabling self-organization of the community is to easily add new members while keeping a suitable control of the intentions of the newly added members.

As shown in Figure 4 if coupled with a rules-oriented process, the mechanism of *cooptation* offers a flexible alternative to overcome the constrains of the two methods decribed above. When a community member suggests to invite a new member to this community, the mechanism of *cooptation* may follow rules based on thresholds for deciding if this suggestion is acceptable for the purpose of the community. For example, one condition for validating the addition of a new member in a

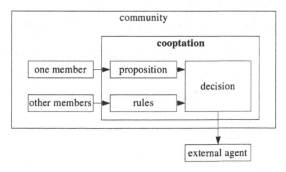

Fig. 4 Flow of events in the mechanism of cooptation

community could be that at least two members grant this request. Every procedure and threshold are entirely specified by the community. Clearly, some parameters might be quite different depending on whether the community is large or small and also all this depends on the specificity of the community. In any case, no external rule must be imposed by the system to the community. Once again, the principle of immanence in AGORA lays on this fundamental assumption.

Since the rules and procedures are specified by the community, there is no single language for specifying the syntax of these rules. Although a language may be recommended, it is most important to leave total freedom for the community to specify the rule language. Once a community has adopted a rule language, there is only the need for a translation service that converts humanly understandable rules into a set of directives and parameters that can be interpreted by an artificial process.

3.3.2 Implication

The mechanism of *implication* provides an estimation of the participation of each member in community activities. The *implication* aims to determine which members are the most involved in a given context and which ones add most value to the community activities. This evaluation must adopt a rating model that satifies the community to keep a high level of motivation of its members by giving them encouragements and rewards from their participation. Also, measurement of this evaluation is dynamic. It takes into account quantitative and qualitative aspects, objective and subjective inputs, as well as their sustainibility in time.

Quantitative aspects can be determined by the attendance of members, the frequency and the time spent accessing the resources, or the amount of contributions to activities. These settings are easily calculable because they can be inferred from the history data provided by the $\boxed{\text{H}}$ service. However, the most significant effect of *implication* can be best obtained from a qualitative evaluation such as assessment of the adequacy of contributions, relevance to the current activities and other subjective parameters such as the reputation of members. This is obtained from other members' feedback.

For this reason, the mechanism of *implication* incorporates a rule-oriented process that converts all inputs (qualitative and quantitative) into a global level of

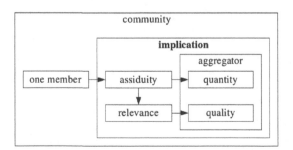

Fig. 5 Flow of events in the mechanism of implication

implication (see Figure 5). In the AGORA model, all rules and parameters of calculation of this level of implication can be specified by the community.

3.3.3   Habilitation

The mechanism of *habilitation* manages the decisions to gradually allocate access rights for community members to services and resources. The goal of *habilitation* is to distribute rights efficiently within communities.

As shown in Figure 6, a decision combines various parameters including the level of implication and the trust that the community has towards this member. The mechanism of habilitation analyzes these parameters to allow members to access services in order to maintain the security and development of the network.

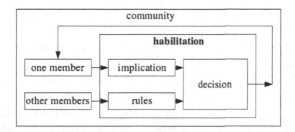

Fig. 6 Flow of events in the mechanism of habilitation

3.3.4   Delegation

The mechanism of delegation consists of allowing one member (the delegator) to transfer part of its rights to another member (the delegate). The major purpose of delegation is continuity of the activities carried out by the community members who cannot always be present to perform their tasks. Delegation may also become useful to motivate members to gain some responsibilities for contributing to the development of the community.

Since the delegate aquires the same responsibilities regarding these tasks, the delegation mechanism must ensure that appropriate rules are specified to prevent

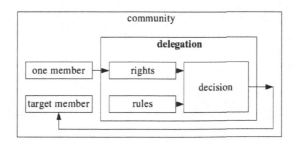

Fig. 7 Flow of events in the mechanism of delegation

risks of failure (see Figure 7). All these rules are specified by the community. The delegation is not permanent, a validity period is specified as well as a condition of revocation at any time.

3.4 *Experimentation on* AGORA

Extensive series of experiments have been performed since 2005 on the AGORA UCS platform, including about one hundred users and a dozen communities in various scientific domains (chemistry, microelectronics, physics, environment as well as human science). Initially, these experiments focused on the usability issues for any user who is not familiar with concepts of computing science. Access is made possible via a simple *stateless thin terminal* such as a web browser. The state of the VCE is maintained at the server side by means of a virtual desktop. An instance of the virtual desktop is dedicated as many times as a user is member of a community. The users have realised the importance of ubiquity since they were able to resume their session in the same state even after switching to another terminal. One advantage of this approach is security since no private information or cookies are stored on the client side. Another advantage of this approach is increase of performance since there is no bottleneck at the client side. Every operation is performed at the server side, therefore the computing resources can be more easily scaled. The aspect of awareness using both asynchronous communication via shared editing or synchronous communicatuion using instant messaging and videoconferencing, allowed an immediate bootstrap of a new VO and the acceptance of the technology was extremely high. This enhanced presence enabled a fast transfer of knowlege in particular for mastering complex computational tools.

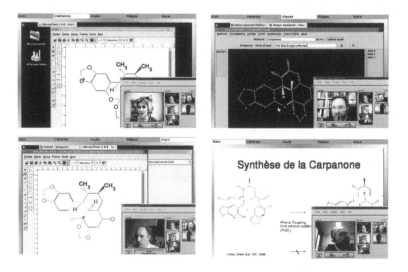

Fig. 8 Experimentation scenario EnCOrE

For instance a scenario called EnCOrE[11] (see Figure 8) has provided the most relevant results [21]. AGORA UCS enabled the visual representation of chemistry models at a distance. Most attention was put by the users on the semantics of their domain rather than solving computing problems. Unskilled users were at ease in their operations. The delegation of rights was important in the absence of some members. The cooptation of new members was also necessary to build a trustful community.

4 Conclusion

Through the description of the AGORA model, immanence in a event-based distributed system appears to be a sound principle to ensure efficient collaboration processes. The main reason is that the flow of events coming from various nodes of a distributed system does not follow a deterministic pattern but is always seeking for self-adaptation until reaching a suitable level of global satisfaction. The principle of immanence brings a strong framework for identifying risks of conceptual limitations in designing event-based distributed systems.

The promise of the GRID infrastructure opens many significant perspectives and gives more strength to the AGORA architecture. Yet, an architecture allowing seamless and secure access to distributed resources remains a real technological challenge since the behavior of a community cannot be forseen in advance. Flexibility of the AGORA UCS is essential to enable a community to freely organise itself. Various situations of collaboration started naturally by using asynchronous collaborative services (common knowledge base, annotation service), then were reinforced with other modalities of interaction by using a synchronous communication interface to facilitate the transfer of knowledge. Discussions in real time, combined with visual representations on a shared desktop, allowed the actors to increase the effectiveness of the collaboration process.

So far, initial experiments have revealed that a user-centric approach is suitable for self-organising communities. The experiments dedicated to validate the mechanisms for immanence started more recently and do not yet cover all the scenarios we are aiming towards. At the moment the objective is focusing on user self-ability to feel at ease in using the environment provided by AGORA UCS and understand the issues related to the self management of the community and the resources provided by the infrastructure. Even at an early stage, this work has already contributed to new ways to approach complex system design where the self-organisation criteria are critical. In many situations, AGORA has achieved results beyond the current state of the art. This progress encourages the belief in a new kind of collaborative system where the *organisation* and the *activity* of communities are mutually related.

---

[11] EnCOrE: Encyclopédie de Chimie Organique Electronique.
Demonstration available at http://agora.lirmm.fr

References

1. Brueckner, S.A., Van Dyke Parunak, H.: Engineering self-organizing applications. In: First IEEE International Conference on Self-Adaptive and Self-Organizing Systems, Tutorial session, SASO 2007, Boston (2007)
2. Cerri, S.A.: Shifting the focus from control to communication: the STReams OBjects Environments model of communicating agents. In: Padget, J. A. (ed.) Collaboration between Human and Artificial Societies 1997. LNCS, vol. 1624, pp. 74–101. Springer, Heidelberg (1999)
3. Cerri., S.A.: Human and Artificial Agent's Conversations on the Grid. In: 1st LEGE-WG Workshop: Towards a European Learning Grid Infrastructure, Lausanne. Educational Models for Grid Based Services (2002),
   http://ewic.bcs.org/conferences/2002/1stlege/session3.htm
4. Cerri, S.A.: An integrated view of Grid services, Agents and Human Learning. In: Towards the Learning Grid: Advances in Human Learning Services, pp. 41–62. IOS Press, Amsterdam (2005),
   http://portal.acm.org/citation.cfm?id=1563266.1563273
5. Dugénie, P.: UCS, Ubiquitous Collaborative Spaces on an infrastructure of distributed resources. PhD thesis. University of Montpellier (2007) (in French), to be downloaded at http://www.lirmm.fr/~dugenie/these
6. Dugénie, P.: Ubiquitous Collaborative Spaces: a step towards collective intelligence. In: European Conference on Computing and Philosophy, ECAP 2008, Montpellier (June 2008)
7. Dugénie, P., Cerri, S.A., Lemoisson, P., Gouaich, A.: Agora UCS Ubiquitous Collaborative Space. In: Woolf, B.P., Aïmeur, E., Nkambou, R., Lajoie, S. (eds.) ITS 2008. LNCS, vol. 5091, pp. 696–698. Springer, Heidelberg (2008)
8. Dugénie, P., Lemoisson, P., Jonquet, C., Crubézy, M., Laurenço, C.: The Grid Shared Desktop: a bootstrapping environment for collaboration. In: Advanced Technology for Learning (ATL), Special issue on Collaborative Learning, vol. 3, pp. 241–249 (2006)
9. Duvert, F., Jonquet, C., Dugénie, P., Cerri, S.A.: Agent-Grid Integration Ontology. In: Meersman, R., Tari, Z., Herrero, P. (eds.) OTM 2006 Workshops. LNCS, vol. 4277, pp. 136–146. Springer, Heidelberg (2006)
10. Ferber, J., Gutknecht, O., Michel, F.: From Agents to Organizations: An Organizational View of Multi-agent Systems. In: Giorgini, P., Müller, J.P., Odell, J. (eds.) AOSE 2003. LNCS, vol. 2935, pp. 214–230. Springer, Heidelberg (2004)
11. Foster, I.: Globus Toolkit Version 4: Software for service-oriented systems. Journal of Computer Science and Technology, 513–520 (2006)
12. Foster, I., Insley, J., Kesselman, C., von Laszewski, G., Thiebaux, M.: Distance visualization: Data exploration on the grid. IEEE Computer Magazine 32(12), 36–43 (1999)
13. Foster, I., Jennings, N.R., Kesselman, C.: Brain meets brawn: why Grid and agents need each other. In: 3rd International Joint Conference on Autonomous Agents and Multiagent Systems, AAMAS 2004 (2004)
14. Foster, I., Kesselman, C., Nick, J., Tuecke, S.: The physiology of the Grid: an Open Grid Services Architecture for distributed systems integration. In: Open Grid Service Infrastructure WG, Global Grid Forum. The Globus Alliance (June 2002)
15. Foster, I., Kesselman, C., Tuecke, S.: The anatomy of the Grid: enabling scalable virtual organizations. Supercomputer Applications 15(3), 200–222 (2001)
16. ITU-T. H.263, infrastructure of audiovisual services, video coding for low bit rate communication. Technical report, International Telecommunication Union (2005)

17. Jonquet, C., Cerri, S.A.: The STROBE model: Dynamic Service Generation on the GRID. Applied Artificial Intelligence Journal 19, 967–1013 (2005)
18. Jonquet, C., Dugénie, P., Cerri, S.A.: Agent-Grid Integration Language. International Journal on Multi-Agent and Grid Systems 4(2), 167–211 (2008), http://hal-lirmm.ccsd.cnrs.fr/lirmm-00139691
19. Jonquet, C., Dugenie, P., Cerri, S.A.: Service-Based Integration of Grid and Multi-Agent Systems Models. In: Kowalczyk, R., Huhns, M.N., Klusch, M., Maamar, Z., Vo, Q.B. (eds.) SOCASE 2008. LNCS, vol. 5006, pp. 56–68. Springer, Heidelberg (2008)
20. Gallop, J.R., Sagar, M., Walton, J.P.R.B., Wood, J.D., Brodlie, K.W., Duce, D.A.: Visualization in grid computing environments. In: Turk, G., Rushmeier, H. J. (eds.) Proceedings of IEEE Visualization, pp. 155–162 (2007), ISBN:0-7803-8788-0
21. Lemoisson, P., Cerri, S.A.: Interactive construction of encore (encyclopédie de chimie organique electronique). Applied Artificial Intelligence Journal Special issue on Learning Grid Services 19, 933–966 (2005)
22. Lévy, P.: L'intelligence collective. Pour une anthropologie du cyberspace. La Découverte, Paris (1994)
23. Luhmann, N.: Social systems. Writing science (1995)
24. Maturana, M.R., Varela., F.J.: Autopoiesis and Cognition. Reidel, Dordrecht (1984)
25. Olson, R.: Access grid hardware specification. Technical report, Argonne National Laboratory (2001)
26. Cerri, S.A., Dimitrakos, T., Gaeta, M., Ritrovato, P., Allison, C., Salerno, S.: Towards the learning grid: advances in human learning services. In: Kok, J.N., Liu, J., Lopez de Mantaras, R., Mizoguchi, R., Musen, M., Zhong, N., Breuker, R.D.J., Guarino, N. (eds.) Frontiers in Artificial Intelligence and Applications, vol. 127, p. X–240. IOS Press, Amsterdam (2005)
27. Pearlman, L., Welch, V., Foster, I., Kesselman, C., Tuecke, S.: A Community Authorization Service for group collaboration. In: 3rd International Workshop on Policies for Distributed Systems and Networks, POLICY 2002, Monterey, CA, USA, pp. 50–59. IEEE Computer Society, Los Alamitos (2002)
28. Ciaschini, V., Frohner, A., Gianoli, A., Lorentey, K., Spataro, F., Alfieri, R., Cecchini, R.: VOMS, an authorization system for virtual organizations. In: Fernández Rivera, F., Bubak, M., Gómez Tato, A., Doallo, R. (eds.) Grid Computing 2003. LNCS, vol. 2970, pp. 33–40. Springer, Heidelberg (2004)
29. Shalf, J., Bethel, E.W.: The grid and future visualization system architectures. IEEE Computer Graphics and Applications 23(2), 6–9 (2004)
30. Gouarderes, G., Cerri, S.A., Nkambou, R.: Learning grid services. Special issue: Applied Artificial Intelligence Journal 19(9-10), 811–1073 (2005)
31. von Bertalanffy, L.: General system theory: foundations, development, applications. Braziller, New York (1968)
32. Weiser, M., Brown, J.S.: Designing calm technology. PowerGrid Journal (July 1996)

Glossary of Terms and Acronyms

AG Access Grid

AGIL Agent-GRID Integration Language

AGR Agent-Group-Role

AGORA UCS AGORA Ubiquitous Collaborative Space is the name of the platform developed for the purpose of experimenting the AGORA principles.

CAS Community Authorization Service

CSCW Computer Suported Collaborative Work

EGEE Enabling GRID for E-Science

ELeGI European Learning Grid Infrastructure

Globus GRID middleware

LDAP Lightweight Directory Access Protocol

MAS Multi-Agent Systems

PCS Persistent Core Services

SOA Service-Oriented Architecture

VO Virtual Organization

VCE Virtual Collaborative Environment

VOMS VO Membership Service

Context-Based Event Processing Systems

Opher Etzion, Yonit Magid, Ella Rabinovich,
Inna Skarbovsky, and Nir Zolotorevsky

Abstract. The concept of context has recently emerged as one of the major abstractions in event processing modeling with presence in event processing products. In this chapter we discuss the notion of context as a first class citizen within event processing modeling, and discuss its implementation in event processing products and models that arise from the current state-of-the practice

Keywords: Context, context-aware computing, context-driven architecture, temporal processing, spatial processing, event processing modeling, event processing products.

1 Introduction: Context in Event Processing Modeling

In real life, many activities are done within a context; we might behave differently within different parts of the day, in different locations, in different states of the weather, these are all types of context. Indeed, context [8] plays the same role in event processing that it plays in real life; a particular event can be processed differently depending on the context in which it occurs, and it may be ignored entirely in some contexts. Contexts have been formalized in some works such as [4], [5], and [11].

There are three main uses of context by event processing applications:

- An event stream is defined [3] as open-ended set of events. If you want to perform an operation on the stream you cannot wait until all these events have been received. Instead you have to divide the stream up into a sequence of context partitions, or *windows*, each of which contains a set of consecutive events. You can then define the operation in terms of its effect on the events in a window. The rule that determines which event instances are admitted into which window is something we call a *temporal context*.

- A collection of events, arriving in one or multiple streams, may contain events that are not particularly connected to one another, even though they might occur close together in a temporal sense. They might, for example,

Opher Etzion · Yonit Magid · Ella Rabinovich · Inna Skarbovsky · Nir Zolotorevsky
IBM Haifa Research Lab, Haifa, Israel
e-mail: {opher,yonit,ellak,inna,nirz}@il.ibm.com

S. Helmer et al.: Reasoning in Event-Based Distributed Systems, SCI 347, pp. 257–278.
springerlink.com   © Springer-Verlag Berlin Heidelberg 2011

refer to occurrences in different locations, or to occurrences involving different entities in the real world. Suppose you were to do some processing of an event stream, such as a simple *aggregate* agent that counts the number of events. By default this would count all the events in the stream, but what if you want to see separate totals for each location where the events occurred? To do this you need to have a separate agent, or at least a separate *instance* of the agent, processing the events for each location. *Spatial* contexts and *Segmentation-oriented* contexts let you assign related events to separate context partitions. You can then have each partition processed by a distinct instance of the event processing agent, so that events in one context partition are processed in isolation from the events in other partitions.

- Context also allows event processing agents to be context sensitive, so that an agent that is active in some contexts may be inactive in others. We refer to this as *state-oriented* context.

In essence, in event processing modeling, an EPA (Event Processing Agent) serves as the basic unit of event processing computation. An EPA may be of three types: filter EPA, transform EPA, and pattern detect EPA (which is the most general one). Context may apply to all of these types; Fig. 1 taken from [8] shows that each EPA is associated with a context, the association is aimed to achieve one or more of the context roles mentioned before.

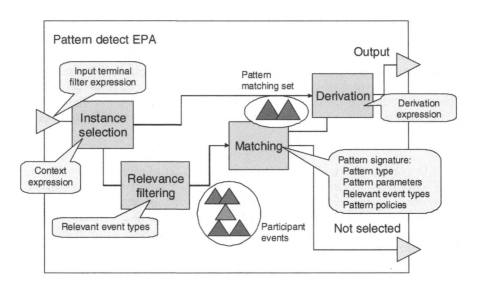

Fig 1. The Event Processing Agent functionality model.

Context affiliation for EPA determines:

- Whether an input event is relevant at all for a specific EPA – this makes the EPA context sensitive; depends on the context an EPA may or may not be involved

- If the EPA has several instances, to which of these instances, an event is relevant for – this serves as grouping of events together according to shared context.

In order to understand the role, let's look at a simple example of segmentation oriented context; events that relate to transactions may be partitioned by customer, since we would like to process the events related to each customer separately.

Within event processing languages, contexts may be explicit, implicit, or partially explicit. Explicit context means that context primitives are first class primitives in the language. Some languages do not support any notion of context, and some support partial notion of contexts. A survey of contexts in various languages will be presented in Section 5, dealing with contexts in practice.

Fig. 2 taken from [8] shows the various context dimensions: temporal, spatial, state oriented, and segmentation oriented that we describe in this section.

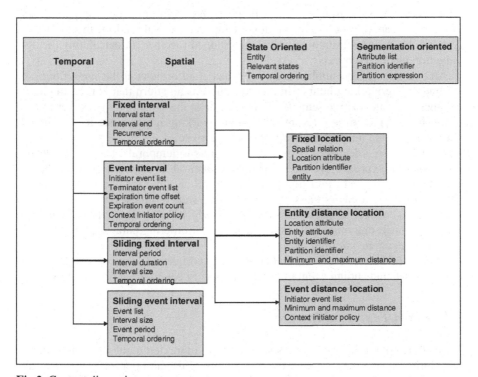

Fig 2. Context dimensions.

The rest of this chapter is structured in the following way: Section 1.2 discusses temporal contexts, Section 1.3 discusses spatial contexts, Section 1.4 discusses segmentation oriented context, Section 1.5 discusses state oriented context, Section 1.6 discusses context composition, Section 1.7 discusses the implementation of context in practice, and Section 1.8 concludes the chapter with discussion on the state of the art, its limitations, and the potential future of contexts in event processing.

2 Temporal Context

Temporal context is aimed to partition the temporal space into time intervals. It is used for two purposes:

- Events grouping: the typical use of temporal contexts is to group events to process them together based on the fact they have occurred during the same interval.
- EPA applicability: Another use of temporal contexts is to indicate that a certain event processing agent is applicable within certain intervals and is not applicable in other intervals. The temporal interval convention is having half-open interval [Ts, Te), such that Ts is the interval starting point, included in the interval, and Te is the interval end-point, not included in the interval.

An example of event grouping is: count all the bids within the auction period. In this case the auction period stands for a temporal interval, and all the events of type bid that occur within this interval are grouped together for calculating the aggregation function (count).

An example of EPA applicability is: detect if all printers at the same floor are off-line during working hours. In this case there is a pattern that is used to check whether all printers in the same floor are off-line at the same time. However monitoring for this pattern is relevant only within working hours, since it is assumed that after working hours there is no technician on-site.

One of the design decisions in associating events to temporal context is to determine whether a specific event falls within the interval. In some cases this is being determined by the detection time, which is the timestamp created by the system [9] when the event enters the system, or by occurrence time which is the time in which the event source specifies as the time in which the event occurred in reality. This decision is determined by using the temporal ordering parameter in the various temporal context definitions

There are four types of temporal contexts: fixed interval, event interval, sliding fixed interval and sliding event interval. We survey briefly each of these types.

2.1 Fixed Temporal Interval

A fixed temporal interval consists of one or more temporal intervals whose boundaries are pre-defined timestamp constants. This can be a one time interval: [July 12 2010 13:30, July 12 2010 17:00) which designates a specific session within a specific conference; it also can be stated as [July 12 2010 10:30, + 3.5 hours), where Te is an offset relative to Ts. A fixed temporal interval can also be recurrent, and in this case the frequency should be specified. Some example of that is [7:00, 9:00) daily, or [Monday 13:30, 14:30) weekly which determines event processing operations that are applicable periodically, for example: some traffic monitoring is done during morning rush hour only, or that during a weekly meeting some monitoring is disabled.

2.2 Event Interval

Event interval is a temporal interval that is being opened or closed when one or more events occur; the meaning of "event occurs" is determined by the temporal ordering parameter and can be interpreted either as the detection time or the occurrence time as explained earlier. The collection of events that open such an interval are called initiator events, and the collection of events that close the temporal interval is called terminator events. An interval may also expire after a certain time offset is reached.

Some examples of event interval are:

- A temporal interval is initiated when a patient is admitted to a hospital, and ends with the release of the same patient from the hospital. Note that here the temporal context is combined with segmentation context, since the temporal context refers to a single patient.
- A temporal interval that starts with a shipment of a package and ends with the delivery of the same package; the temporal interval expires after 3 days, which is the delivery designated time, even if the package has not been shipped, thus there are two ways in which this interval can terminate: the desired way (delivery) and the time-out way (3 days have passed).

2.3 Sliding Fixed Interval

In a sliding fixed interval context, new windows are opened at regular intervals. Unlike the non-sliding fixed interval context these windows are not tied to particular times, instead each window is opened at a specified time after its predecessor. Each window has a fixed size, specified either as a time interval or a count of event instances.

A sliding fixed interval is specified by the interval period which designates the frequency in which new windows are opened, the interval duration, which specifies how long this interval spans, and the interval size, which determines how many events are included in a single window. The specification must include an *interval period* parameter and either an *interval duration* or *interval size* (or both, in which case it may end earlier than the duration if the event count is exceeded). Sliding intervals may be overlapping or non-overlapping; they are overlapping if and only if the *interval period* < *interval duration*.

Some examples are as follows:

- A temporal interval starts every hour, with the duration of one hour. In this case we partition the time into time windows of one hour each.
- A temporal interval starts every day, and ends when 50 orders have been placed in the system, or at the end of the day. In this case the interval partitions the temporal space into time windows of one day; however, the daily interval can terminate earlier if there are 50 orders.
- A temporal interval starts every 10 minutes, and lasts for an hour. In this case, each event (starting from the second hour of operation) is associated with 6 different windows that are active in parallel. This type of context can be used for trend seeking windows in time series' events.

2.4   Sliding Event Interval

The *sliding event interval* context is similar to the sliding fixed interval context. The difference is that the criterion for opening a new window is specified as a count of events, rather than as a time period. A sliding event interval is specified by list of events (possibly with a predicate for each event), interval size (in event count), and event period.   The event list designates event types whose instances can start the temporal window; the interval size determines the event count that closes the interval, while the event period (which defaults to the interval size) determines the event count to open an additional window.

An example is the following:

- A sensor measures the temperature; an interval consists of five consecutive measurements and starts with every single measurement. Each measurement is associated with five different intervals.

3   Spatial Context

A spatial context groups event instances according to their geospatial characteristics. This type of context assumes that an event has a spatial attribute designated its location. Location can be represented in three ways [8]: point in space, line or polyline, and area (polygon). As shown in Fig.3, there are three types of spatial contexts: fixed location, entity distance location and event distance location.

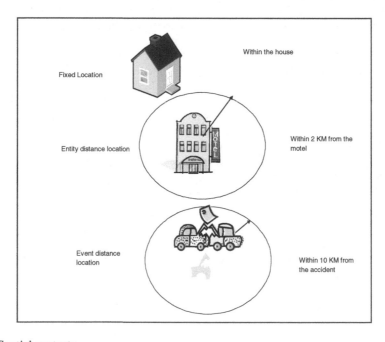

Fig 3. Spatial contexts.

3.1 Fixed Location

A *fixed location* context has one or more context partitions, each of which are associated with the location of a reference entity, sometimes referred to as a *geofence*. An event instance is classified into a context partition if its location's attribute correlates with the spatial entity in some way. Like the event location, the location of the reference entity can be of any of the three types: point, line, area, thus there can be several relations between the event's location and the reference location as shown in Fig.4.

The *contained in* relation is true if the entity location is completely enclosed within the event location.

The *contains* relation is true if the event location is completely enclosed within the entity location.

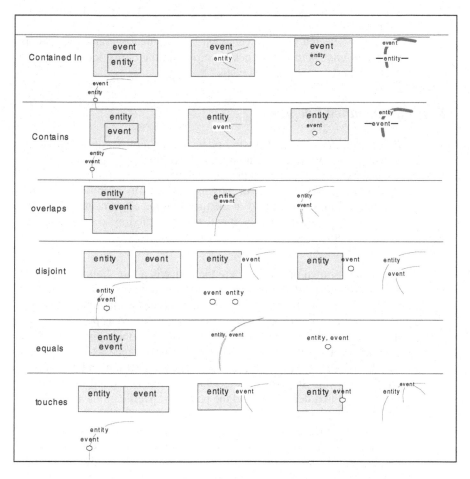

Fig 4. Fixed location relations - taken from [8].

The *overlaps* relation is true if there is some overlap between the entity location and event location, but neither of them is contained within the other.

The *disjoint* relation is true if there is no overlap between the entity location and event location.

The *equals* relation is true if the entity location and event location are identical.

The *touches* relation is true if the event location borders the entity location. Depending on the implementation, some of these relations may have some tolerance, e.g. two points can be considered equal if they are no more than some distance apart.

Not every combination of entity location type and event location type is valid, as seen in Fig. 4, for example: a point cannot be contained in another point, an area cannot be equal to a line, and a point cannot overlap an area.

Here are some examples:

- A car fleet management application follows cars that are driving on a certain highway. The car location, obtained by the GPS, is represented as a point, the highway is represented by a polyline, and the relationship is "contained in".

- A plague is detected within a geographical area that overlaps the borders of a certain city, both the event's location and the entity location are areas, and the relationship is "overlaps."

3.2 Entity Distance Location

An *entity distance location* context gives rise to one or more context partitions, based on the distance between the event's location attribute and some other entity. The word distance refers to the shortest distance between two spatial entities.

This entity may be either stationary or moving. If the entity is moving, the distance relates to the location of the entity at the time that the event occurred (its *occurrence time*). The entity may either be one that is specified by another attribute of the event, or one that is specified in the context definition.

Some examples are as follows:

- The example shown in Fig.3, where the entity is a motel and the context is used to track fire alarms within 2 km of the motel, in order to alert the motel manager of possible danger. This context has a single partition, and both the event location and the entity location are given as points.

- Vehicle breakdown events partitioned according to their distance from a particular service center. They are grouped by distance as follows: less than 10 km, between 10 km and 30 km, between 30 km and 60 km and more than 60 km. In this case the service center is a fixed entity specified in the context definition, and the partition specifications are <10 km; ≥ 10 km and < 30 km; ≥ 30 km and < 60 km; ≥60 km.

- An alerting service at a big conference, which attendees can use to receive alerts when they are in the proximity of another person. Attendees send periodic events giving their own location and these events include the name of the person they are interested in meeting. This example shows the use of an entity distance location context where the entity is itself moving. The context specification has an entity attribute which tells the context which attribute from the event gives the name of the person being tracked and an explicit partition with a maximum distance of 100m. This means that the alert generation event processing agent is only invoked if the two people are less that 100m apart.

3.3 Event Distance Location

This type of context specifies an event type and a matching expression predicate. A new partition is created if an event occurrence is detected that matches this predicate. Subsequent events are then included in the context partition if they occurred within a specific distance of the initiating event. Distance is defined as in the entity distance location context.

Some examples are as follows:

- Detecting the presence of an aircraft within a 10 km radius of a volcanic ash
- Detecting a case of scarlet fever within a distance of 100km from a previous outbreak of this disease (*event type* is disease report, predicate is disease="scarlet fever", *maximum distance* is 100 km).

4 Segmentation Oriented Context

A *segmentation-oriented* context is used to group event instances into context partitions based on the value of some attribute or collection of attributes in the instances themselves. As a simple example consider an event processing agent that takes a single stream of input events, in which each event contains a customer identifier attribute. The value of this attribute can be used to group events so that there's a separate context partition for each customer. Each context partition only contains events related to that customer, so that the behavior of each customer can be tracked independently of the other customers.

In this kind of segmentation-oriented context, the context specifies just the attribute (or attributes) to be used, and this implicitly defines a context partition for every possible value of that attribute. Alternatively, a segmentation-oriented context definition can list its context partitions explicitly. Each partition specification includes one or more predicate expressions involving one or more of the event attributes; an event is assigned to a context partition if one of its predicates evaluates to true.

Partitions can be specified:

- By attribute value: e.g. there is a context partition for each customer, for all events related to this customer
- By combination of attributes: e.g. all car accidents are grouped by car type and driver license type.
- By explicit predicate: e.g. there is a context partition by an employee's band. One partition for band < 5, one partition for band = 6 or 7, one partition for band 8, one partition for band 9, and one partition for band > 9.

5   State-Oriented Context

State-oriented context differs from the other dimensions in that the context is determined by the state of some entity that is external to the event processing system. This is best illustrated with some examples:

- An airport security system could have a threat level status taking values green, blue, yellow, orange, or red. Some events may need to be monitored only when the threat level is orange or above, while other events may be processed differently in different threat levels.
- Traffic in a certain highway has several status values: traffic flowing, traffic slow, traffic stationary.

Each of these states issues a context partition.

6   Context Composition

Event processing applications often use combinations of two or more of the simple context types that we have discussed so far. In particular a temporal context type is frequently used in combination with a segmentation-oriented context, and we show and example of this in Fig. 5.

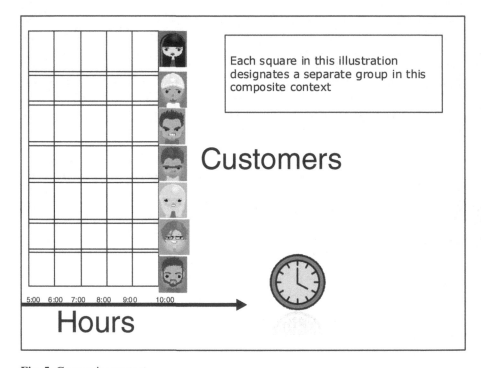

Fig 5. Composite context.

Fig. 5 is showing composition of segmentation-oriented context and temporal context. Each square is a separate context partition and designates the combination of an hour-long interval and a specific customer.

Composition of context is typically an intersection between two context dimensions, i.e. an event belongs to the composite context if it belongs to each of the member contexts; other possible set operations for composition union and set difference. In these cases typical uses are cases in which the member contexts are of the same context dimension (example: union of temporal intervals, union of spatial fixed areas). It should be noted that the definition of composite context in any case may relate both to member contexts of the same dimension or of different dimension.

These context dimensions generalize the functionality that exists within various products and models of event processing. Next, we move on to context implementation in five event processing products that samples that implementation context in various products.

7 Context Realization within Current Event Processing Products

In the following sections we introduce context support within a sample of existing event processing commercial platforms. We classify this context support according to the four context dimensions defined in Section 4: temporal context, segmentation context, spatial context and state-oriented context. The five languages we survey are: StreamBase StreamSQL, Oracle EPL, Rulecore Reakt, Sybase's CCL (formerly Aleri/Coral8) and IBM's Websphere Business Events.

7.1 Stream Base StreamSQL

StreamSQL [17] is a SQL based Event Processing Language (EPL) focusing on event stream processing, extending the SQL semantics and adding stream-oriented operators as well as windowing constructs for subdividing a potentially infinite stream of data into analyzable segments.

The windowing constructs correspond to the temporal context dimension, where the windows are defined based on time, or based on number of events, or based on an attribute value within an event.

Additional constructs in StreamSQL, such as GroupBy and PartitionBy clauses correspond to the segmentation context dimension.

The temporal context supported is the sliding fixed interval called in StreamSQL **sliding fixed window** and sliding event interval called **sliding event window**. Default evaluation policy for the operators is deferred upon reaching the specified window size, however this can be changed in the operator specifications to immediate or semi-immediate (performing the operation once every number of units even if window size is not reached).

7.1.1 Sliding Fixed Window

A window construct of type "time" supports the notion of sliding fixed interval. New windows are opened at regular intervals relative to each other, as specified in the ADVANCE parameter; the windows remain open for a specified period of time, as stated in the SIZE parameter. Additional parameters further influence the behavior of the window construct, such as OFFSET which indicates a value by which to offset the start of the window.

For example, Listing 1 creates sliding overlapping windows which are opened every second and remain open for 10 seconds

---

CREATE WINDOW *tenSecondsInterval*(
SIZE *10* ADVANCE *1* {TIME});

Listing 1. StreamSQL sliding fixed window example.

7.1.2 Event Interval and Sliding Event Interval

A window construct of type "tuple" supports the notion of sliding event intervals, where the window size is defined based on the event number, and the repeating intervals are defined in terms of events as well.

```
CREATE WINDOW eventsInterval (
    SIZE 10 ADVANCE 1 {TUPLE} TIMEOUT 20);
```

Listing 2. StreamSQL sliding event window example.

The repetition is optional, a policy determines whether it is necessary to open new windows on event arrival or not, therefore this supports the notion of event interval.

For example, Listing 2 defines a context where a new window is opened upon arrival of a new tuple (event).

The windows are closed either after accumulating 5 events or when a timeout of 20 seconds from window opening expires.

Options exist to use the temporal and segmentation context dimensions in conjunction thus introducing a composite context consisting of multiple dimensions.

7.2 Oracle EPL

The Oracle CEP [27] offering is a Java-based solution targeting event stream processing applications. Oracle EPL supports both the segmentation and temporal context dimensions [18]. The temporal context supports four types of sliding windows: row-based, time-based, batched row and time-based, which correspond to sliding event and sliding fixed interval in this dimension with different initiation policies.

The sliding row-based window supports retention of last N event entries. For example, the clause shown in Listing 3 retains the last five events of the Withdrawal stream, when the oldest event is "pushed out" of the sliding window, providing a form of sliding event interval.

```
SELECT  *  FROM  Withdrawal  RETAIN  5
```

Listing 3. Oracle EPL sliding raw-based window example.

Fig. 6 demonstrates the flow of events through the event stream and the content of the window for this statement.

The time based sliding window is a moving window aggregating events for a specified time, implementing the case of sliding fixed temporal interval. For example, the clause shown in Listing 4 defines a window holding a view into an event stream of all events received in the last 4 seconds.

Fig. 7 demonstrates the time-based window behavior over a period of time

An example of a batched time windows is shown in the clause presented in Listing 5.

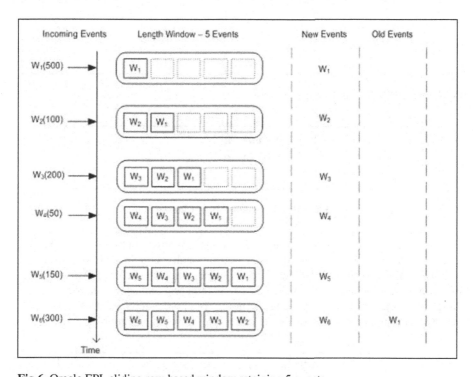

Fig 6. Oracle EPL sliding row-based window retaining 5 events.

```
SELECT * FROM Withdrawal RETAIN 4 SECONDS
```

Listing 4. Oracle EPL sliding time-based window example.

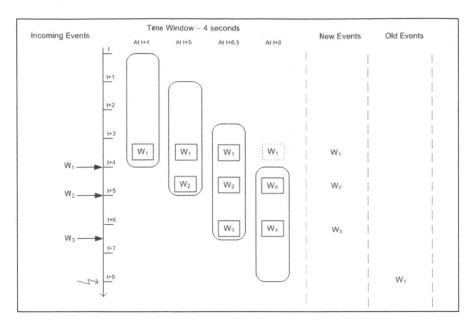

Fig 7. Oracle EPL Sliding time-based window.

SELECT * FROM *Withdrawal*
RETAIN BATCH OF *4* SECONDS

Listing 5. Oracle EPL batched time window example.

The windows are non overlapping windows, and a new window is opened only when its predecessor window is closed. Additionally Oracle EPL supports the GROUP BY clause which enables further segmentation of the event stream. The EPL language supports the composition of the temporal and segmentation context dimensions together thus supporting the notion of composite context. An example of the syntax supporting the composite context is shown in Listing 6.

INSERT INTO *TicksPerSecond*
SELECT   *feed*,   COUNT(*)   AS   *cnt*   FROM
MarketDataEvent
RETAIN BATCH OF 1 SECOND
GROUP BY *feed*

Listing 6. Oracle EPL composite context example.

The composite context that is shown in Listing 6 is composed of a temporal context of a sliding window of 1 second, and a segmentation context by feed.

7.3 RuleCore Reakt

RuleCore [20] executes ECA rules defined using the Reakt language. The main language element in Reakt is the rule. RuleCore uses a notion of "views" as execution context for the rule. Each rule instance maintains its own virtual view into the incoming event stream. The view provides a window into the incoming event stream providing context for rule evaluation. Reakt supports segmentation, temporal, spatial and state contexts.

The temporal context is implemented using the MaxAge view property which defines the maximum age for each event in the view.

Example: Listing 7 shows a definition that limits every rule instance to view only events from the past 10 minutes.

```
<MaxAge>00:10:00</MaxAge>
```

Listing 7. RuleCore Reakt fixed window example.

Segmentation context is implemented using the [?] match view property which is very flexible and can select values from the event bodies using the full expressive power of XPath. This allows for powerful semantic matching even of complex event structures. An example is shown in Listing 8.

```
<Match>
 <Value>
<Event><base:XPath>EventDef[eventType="Warning]</bas
e:XPath></Even>
 <Field><base:XPath xmlns:base="base">base:
ServerIP base:XPath></Field>
 </Value>
<Match>
```

Listing 8. RuleCore Reakt match view.

Spatial context is implemented using the Type view property by defining a new geographic zone entity type defined by its coordinates and Match view property. An event is then associated with a specific entity which may have a set of properties such as position or a Zone. Spatial Evaluation is done by applying Match property on specific Entity type with geographic location and to specific Zone that specifies the reference geo area, all events of specific entity type and which match Zone property will be classified to this specific context partition.

Example: Listing 9 illustrates the association of specific events that are contained in a location Zone. For this rule we define ZoneEntry view, which will contain events from "vehicle" entity type (Match/vehicle) and only from specific zone location (Match/Zone).

```
<ViewDef name="ZoneEntry">

<Propertie><!– All events in the view must be of types InsideZone or OutsideZone –>

   <Type><Event><XPath>EventDef[@eventType="InsideZone"]</XPath></Event>

      <Event><XPath>EventDef[@eventType="OutsideZone"]</XPath> </Event>

</Type>

   <!– All events in the view must be from the same vehicle –>

   <Match name="vehicle">

      <Value><Event><XPath>EventDef[@eventType="OutsideZone"]</XPath></Event>

       <Property name="Vehicle"></Value>

      <Value> <Event> <XPath>EventDef[@eventType="InsideZone"]</XPath></Event>

<Property name="Vehicle"> </Value>

   </Match><!– All events in the view must be from the same zone –>

   <Match name="zone">

    <Value> <Event> <XPath>EventDef[@eventType="InsideZone"]</XPath> <Event>

       <Property name="Zone"> </Value>

      <Value> <Event><XPath>EventDef[@eventType="OutsideZone"]</XPath> </Event>

       <Property name="Zone"> </Value>

    </Match>

  </Properties>

</ViewDef>
```

Listing 9. RuleCore Reakt - spatial context example.

State contexts are implemented in a similar way to the way in which   spatial contexts are implemented, using Type view property where each event entity has specific properties. State evaluation is done by applying Match property on specific Entity type property, all events of specific entity type and which match specific property will be in the context.

The sliding event-based MaxCount view supports retention of last N event entries, for example the statement guarantees that an event view never contains more than 100 events.

```
<MaxCount>100</MaxCount>
```

Listing 10. RuleCore Reakt - sliding event interval example.

The Reakt language allows the composition of the temporal, segmentation, spatial and state context dimensions together thus supporting the notion of composite context.

7.4 Sybase Aleri -Coral8 and CCL

The Sybase Aleri CEP offering consists of Aleri Streaming Platform and Coral8 engine. Coral8's authoring language[22] [23] CCL (Continuous Computation Language) is a SQL-based event processing language extending SQL in a different way relative to the Streambase and Oracle examples presented previously. CCL supports different combinations of window types: sliding and jumping windows, either row-based (event-based) or time-based.

Examples include:

- Time intervals ("keep one hour's worth of data").
- Row counts ("keep the last 1000 rows").
- Value groups ("keep the last row for each stock symbol").

Relative to the context dimension classification defined in section 4, CCL supports both the temporal and segmentation context dimensions. In the temporal dimension, the sliding fixed interval, sliding event interval and event interval are supported. The difference in CCL terminology between sliding and jumping windows is the way they treat the interval periods and the window initiation policies.

```
    INSERT INTO MaxPrice
SELECT Trades.Symbol, MAX(Trades.Price)
FROM Trades KEEP 15 MINUTES
WHERE Trades.Symbol = 'IBM';
```

Listing 11. Sybase CCL - Sliding fixed interval example.

Listing 11 illustrates a sliding fixed interval representing a 15 minute window into the Trades event stream, and the aggregation is determining the highest trade in the last 15 minutes. The interval period is once every minute, and the interval duration is 15 minutes. The initiation policy is "refresh" – the previous window is closed once the new one is created.

A similar example illustrating the sliding event interval is deducing the highest price in the last 15 trades instead of last 15 minutes. This example is shown in Listing 12.

```
INSERT INTO MaxPrice
SELECT Trades.Symbol, MAX(Trades.Price)
FROM Trades KEEP 15 ROWS
WHERE Trades.Symbol = 'IBM';
```

Listing 12. Sybase CCL - Sliding event interval.

The window can be also defined implicitly in pattern matching queries, such as the following sliding event interval definition, in Listing 13, describing a period which starts with each event A arrival, and ends 10 seconds later; during this period event A should occur, followed by event B, and event C should not occur between the time in which B occurs and the end of the temporal window.

```
[10 SECONDS: A, B, !C]
```

Listing 13. Sybase CCL - pattern embedded context.

The GROUP BY clause supports the segmentation context implementation.
A combination of the temporal and segmentation context enables the expression of composite contexts. In the example shown in Listing 14, the sliding fixed interval window is further segmented according the stock's symbol. This query determines the maximum price for each stock type in the last 15 minutes.

```
INSERT INTO MaxPrice
SELECT Trades.Symbol, MAX(Trades.Price)
FROM Trades KEEP 15 MINUTES
GROUP BY Trades.Symbol;
```

Listing 14. Sybase CCL - composite context example.

7.5 IBM Websphere Business Events

IBM WBE [21] is a business event processing system, focusing on the business user and his needs. A segmentation context is easily expressed in WBE using the "Related By" construct, shown in Fig. 8, which supports the segmentation group of events based on event field value or a composition of field values [19].

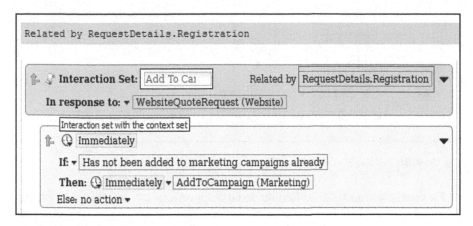

Fig 8. WBE definition of segmentation context.

Fig. 8 shows an example in which the events are related to each other by the same Registration, making it a segmentation context. The temporal context capability is supported using deferred evaluation mode (Fig. 9). Events can be evaluated using time delays from the triggering event. These time delays can be for a fixed amount of time or until a certain amount of time has elapsed after another event has occurred, or until a specific date.

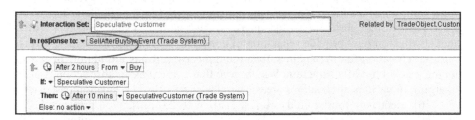

Fig 9. WBE definition of temporal context.

WBE also supports temporal event evaluation within context applying filtering conditions to the rule. This is done using temporal functions such as: **Follows After, Follows before, and Follows Within**. These functions determine if the specified event in the associated scope will follow the event or action defined in the filter after/before/within certain amount of time. Functions such as**: Is Present** before, **Is Present After, Is Present Within**, determine if the event specified in the associated scope has previously occurred in the same context after/before/within the date/time period defined in the filter. An example of the Follows By construct is shown in Fig. 10.

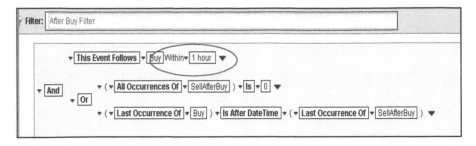

Fig 10. WBE Follows By construct.

To conclude, most CEP offerings today support one or another variation on temporal and segmentation context dimensions. While in most cases state-oriented context can be implemented in those solutions indirectly, using event data and filtering on this data, out-of-the-box geospatial capabilities are lacking in most of the mentioned products.

8 Conclusion

This chapter has discussed various aspects of contexts in event processing. We have presented the general notion of context, a general view of context in event processing, and a sample of implementations of this notion within various products.   As can be seen, most of the current languages has many of the context functions, but part of all these functions are not implemented as separate abstraction, but as part of other abstractions (like queries); the benefit of using context as a distinct abstraction may provide separation of concerns, and higher flexibility, and is consistent with current trends in software engineering; it has a trade-off of introducing a new kind of abstraction. We observe that context as an abstraction is also emerging in various applications areas: wireless networks [1], location based services [6], electronic tourist guidance [7], [15], Web 2.0 [16], Web presence [10], Business process management and design [13], [14], healthcare monitoring [12], telecommunication [25] and payment systems [26]. These works on context are expected to merge with the work existing in event processing and; Furthermore, We believe that the next generation of event processing as well as computing will view context as a separate abstraction. The experience gained in applying context in event processing in event processing will assist with the standardization of the semantics of context for the various usages of context.

References

1. Abowd, G.D., Atkeson, C.G., Hong, J., Long, S., Kooper, R., Pinkerton, M.: Cyberguide: A mobile context-aware tour guide. Wireless Networks: special issue on mobile computing and networking: selected papers from MobiCom 3(5), 421–433 (1997)

2. Adi, A., Biger, A., Botzer, D., Etzion, O., Sommer, Z.: Context Awareness in Amit. In: 5th Annual International Workshop on Active Middleware Services (AMS 2003), pp. 160–167 (2003)
3. Arasu, A., Babcock, B., Babu, S., Datar, M., Ito, K., Nishizawa, I., Rosenstein, J., Widom, J.: STREAM: The Stanford Stream Data Manager. In: SIGMOD Conference 2003, p. 666 (2003)
4. Buvac, S., Mason, I.A.: Propositional logic of context. In: Proc. of the 11th National Conference on Artificial Intelligence (1993)
5. Buvac, S.: Quantificational Logic of Context. In: Proc. of the 13th National Conference on Artificial Intelligence (1996)
6. Broadbent, J., Marti, P.: Location-Aware Mobile Interactive Guides: Usability Issues. In: Proc. Inter. Cultural Heritage Informatics Meeting, Paris, France (1997)
7. Heverst, K., Davies, N., Mitchell, K., Friday, A., Efstratiou, C.: Developing a Context-Aware Electronic Tourist Guide: some issues and experiences. In: Proc. CHI, The Hague, Netherlands (2000)
8. Etzion, O., Niblett, P.: Event Processing in Action. Manning Publications (2010)
9. Jensen, C.S., et al.: The Consensus Glossary of Temporal Database Concepts - February 1998 Version. In: Etzion, O., Jajodia, S., Sripada, S. (eds.) Temporal Databases Research and Practice, pp. 367–405. Springer, Heidelberg (1998)
10. Kindberg, T., et al.: People, Places, Things: Web Presence for the Real World. In: Proc. 3rd Annual Wireless and Mobile Computer Systems and Applications, Monterey, CA (2000)
11. Luciano, S., Fausto, G.: ML Systems: A Proof Theory for Contexts. Journal of Logic, Language and Information 11, 471–518 (2002)
12. Mohomed, I., Misra, A., Ebling, M., Jerome, W.: Context-aware and personalized event filtering for low-overhead continuous remote health monitoring. In: International Symposium on a World of Wireless, Mobile and Multimedia Networks (2008)
13. Rosemann, M., Recker, J., Flender, C., Ansell, P.: Context-Awareness in Business Process Design. In: 17th Australasian Conference on Information Systems, Adelaide, Australia (2006)
14. Rosemann, M., Recker, J., Flender, C.: Contextualization of Business Processes. International Journal of Business Process Integration and Management 3, 47–60 (2008)
15. Woodruff, A., Aoki, P.M., Hurst, A., Szymanski, M.H.: Electronic Guidebooks and Visitor Attention. In: Proc. 6th Int. Cultural Heritage Informatics Meeting, Milan, Italy, pp. 437–454 (2001)
16. Context in Web 2.0 realm, http://blog.roundarch.com/2010/03/25/in-the-realm-of-web-2-0-context-is-king-part-one/
17. StreamSQL-Streambase EP language, http://dev.streambase.com/developers/docs/sb37/streamsql/index.html
18. Oracle EPL-Oracle EP language, http://download.oracle.com/docs/cd/E13157_01/wlevs/docs30/epl_guide/overview.html
19. IBM Websphere Business Events Context configuration, http://publib.boulder.ibm.com/infocenter/wbevents/v7r0m0/index.jsp?topic=/com.ibm.wbe.install.doc/doc/configuringtheruntime.html
20. RuleCore CEP Server, http://rulecore.com/content/view/35/52/
21. IBM Websphere Business Events Overview, http://www01.ibm.com/software/integration/wbe/features/?S_CMP=wspace

22. Sybase-Aleri Continuous Computation Language (CCL),
    `http://www.aleri.com/products/aleri-cep/coral8-engine/ccl`
23. Sybase-Aleri CCL context definition, `http://www.aleri.com/WebHelp/`
    `programming/programmers/c8pg_using_windows.html`
24. Context Delivery Architecture by Gartner,
    `http://www.gartner.com/DisplayDocument?id=535313`
25. Context aware mobile communications system, `http://www.geovector.com/`
26. Context in mobile payment systems, `http://www.nttdocomo.com/`
27. Oracle CEP overview, `http://www.oracle.com/technologies/`
    `soa/complex-event-processing.html`

Event Processing over Uncertain Data

Avigdor Gal, Segev Wasserkrug, and Opher Etzion

Abstract. Events are the main input of event-based systems. Some events are generated externally and flow across distributed systems, while other events and their content need to be inferred by the event-based system itself. Such inference has a clear trade-off between inferring events with certainty, using full and complete information, and the need to provide a quick notification of newly revealed events. Timely event inference is therefore hampered by the gap between the actual occurrences of events, to which the system must respond, and the ability of event-based systems to accurately infer these events. This gap results in uncertainty and may be attributed to unreliable data sources (*e.g.*, an inaccurate sensor reading), unreliable networks (*e.g.*, packet drop at routers), the use of fuzzy terminology in reports (*e.g.*, normal temperature) or the inability to determine with certainty whether a phenomenon has occurred (*e.g.*, declaring an epidemic). In this chapter we present the state-of-the-art in event processing over uncertain data. We provide a classification of uncertainty in event-based systems, define a model for event processing over uncertain data, and propose algorithmic solutions for handling uncertainty. We also define, for demonstration purposes, a simple pattern language that supports uncertainty and detail open issues and challenges in this research area.

1 Introduction

This chapter covers a topic of increasing interest in the event processing research and practice communities. Event processing typically refers to an approach to software systems that is based on event delivery, and that includes specific logic to filter,

Avigdor Gal
Technion – Israel Institute of Technology, Faculty of Industrial Engineering & Management, Technion City, 32000 Haifa, Israel
e-mail: avigal@ie.technion.ac.il

Segev Wasserkrug
IBM Haifa Research Lab, Haifa, Israel
e-mail: segevw@il.ibm.com

S. Helmer et al.: Reasoning in Event-Based Distributed Systems, SCI 347, pp. 279–304.
springerlink.com   © Springer-Verlag Berlin Heidelberg 2011

transform, or detect patterns in events as they occur. A first generation of event processing platforms is diversified into products with various approaches towards event processing, including the stream oriented approach, such as: StreamSQL Event-Flow (StreamBase),[30] CCL (Sybase),[31] Oracle CEP (Oracle),[24] and Stream Processing Language (IBM) [29]; the rule oriented approach that is implemented in products such as: AutoPilot M6 (Nastel),[4] Reakt (ruleCore),[28] TIBCO BusinessEvents (TIBCO),[33] and Websphere Business Events (IBM) [38]); the imperative approach is implemented in products such as Apama (Progress Software) [3] and Netcool Impact Policy Language (IBM) [23]; and, finally, the publish-subscribe approach that is part of Rendezvous (TIBCO),[32] Websphere Message Queue (IBM),[39] and RTI Data Distribution System (RTI).[27] As a common denominator, all of these approaches assume that all relevant events are consumed by the event processing system, all events reported to the system have occurred, and processing of event processing systems can be done in a deterministic fashion. These assumptions hold in many of the applications first generation platforms support.

Moving to the second generation of event processing platforms, one of the required characteristics is the extension of the range of applications that employ event processing platforms beyond the early adopters. Some of the target applications deviate from the basic assumptions underlying the first generation of event processing platforms in both functional and non-functional aspects. In this chapter we focus on a specific functional aspect, the ability to deal with inexact reasoning [11]. We motivate this requirement with two examples.

Example 1. An increasing number of event processing systems are based on Twitter feeds as raw events, by using either structured Twitter feeds, or using tags, *e.g.,* a bus notiyfing Twitter every time it arrives at a station;[1] automated trading decision applications, based on analysis of Tweets about traded companies,[2] *etc..* One cannot assume that the collection of events sent to Twitter is complete. Furthermore, one cannot assume it is accurate, as some tweets may be of rumor types, some may contain inaccurate information, and some are even sent by malicious sources. Yet, in the highly competitive world of trading, Twitter events are considered to be a good source of information, given that the processing can take into account all these possible inaccuracies.

Example 2. A service provider is interested in detecting the frustration of a valued customer in order to mitigate the risk of the customer deserting. In some cases there is an explicit event where a customer calls and loudly expresses dissatisfaction, however in most cases this is inferred from detecting some patterns over the event history. Assume that practice (or applying some machine learning techniques) concluded that if a customer approaches the customer service center three times within a single day about the same topic, this customer is frustrated. The fact that the customer is frustrated is the business situation that the service provider wants

---

[1] http://twitter.com/hursleyminibus

[2] http://www.wallstreetandtech.com/data-management/
showArticle.jhtml?articleID=218101018

to identify, while the pattern detected on the customer interaction event is only an approximation of this situation.

In the absence of support for inexact reasoning, applications as those described above may suffer, either directly or indirectly, from incorrect situation detection. Using current technology, there are four different ways to support such situations of uncertainty. First, situations of uncertainty can simply be ignored. Such a solution may be cost-effective if situations of uncertainty are relatively infrequent, and the damage of not handling them is not substantial. For example, in network management systems problem indications such as device time-out may be lost, but such events are not critical since they are issued on a recurring basis.

A second solution for handling uncertainty requires situations to be created only when the event pattern is a necessary and sufficient condition to detect the situation in a deterministic way. This is the case in many current systems. According to a third solution, the system is designed so that some detected situations require reinforcement from multiple indications. For example, in a fraud detection system, often a fraud suspicion requires reinforcement from multiple patterns, and possibly within the context of a customer's history. This is useful when false positives should be minimized, and it comes at the cost of false negatives.

Finally, it is possible to notify the result of each situation detection to a human observer that needs to decide whether an action should be taken. Again, this method is aimed at minimizing false positives.

The applications motivating a second generation of event processing platforms include applications in which false positives and false negatives may be relatively frequent, and the damage inflicted by these cases may be substantial, sometimes critical. For example, a stock value of a company may collapse if automated trading decisions are based on false rumors. While the motivation exists, the handling of inexactness in event processing in general is still a challenge in the current state-of-the-art and in this chapter we explore quantitative methods to manage inexactness.

The rest of the chapter is structured in the following way: Section 2 provides basic terminology to be used in this chapter; Section 3 provides a taxonomy of uncertainty cases and the handling of uncertain events is discussed in Section 4; Section 5 discusses the handling of uncertain situations and Section 6 describes an algorithm for uncertain derivation of events; We conclude with research challenges in Section 7.

2 Preliminaries

In this section we provide some basic concepts in event processing (Section 2.1) and uncertainty handling (Section 2.2). Throughout this chapter we shall use the following two examples for demonstration.

Example 3. A thermometer generates an event whenever the temperature rises above 37.5°C. However, the thermometer is known to be accurate to within ±0.2°C. Therefore, when the temperature measured by the thermometer is 37.6°C, there is some uncertainty regarding whether the event has actually occurred.

Example 4. Consider an e-Trading Web site, where customers can buy and sell stocks, check their portfolio and receive information regarding the current price of any stock. We would like to identify a variety of events, including that of speculative customers (illegal trading events) and customers becoming dissatisfied (CRM – Customer Relationship Management – related events).

2.1 Event Processing

We base our description of basic concepts in event processing on the model proposed by Etzion and Niblett [11].

An *event* is an occurrence within a particular system or domain; it is something that has happened, or is contemplated as having happened in that domain. The word event is also used to mean a programming entity that represents such an occurrence in a computing system. We classify events as either raw or derived events. A *raw event* is an event that is introduced into an event processing system by an external event producer. A *derived event* is an event that is generated as a result of event processing that takes place inside the event processing system. A derived event can be generated either by event transformation or event pattern matching. An event transformation transforms one or more event inputs into one or more event output by translation, enrichment from external data source, aggregation, composition or splitting. Example of event transformation is an aggregation that finds the average of a temperature measurement over a one hour shifting window.

An *event pattern* is a template, specifying one or more combinations of events. Given any collection of events one may be able to find one or more subsets of those events that match a particular pattern. We say that such a subset *satisfies* the pattern. An example of an event pattern is a sequence of events of type "buy stock" followed by an event of type "sell stock." The pattern matching process in this case creates pairs of events of these two types that match this pattern, satisfying all matching conditions (same stock, same customer, same day), defined to be the *pattern matching set*. A pattern matching set can serve as a *composite event*, composed of all member events in the event set, *e.g.*, the two matched event {buy stock, sell stock}. Alternatively, it can yield an event, created as some transformation of the matching set. For example, the event can be a newly created event type, containing some information from the payloads of both events, such as: ⟨customerid, buyamount, sellamount⟩.

A *situation* is an event occurrence that might require a reaction and is consumed by an external event consumer. In current systems, a situation can either be a raw event or a derived event, however, in some cases a derived event is merely an approximation of a situation. An example of a situation may be the detection of a speculative customer, with the condition that a speculative customer is a customer that buys and then sells the same stock on the same day, both transactions exceeding $1M.

The linkage between an event pattern and a situation might suffer from false positive or false negative phenomena. *False negative situation detection* refers to cases in which a situation occurred in reality, but the event representing this situation was not emitted by an event processing system. In the example discussed above, the amount paid for a stock was recorded to be slightly less than \$1M, and yet the customer was indeed a speculative customer. In this case, the event processing system did not detect it. *False positive situation detection* refers to cases in which an event representing a situation was emitted by an event processing system, but the situation did not occur in reality. In our example, the purchase was recorded erroneously, and was later corrected, yet the event system declared the customer to be speculative. Two of the main goals of an uncertainty handling mechanism for events are a) making explicit, and b) quantifying, the knowledge gap between an event pattern and the corresponding situation.

A variety of data can be associated with the occurrence of an event. Two examples of such data are: The point in time at which an event occurred, and the new price of a stock in an event describing a change in the price of a specific stock. Some data are common to all events (*e.g.*, their time of occurrence), while others are specific only to some events (*e.g.*, data describing stock prices are relevant only for stock-related events). The data items associated with an event are termed *attributes*. In what follows, *e.attributeName* denotes the value of a specific attribute of a specific event e. For example, $e_1.occT$ refers to the occurrence time of event e_1. In addition, the type of event e is denoted by $e \in type$.

2.2 Uncertainty Management Mechanisms

There is a rich literature on mechanisms for uncertainty handling, including, among others, *Lower and Upper Probabilities*, *Dempster-Shafer Belief Functions*, *Possibility Measures* (see [17]), *Fuzzy Sets* and *Fuzzy Logic* [41]. We next present, in more details, three common mechanisms that were applied in the context of event processing, namely probability theory, fuzzy set theory, and possibility theory.

2.2.1 Probability Theory

The most well known and widely used framework for quantitative representation and reasoning about uncertainty is probability theory. An intuitively appealing way to define this probability space involves possible world semantics [13]. Using such a definition, a probability space is a triple $pred = (W, F, \mu)$ such that:

- W is a set of possible worlds, with each possible world corresponding to a specific set of event occurrence that is considered possible. A typical assumption is that the real world is one of the possible worlds.
- $F \subseteq 2^{|W|}$ is a σ-algebra over W. σ-algebra, in general, and in particular F, is a nonempty collection of sets of possible worlds that is closed under complementation and countable unions. These properties of σ-algebra enable the definition of a probability space over F.
- $\mu : F \to [0, 1]$ is a probability measure over F.

We call the above representation of the probability space the *possible world representation*. One problem with possible worlds semantics is performance. Performing operations on possible worlds can lead to an exponential growth of alternatives. In

Section 5.1 we present an alternative representation to the possible worlds semantics and in Section 6 we use this alternative representation to compute efficiently probabilities of complex events.

The most common approach for quantifying probabilities are Bayesian (or belief) networks [25]. However, Bayesian networks are only adequate for representing propositional probabilistic relationships between entities. In addition, standard Bayesian networks cannot explicitly model temporal relationships. To overcome these limitations, several extensions to Bayesian networks have been defined, including Dynamic Belief Networks [19] , Time Nets [18], Modifiable Temporal Belief Networks [8] and Temporal Nodes Bayesian Networks [14]. Although these extensions are more expressive than classical Bayesian networks, they nonetheless lack the expressive power of first-order logic. In addition, some of these extensions allow more expressive power at the expense of efficient calculation.

Another formal approach to reasoning about probabilities involves probabilistic logics (*e.g.*, [5] and [16]). These enable assigning probabilities to statements in first-order logic, as well as inferring new statements based on some axiomatic system. However, they are less suitable as mechanisms for the calculation of probabilities in a given probability space.

A third paradigm for dealing with uncertainty using probabilities is the KBMC (Knowledge Based Model Construction) paradigm [7]. This approach combines the representational strength of probabilistic logics with the computational advantages of Bayesian networks. In this paradigm, separate models exist for probabilistic knowledge specification and probabilistic inference. Probabilistic knowledge is represented in some knowledge model (usually a specific probabilistic logic), and whenever an inference is carried out, an inference model would be constructed based on this knowledge.

2.2.2 Fuzzy Set Theory

The background on fuzzy set theory is based on [41, 12, 22]. A fuzzy set M on a universe set U is a set that specifies for each element $x \in U$ a degree of membership using a membership function

$$\mu_M : U \to [0,1]$$

For example, considering Example 3, the membership function that assigns a value to the reading of a thermometer can be represented as a bell shape over the range $[37.3°C, 37.7°C]$, with higher membership value in the center ($37.5°C$), slowly decreasing to 0 on both sides. It is worth noting that, unlike probability theory, the area under the curve does not necessarily sum to 1.

2.2.3 Possibility Theory

Possibility theory formalizes users' subjective uncertainty of a given state of the world [10]. Therefore, an event "customer x is frustrated" may be associated with a confidence measure $\pi_{\text{frustrated}}(x)$. Both fuzzy set and possibility theories use a numerical measure, yet they express different uncertainties. Fuzzy set theory is more suitable to represent vague description of an object (*e.g.*, value of a temperature reading) and possibility measures define the subjective confidence of the state in the world (*e.g.*, the occurrence of an event).

In [22], two measures were defined to describe the matching of a subscription to a publication and can be easily adopted to event processing under uncertainty. The *possibility measure* (Π) expresses the plausability of an event occurrence. The *necessity measure* (N) expresses the necessity of occurrence of an event e or, formulated differently, the impossibility of the complement of e. If it is completely possible to have occurred then possibility is $\Pi(e) = 1$. If it is impossible then the possibility is $\Pi(e) = 0$. Intermediate numbers in $[0, 1]$ represent an intermediate belief in event occurrence. A necessity measure is introduced to complement the information available about the state described by the attribute. The relationship between possibility and necessity satisfies:

$$N(e) = 1 - \Pi(\bar{e})$$
$$\forall e, \Pi(e) \geq N(e)$$

where $\bar{e}$ represents the complement of e. It is worth noting that if $\Pi(e)$ is a probability distribution, then $\Pi(e) = N(e)$.

2.2.4   Discussion

The literature carries heated debates about the role of fuzzy sets framework and probabilistic methods. A probabilistic-based approach assumes that one has an incomplete knowledge on the portion of the real world being modeled. However, this knowledge can be encoded as probabilities about events. The fuzzy approach, on the other hand, aims at modeling the intrinsic imprecision of features of the modeled reality. Therefore, the amount of knowledge at the user's disposal is of little concern. In addition to philosophical reasoning, the debate also relates to pragmatics. Probabilistic reasoning typically relies on event independence assumptions, making correlated events harder to assess. Results presented in [9] show a comparative study of the capabilities of probability and fuzzy methods. This study shows that probabilistic analysis is intrinsically more expressive than fuzzy sets. However, fuzzy methods demonstrate higher computational efficiency.

3   Taxonomy of Event Uncertainty

This section provides a taxonomy of event uncertainty. Section 3.1 defines two dimensions to classify uncertainties as relating to events and Section 3.2 describes the causes of event uncertainties for these dimensions.

3.1 Dimensions of Event Uncertainty

We classify the uncertainty according to two orthogonal dimensions: *Element Uncertainty* and *Origin Uncertainty*. The first dimension, *Element Uncertainty*, refers to the fact that event-related uncertainty may involve one of two elements:

Uncertainty regarding event occurrence: Such uncertainty is associated with the fact that although the actual event occurrence is atomic, the system does not know whether or not this event has in fact occurred. One example of such an event is the thermometer reading event from Example 3. Another example is money laundering, where at any point in time, money laundering may have been carried out by some customer. However, a Complex Event Processing (CEP) system can probably never be certain whether money laundering actually took place.

Uncertainty regarding event attributes: Even in cases in which the event is known to have occurred, there may be uncertainty associated with its attributes. For example, while it may be known that an event has occurred, its time of occurrence may not be precisely known. As another example, an event may be associated with a fuzzy domain, stating that a temperature is mild, in which case there is uncertainty regarding the exact temperature.

The second dimension, *Origin Uncertainty*, pertains to the fact that in a CEP system, there may be two types of events, raw events, signalled by event sources, and derived events, which are inferred based on other events. In the following, events which serve as the basis for the inference of other events will be termed *evidence events*, be they raw or derived events. Therefore, there are two possible origins for uncertainty:

Uncertainty originating at the event source: For raw events, there may be uncertainty associated either with the event occurrence itself, or the event's attributes, due to a feature of the event source. Example 3, in which uncertainty regarding an event occurrence is caused by the limited measuring accuracy of a thermometer, illustrates such a case.

Uncertainty resulting from event inference: Derived events are based on other events and uncertainty can propagate to the derived events. This is demonstrated by Example 2, in which uncertainty regarding measures of frustration of a customer propagates to the uncertainty of a customer deserting event.

Two additional examples are given next. In Example 5, the uncertainty of an event originates from the source, but is limited to its attributes, rather than to its occurrence. Example 6 shows uncertainty regarding an event's attributes resulting from event inference.

Example 5. Consider a case in which an event is generated whenever the temperature reading changes. Assume that the thermometer under discussion has the same accuracy as the one defined in Example 3. Furthermore, assume that the new temperature is an attribute of this event. In this case, there is no uncertainty regarding the actual occurrence. There is only uncertainty regarding this new temperature attribute.

Table 1 Uncertainty classification

	Origin: Event Source	Origin: Event Inference
Uncertainty Regarding Event Occurrence	Unreliable Source Imprecise Source Problematic Communication Medium Estimates	Propagation of Uncertainty Uncertain Rules
Uncertainty Regarding Event Attributes	Unreliable Source Imprecise Source Problematic Communication Medium Estimates Time Synchronization in Distributed Systems	Propagation of Uncertainty

Example 6. Consider a case in which an event e_3 should be inferred whenever an event e_2 occurs after an event of type e_1. Assume that the inferred event e_3 has an attribute a_1^3 whose value is the sum of the value of the attribute a_1^1 of event e_1 and the value of the attribute a_1^2 of event e_2. Now assume that both e_1 and e_2 are known to have occurred with certainty, but there is uncertainty regarding the value of attribute a_1^1. In this case there is uncertainty only regarding the value of attribute a_1^3, and this uncertainty results from event inference.

The above examples demonstrate that the two dimensions are indeed orthogonal. Therefore, uncertainty associated with events could be mapped into one of four quadrants, as shown in Table 1. In addition, due to the orthogonality of these dimensions, we define four types of event uncertainty: *Uncertainty regarding event occurrence originating at an event source, uncertainty regarding event occurrence resulting from inference, uncertainty regarding event attributes originating at an event source*, and *uncertainty regarding event attributes resulting from event inference*.

3.2 Causes of Event Uncertainty

This section describes, at a high level, the various causes of event uncertainty, according to the dimensions defined in Section 3.1. Table 1 summarizes the causes of uncertainty.

3.2.1 Causes of Uncertainty Originating at the Source

Uncertainty regarding event *occurrence* originating at an event source is caused by one of the following:

An unreliable source: An event source may malfunction and indicate that an event has occurred even if it has not. Similarly, the event source may fail to signal the occurrence of an event which has, in fact, occurred.

An imprecise event source: An event source may operate correctly, but still fail to signal the occurrence of events due to limited precision (or may signal events that did not occur). This is illustrated by Example 3.

Problematic communication medium: Even if the event source has full precision, and operates correctly 100% of the time, the communication medium between the source and the active system may drop indications of an event's occurrence, or generate indications of events that did not occur.

Uncertainty due to estimates: In some cases, the event itself may be a result of a statistical estimate. For example, it may be beneficial to generate an event whenever a network Denial of Service (DoS) event occurs, where the occurrence of such a DoS event is generated based on some mathematical model. However, the reasoner that deduce the event occurrence may produce a false positive type of error and hence this event also has uncertainty associated with it.

Uncertainty regarding the *attributes* originating at the event source can also be caused by any of the above reasons. An unreliable or imprecise source may be unreliable or imprecise regarding just the attribute values. Similarly, the communication medium may garble just the values of attributes, rather than messing with event occurrence. Finally, estimates or fuzzy values may also result in uncertainty regarding event attributes.

In distributed systems, there exists an additional cause for uncertainty regarding the special attribute capturing the occurrence time of the event. This is due to the fact that in distributed systems, the clocks of various nodes are usually only guaranteed to be synchronized to within some interval of a global system clock [21]. Therefore, there is uncertainty regarding the occurrence time of events as measured according to this global system clock.

It is worth noting that in both of the above cases, uncertainty regarding a specific event may be caused by a combination of factors. For example, it is possible that both the event source itself and the communication medium simultaneously corrupt information sent regarding the same event.

3.2.2 Causes of Inferred Uncertainty

Uncertainty regarding event *occurrence* resulting from inference has the following two possible causes:

1. **Propagation of Uncertainty:** A derived event can be a result of a deterministic pattern. However, when there is uncertainty regarding the events that are used for the derivation, there is also uncertainty regarding the derived event.
2. **Uncertain Patterns:** The pattern itself may be defined in an uncertain manner, whenever an event cannot be inferred with absolute certainty based on other events. An example of this is money laundering, where events denoting suspicious transactions only serve to indicate the possible occurrence of a money laundering event. Usually, a money laundering event cannot be inferred with certainty based on such suspicious transactions.

Note that these two causes may be combined. That is, it may happen that not only is the inference of an event based on an uncertain pattern, but also uncertainty exists regarding the occurrence (or attribute values) of one of the events which serve as evidence for this inference.

Regarding uncertainty of derived event *attributes*, the possible causes depend on how these attributes are calculated from the attributes of the evidence events. The most intuitive way to calculate such derived attributes is using deterministic functions defined over the attributes of the evidence events. In such a case, the only cause of uncertainty in the derived attributes is the propagation of uncertainty from the attributes of the evidence events. This is because the uncertainty regarding the event attributes is defined to be the uncertainty regarding the attribute values given that the event occurred. Therefore, the pattern cannot induce uncertainty regarding the attribute values of the derived events. However, if the inference system makes it possible to define the attribute values of the derived events in a non-deterministic (*e.g.*, fuzzy) manner, uncertain patterns may also be a cause of derived attribute uncertainty.

4 Handling Uncertainty at the Source

In this section we present two models for representing uncertainty at the source.

4.1 Probability Theory-Based Representation

The model presented next provides a probability theory-based representation for event uncertainty, based on [35]. A similar model was presented later by Balazinska *et al.* in [6] for RFID data. Arguing in favor of probability theory includes following reasons:

- Probability theory has widespread acceptance.
- Probability theory is a well-understood and powerful tool.
- Many technical results that facilitate its use have been shown formally.
- Under certain assumptions, probability is the only "rational" way to represent uncertainty (see [17]).
- There are well-known and accepted methodologies for carrying out inferences based on probability theory, involving structures such as Bayesian networks.
- Probability theory can be used together with utility theory (see [26]) for automatic decision making. Such automatic decision making would facilitate the implementation of automatic actions by complex event processing systems.

To apply probability theory to the handling of event processing in the context of uncertainty the notion of an event has to be extended, to allow the specification of uncertainty associated with a specific event. Also, deriving events in the context of uncertainty needs to be defined.

We represent the information a composite event system holds about each event instance with a data structure we term *Event Instance Data* (*EID*). *EID* incorporates all relevant data about an event, including its type, time of occurrence, *etc*. In event composition systems with no uncertainty, each event can be represented by a single tuple of values $Val = \langle val_1, \ldots, val_n \rangle$, one value for each attribute associated with the event (see Section 2.1 for the introduction of event attributes). In our case, to capture the uncertainty associated with an event instance, the *EID* of each event instance is a Random Variable (*RV*). The possible values of *EID* are taken from the domain $V = \{notOccurred\} \cup V'$, where V' is a **set** of tuples of the form $\langle val_1, \ldots, val_n \rangle$.

The semantics of a value of E (encoded as an *EID*), representing the information the system has about event e are as follows: The probability that the value of E belongs to a subset $S \subseteq V \setminus \{notOccurred\}$ is the probability that event e has occurred, and that the value of its attributes is some tuple of the form $\langle val_1, \ldots, val_n \rangle$, where $\{\langle val_1, \ldots, val_n \rangle\} \subseteq S$. Similarly, the probability associated with the value $\{notOccurred\}$ is the probability that the event did not occur.

Example 7. Consider an event that quotes a price of \$100 for a share of IBM stock at time 10:45. Say that the system considers the following possible: The event did not occur at all; the event occurred at time 10:34 and the price of the stock was \$105; and the event occurred at time 10:45 and the price was \$100. In addition, say that the system considers the probabilities of these possibilities to be 0.3, 0.3 and 0.4, respectively. Then, the event can be represented by an RV E whose possible values are $\{notOccurred\}, \{10{:}34, IBM, 105\}$, and $\{10{:}45, IBM, 100\}$. Also, $\Pr(E = \{notOccurred\})$ and $\Pr\{E = \{10:34, IBM, 105\}\}$ are both 0.3 and $\Pr\{E \in \{\{10:45, IBM, 100\}, \{10:34, IBM, 105\}\}\} = 0.7$.

Example 8. In case of the thermometer related event appearing in Example 3, assume the following: the conditional probability that the temperature is 37.3°C, given that the thermometer reads 37.5°C, is 0.1, the probability that the temperature is 37.4°C is 0.15, the probability that the temperature is 37.5°C is 0.5, the probability that the temperature is 37.6°C is 0.15, and the probability that the temperature is 37.7°C is 0.1. The probability that the event did not occur is 0.3 and the probability that the event did occur is 0.7. Moreover, assume that whenever the event does occur, the temperature is an attribute of this event. Assume that at time 5, the thermometer registers a reading of 37.5°C. This information would be represented by an EID E with the following set of values: $\{notOccurred\}$ - indicating that the event did not occur, and 5 value sets of the form $\{5, 37.X°C\}$ - indicating that the event occurred at time 5 with temperature 37.X°C, and X stands for 1, 2, 3, 4 or 5. Examples of probabilities defined over E are $\Pr(E = \{notOccurred\}) = 0.3$ and $\Pr(E = \{5, 37.5°C\}) = 0.5 \cdot 0.7 = 0.35$ (due to the removal of the conditioning).

The set of possible values of the EID RVs contains information regarding both the occurrence of the event and its attributes. Therefore, this representation is sufficient to represent the uncertainty regarding both occurrence and attributes.

An additional concept, relevant in the context of event composition systems, is that of *Event History* (*EH*). An event history $EH_{t_1}^{t_2}$ is the set of all events (of interest to the system), as well as their associated data, whose occurrence time falls between t_1 and t_2. For example, consider the following events: an event e_1, at time 10:30, quoting the value of an IBM share as \$100; event e_2, at time 10:45, quoting the value of an IBM share as \$105; event e_3 at time 11:00, quoting the value of an IBM share as \$103. Using the notation described above, the events e_1, e_2, e_3 can be described by the tuples $\{10:30, IBM, 100\}$, $\{10:45, IBM, 105\}$, and $\{11:00, IBM, 103\}$. Examples of event histories defined on these events are the following: $EH_{10:30}^{10:45} = \{e_1, e_2\}$, $EH_{10:45}^{11:00} = \{e_2, e_3\}$, $EH_{10:30}^{10:35} = \{e_1\}$ and $EH_{10:30}^{11:00} = \{e_1, e_2, e_3\}$. It is worth noting that there does not exist an event history that consists of both e_1 and e_3, and that does not include e_2.

The actual event history is not necessarily equivalent to the information regarding the event history possessed by the system. For example, a thermometer reading of $37.5°C$ is not necessarily the "true" one, as illustrated in example 8. We will therefore make the distinction by denoting the event history possessed by the system by *system event history*.

4.2 Fuzzy Set Theory-Based Representation

In [22], a model for representing basic fuzzy events, both in subscriptions and in publications, is given. An event will be of the form "x is $\tilde{A}$" where x is an attribute of an event and $\tilde{A}$ is a fuzzy value, associated with a membership function. For example, following Example 8, an event of the form "*temperature* of sensor s is *normal*" can be defined with the following membership definition of the fuzzy term *normal*:

$$\mu_{normal}(x) = \begin{cases} 0 & \text{if } x < 37.3°C \\ 0.1 & \text{if } x = 37.3°C \\ 0.15 & \text{if } x = 37.4°C \\ 0.5 & \text{if } x = 37.5°C \\ 0.15 & \text{if } x = 37.6°C \\ 0.1 & \text{if } x = 37.7°C \\ 0 & \text{if } x > 37.7°C \end{cases}$$

It is worth highlighting the differences between the two approaches presented in this section. The example above provides an interpretation of a fuzzy term using a distribution of values. In Example 8, we provided an interpretation of possible values, given an exact reading. Given the different settings, there is a room for both representations, and investigating the best method for combining them is a topic for future research.

5 Handling Inference Uncertainty

Inference uncertainty is handled by extending the notion of an event pattern, used in deterministic event processing systems, to represent the uncertainty associated with event derivation. This is described in this section, based on [35]. Although the description uses probability theory as a basis for the extension, this framework can be adapted to the use of fuzzy set theory or possibility theory. In the next section we discuss methods for precise and efficient probability computation, as well as methods for evaluating complex fuzzy events.

In this chapter, we follow the KBMC approach, as introduced in Section 2.2: Knowledge is represented as probabilistic patterns (see Section 5.1), while probability calculation is carried out by constructing a Bayesian network based inference model (see Section 6).

Contemporary deterministic event processing systems use pattern matching for situation detection. In such systems, a situation is said to have occurred if the stream of incoming events match some pattern. At each point in time t, the existence of relevant patterns can be checked in the event histories that are known at that time. A pattern p is given in the form $\langle sel_p^n, pred_p^n, eventType_p, mappingExpressions_p, prob_p \rangle$ where:

sel_p^n is a deterministic predicate returning a subset of an event history of size less than or equal to n (for some integer n). If the returned subset has strictly less then n elements, no further evaluation of the pattern is carried out. A possible selection expression is "the first two events of type $stockQuote$." Therefore, for the event history e_1, e_2, e_3, if only e_1 is of type $stockQuote$ the pattern is not triggered. However, if both e_1 and e_3 are of type $stockQuote$, then the subset $\{e_1, e_3\}$ is selected, and evaluation of the pattern continues.

$pred_p^n$ is a predicate of arity n over event instances (note that this is the same n appearing in sel_p^n). This predicate is applied to the n event instances selected by sel_p^n. An example of $pred_p^n$ is "the events e_1, e_2, e_3 have occurred in the order e_1, e_2, e_3, all three events are events regarding the same stock, and event e_3 occurred no later than 5 minutes after event e_1."

$eventType_p$ is the type of event inferred by this pattern. It can be, for example, an event of type $speculativeCustomer$.

$mappingExpressions_p$ is a set of functions, mapping the attribute values of the events that triggered this pattern to the attribute values of the derived event.

$prob_p \in [0, 1]$ is the probability of inferring the derived event given that the pattern has occurred. The exact semantics of this probability are defined in Section 5.1.

By definition, the predicates defined by sel_p^n and $pred_p^n$ are deterministic, as are the functions $mappingExpressions_p$. Therefore, the only uncertainty present in the pattern is represented by the quantity $prob_p$. Indeed, many deterministic composite event languages, $e.g.$, the Situation Manager pattern Language [2] can be viewed as defining a set of patterns P such that each pattern $p \in P$ is of the form $\langle sel_p^n, pred_p^n, eventType_p, mappingExpressions_p \rangle$.

We assume that an event type is either explicit or inferred by a single pattern. This is done for simplicity. If there is more than one source of information for an event type ($e.g.$, two patterns), the probabilities supplied by the separate sources (patterns) must be combined to create a well-defined probability space (see Section 5.1).

We conclude this section with the definition of a specific language, instantiating each part of the rule, including the allowable syntax and semantics of s_r, p_r, etc. In this language, we have the following:

- sel_p^n is of the form $\langle selExpression_1, \ldots, selExpression_n \rangle$, where $selExpression_i$ is a selection expression of the form $\varepsilon_i \in eventType_i$, with $eventType$ being a valid event type. Given an event history, $selExpression_i$ will select a single event, ε_i. The event ε_i selected by $selExpression_i$ is the first event in the event history of type $eventType_i$ that was not selected by a selection expression $selExpression_j$ such that $j < i$.

- $pred_p^n$ is a conjunctive predicate defined over the events $\varepsilon_1, \ldots, \varepsilon_n$ selected by sel_p^n, of the form $\bigwedge_{i=1}^{m} predicate_i$. $predicate_i$ is either a temporal predicate, or an equality relation between attributes. If $predicate_i$ is an equality predicate, it is of the form $\varepsilon_k.attribute_l = \varepsilon_j.attribute_m$ for $k \neq j$. This specifies that the value of $attribute_l$ of event ε_k must be the same as $attribute_m$ of event ε_j. A temporal predicate $predicate_i$ takes one of the following forms:

  - $a \leq \varepsilon_k.occT \leq b$, where a and b are temporal constants denoting time points in the range $[0, \infty]$. This predicate specifies that the event has occurred within the interval $[a, b]$.
  - $\varepsilon_j.occT < \varepsilon_k.occT$ for $k \neq j$. This predicate defines a partial order over subsets of events.
  - $\varepsilon_j.occT \leq \varepsilon_k.occT \leq \varepsilon_j.occT + c$ for $k \neq j$, where c is a temporal constant such that $c > 0$. This predicate specifies that an event has happened within a specified interval relative to another event.

- Regarding $mappingExpression_r$ the occurrence time of the derived event is always determined to be the point in time at which the inference was carried out. As for other attributes, two types of functions are allowed. The first is a function, mapping a specific attribute value of a specific event participating in $pattern_r$ to an attribute of the derived event. The second is the mapping of a constant value to a derived attribute.

5.1   Event Inferencing Using Probability Theory

A valid probability space needs to be defined to quantify the probabilities associated with each event derived using an uncertain pattern. Therefore, pattern reasoning facilities need to be able to compute at any point in time t the probability that an event e, with specific data, occurred at some time $t' \leq t$. In addition, the only evidence that can be taken into account is that which is known to the system at time t. Therefore, a (possibly different) probability space is defined for each t. Extending our discussion in Section 2.2.1, the probability space at time t is a triple $pred_t = (W_t, F_t, \mu_t)$ such that:

- W_t is a set of possible worlds, with each possible world corresponding to a specific event history that is considered possible at time t. An assumption that holds in all practical applications is that the number of events in each event history, as well as the overall number of events, is finite. This is because an actual system cannot consider an infinite number of events in a finite time period. Therefore, each possible world corresponds to an event history that is finite in size. In addition, we assume that the real world is one of the possible worlds.

- $F_t \subseteq 2^{|W_t|}$ is a σ-algebra over W_t.
- $\mu_t : F_t \to [0,1]$ is a probability measure over F_t.

A less intuitive, yet more computationally useful way to define the probability space is as follows. Let $E_1, E_2, \ldots$ be the set of EIDs representing the information about all events of interest. It is clear that each finite event history can be represented by a finite number of values $e_1, \ldots, e_n$, such that there exists a finite number of EIDs $E_1, \ldots, E_n$ where e_i is a possible value of E_i. Therefore, each possible world $w_t \in W_t$ can be represented by such a finite number of values. In addition, as the overall number of events is finite, there is a finite number of events $E_1, \ldots, E_m$ such that E_i could have occurred in some $w_t \in W_t$. Finally, if $|W_t|$ is finite, each E_i can only have a finite number of associated values (one for each world in W_t) in which it appears. Note that in such a case, each possible w_t can be represented by a finite number of values $Val_1, \ldots, Val_m$, where the value $Val_1, \ldots, Val_n$ for some $n \leq m$ is a set of values, each such set representing the values of one of the n events that occurred in w_t, and $Val_{n+1}, \ldots Val_m$ are all $\{notOccurred\}$. From this it follows that if the probability space $pred_t$ represents the knowledge of the composite event system at time t, this knowledge can be represented by a set of m EIDs - $E_1, \ldots, E_m$.

Therefore, in the case where $|W_t|$ is finite, it is possible to define the probability space $pred_t$ as (Ω_t, F_t, μ_t') where:

- $\Omega_t = \{Val_1, \ldots, Val_m\}$ such that the tuple $Val_1, \ldots, Val_m$ is the set of values corresponding to an event history, where this event history is a possible world $w_t \in W_t$ as described above. Obviously, $|\Omega_t|$ is finite.
- $F_t = 2^{|\Omega_t|}$
- $\mu_t'(\{Val_1, \ldots, Val_m\}) = \mu_t(w_t)$ such that w_t is the world represented by $\{Val_1, \ldots, Val_m\}$

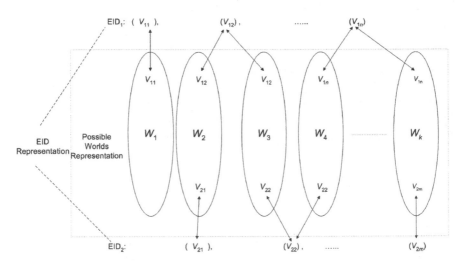

Fig. 1 Alternative probability space representation

This representation is termed the *EID representation*.

Figure 1 provides an illustration of two equivalent representations of the probability space. A possible world is marked as ovals in Figure 1. The figure presents two EIDs, one at the top and the other at the bottom. The participation of an event with a concrete value is marked by an arrow from the specific value to the possible worlds in which it participates. It is worth noting that the bottom event does not participate in world W_1, and therefore its value there is $\{notOccurred\}$.

As a concluding remark, note that each possible set of values of EIDs, $\{Val_1, \ldots, Val_m\}$ corresponds to some event history. Therefore, given an EID representation of $pred_t$, where $|\Omega_t|$ is finite, it is obviously possible to create the corresponding finite-size possible worlds representation by defining a possible world $w_t \in W_t$ for each distinct set of values $\{Val_1, \ldots, Val_m\}$.

We now define the semantics of patterns in the probability space discussed above. Intuitively, in such a probability space the semantics of each pattern p are as follows: Let $EH_{t_1}^{t_2}$ be an event history. If the pattern p is applied at some time $t \geq t_2$, and the set of events selected by sel_p^n from $EH_{t_1}^{t_2}$ is of size n and is such that $pred_p^n$ on this event is true, then the event inferred by pattern p occurred with probability $prob_p$. In addition, in such a case, the value of its corresponding attributes is the value defined by $mappingExpressions_p$. Otherwise, the event cannot be inferred.

Formally, let $sel_p^n(EH_{t_1}^{t_2})$ denote the set of events selected by sel_p^n from $EH_{t_1}^{t_2}$, and let $pred_p^n(sel_p^n(EH_{t_1}^{t_2}))$ denote the value of the predicate $pred_p^n$ on $sel_p^n(EH_{t_1}^{t_2})$ (recall that if $|sel_p^n(EH_{t_1}^{t_2})| < n$ then the pattern is not applied). In addition, let $val_1, \ldots, val_n$ denote the value of the attributes of the inferred event e_p as defined by $mappingExpressions_p$. Then, if the specific event history is known, and denoting by E_p the EID corresponding to e_p, we have the following:

$$\Pr(E_p = \{occurred, val_1, \ldots, val_n\} \mid EH_{t_1}^{t_2}) = prob_p \quad if\ pred_p^n(SEL_p^n(EH_{t_1}^{t_2})) = true \tag{1}$$

$$\Pr(E_p = \{notOccurred\} \mid EH_{t_1}^{t_2}) = (1 - prob_p) \quad if\ pred_p^n(SEL_p^n(EH_{t_1}^{t_2})) = true \tag{2}$$

$$\Pr(E_p = \{notOccurred\} \mid EH_{t_1}^{t_2}) = 1 \quad if\ pred_p^n(SEL_p^n(EH_{t_1}^{t_2})) = false \tag{3}$$

Recall from Section 5.1 that if $|W_t|$ is finite, the probability space can be represented by a finite set of random variables, each with a finite set of values. In addition, note that the pattern semantics defined above specify that the probability of the derived event does not depend on the entire event history, but rather on the events selected by sel_p^n. Therefore, let us denote by $E_1, \ldots, E_m$ the set of EIDs that describe knowledge regarding the event history, and assume, without loss of generality that $\{E_1, \ldots E_l\}$ describe the subset of $\{E_1, \ldots, E_m\}$ that are candidates for selection by sel_p^n (note that $l \geq n$, as sel_p^n must choose the first n events that have **actually occurred**). An EID E is a candidate for selection if there is a possible event history in the probability space $pred_t$ such that there is a set of n events which will be chosen by sel_p^n from this event history, and the event e corresponding to E is in this set. Then for all sets of values $\{Val_1, \ldots, Val_m\}$ such that $E_i = Val_i$, we have that

$$Pr(E_p|E_1, \ldots E_l) = Pr(E_p|E_1, \ldots, E_l, E_{l+1}, \ldots E_m) \tag{4}$$

i.e., E_p is conditionally independent of $\{E_{l+1}, \ldots, E_m\}$ given $\{E_1, \ldots E_l\}$. Now let $Val_1, \ldots, Val_l$ denote a specific set of values of $E_1, \ldots E_l$. Given such a set of specific values, the subset $\{e'_{j_1}, \ldots, e'_{j_n}\}$ selected by sel_p^n is well defined. Therefore, we have from the above equations that:

$$Pr(E_p = \{occurred, val_1, \ldots, val_n\}|Val_1, \ldots, Val_l) = prob_p \quad if \; pred_p^n(e'_{j_1}, \ldots, e'_{j_n}) = true \tag{5}$$

$$Pr(E_p = \{notOccurred\}|Val_1, \ldots, Val_l) = (1 - prob_p) \quad if \; pred_p^n(e'_{j_1}, \ldots, e'_{j_n}) = true \tag{6}$$

$$Pr(E_p = \{notOccurred\}| e_1, \ldots, e_l) = 1 \quad if \; pred_p^n(e'_{j_1}, \ldots, e'_{j_n}) = false \tag{7}$$

Eqs. 5-7 state that the inferred event and its values are conditionally probabilistically independent of all events prior to the inference, given exact information regarding the events that are candidates for selection by the corresponding selection expression.

As a concluding remark, we note that the mechanism of this section can also be used to predict future events, i.e., at each point in time t, events occurring at any point in time t' could be inferred, based on a set of probabilistic patterns described at the beginning of the chapter. In order to enable such prediction, however, prediction patterns should be defined that, at time t, given a set of events, infer events whose occurrence time t' is such that $t' > t$.

5.2   Event Inferencing Using Fuzzy Set Theory

Recall that in Section 4.2, we have defined events of the form "x is $\tilde{A}$," associated with a membership function $\mu_A(x)$. Let R be a relation, describing how complex events are evaluated. The membership function of a complex event s over a set of m events $(x_1, x_2, \ldots, x_m)$ is defined as follows:

$$M(x_1, x_2, \ldots, x_m) = R(\mu_{A_1}(x_1), \mu_{A_2}(x_2), \ldots, \mu_{A_m}(x_m))$$

R determines the method according to which membership values of different fuzzy sets can be combined. For example, consider the event "*temperature* of sensor s is *normal*" presented above, and assume another fuzzy event "*day t* is *hot*," with the following membership function:

$$\mu_{hot}(x) = \begin{cases} 0 & \text{if } x < 20°C \\ 0.1 & \text{if } 20°C \leq x < 25°C \\ 0.15 & \text{if } 25°C \leq x < 30°C \\ 0.75 & \text{if } 30°C \leq x \end{cases}$$

We can define a complex event "*temperature* of sensor s is *normal* and *day* t is *hot*" with the membership function

$$M(x_1, x_2) = \min\left(\mu_{normal}(x_1), \mu_{hot}(x_2)\right)$$

The min operator is the most well-known representative of a large family of operators called *triangular norms* (t-norms, for short), routinely deployed as interpretations of fuzzy conjunctions (see, for example, the monographs [20, 15]).

A *triangular norm* $T : [0,1] \times [0,1] \to [0,1]$ is a binary operator on the unit interval satisfying the following axioms for all $x, y, z \in [0,1]$:

$$T(x, 1) = x \text{ (boundary condition)}$$
$$x \leq y \text{ implies } T(x, z) \leq T(y, z) \text{ (monotonicity)}$$
$$T(x, y) = T(y, x) \text{ (commutativity)}$$
$$T(x, T(y, z)) = T(T(x, y), z) \text{ (associativity)}$$

The following t-norm examples are typically used as interpretations of fuzzy conjunctions:

$$Tm(x, y) = \min(x, y) \text{ (minimum t-norm)}$$
$$Tp(x, y) = x \cdot y \text{ (product t-norm)}$$
$$Tl(x, y) = \max(x + y - 1, 0) \text{ (Lukasiewicz t-norm)}$$

All t-norms over the unit interval can be represented as a combination of the triplet (Tm, Tp, Tl) (see [15] for a formal presentation of this statement). For example, the Dubois-Prade family of t-norms T^{dp}, also used often in fuzzy set theory and fuzzy logic, is defined using Tm, Tp and Tl as:

$$T^{dp}(x, y) = \begin{cases} \lambda \cdot Tp(\frac{x}{\lambda}, \frac{y}{\lambda}) & (x, y) \in [0, \lambda]^2 \\ Tm(x, y) & \text{otherwise} \end{cases}$$

The *average* operator belongs to another large family of operators termed *fuzzy aggregate operators* [20]. A fuzzy aggregate operator $H : [0,1]^n \to [0,1]$ satisfy the following axioms for every $x_1, \ldots, x_n \in [0,1]$:

$$H(x_1, x_1, \ldots, x_1) = x_1 \text{ (idempotency)} \qquad (8)$$

for every $y_1, y_2, \ldots, y_n \in [0,1]$ such that $x_i \leq y_i$,

$$H(x_1, x_2, \ldots, x_n) \leq H(y_1, y_2, \ldots, y_n) \text{ (increasing monotonicity)} \qquad (9)$$

H is a continuous function $\qquad (10)$

Let $\bar{x} = (x_1, \ldots, x_n)$ be a vector such that for all $1 \leq i \leq n$, $x_i \in [0,1]$ and let $\bar{\omega} = (\omega_1, \ldots, \omega_n)$ be a weight vector that sums to unity. Examples of fuzzy aggregate operators include the *average* operator $Ha(\bar{x}) = \frac{1}{n}\sum_1^n x_i$ and the *weighted average* operator $Hwa(\bar{x}, \bar{\omega}) = \bar{x} \cdot \bar{\omega}$. Clearly, *average* is a special case of the *weighted average* operator, where $\omega_1 = \cdots = \omega_n = \frac{1}{n}$. It is worth noting that Tm (the min t-norm) is also a fuzzy aggregate operator, due to its idempotency (its associative property provides a way of defining it over any number of arguments). However, Tp and Tl are not fuzzy aggregate operators.

T-norms and fuzzy aggregate operators are comparable, using the following inequality:

$$\min(x_1, \ldots, x_n) \leq H(x_1, \ldots, x_n)$$

for all $x_1, \ldots, x_n \in [0,1]$ and function H satisfying axioms 8-10.

Tm is the only idempotent t-norm. That is, $Tm(x,x) = x$.[3] This becomes handy when comparing t-norms with fuzzy aggregate operators. It can be easily proven (see [15]) that

$$Tl(x,y) \leq Tp(x,y) \leq Tm(x,y) \tag{11}$$

for all $x, y \in [0,1]$.

Following this discussion, it becomes clear that the space of possible computation methods for the similarity of complex events is large. Additional research is required to identify the best fuzzy operator for a given complex event.

6 Algorithms for Uncertain Inferencing

This section is devoted to the efficient inferencing in a setting of uncertainty. We provide an example of an algorithm for uncertain inferencing of events, following [36]. In a nutshell, the proposed algorithm works as follows: Given a set of rules and a set of EIDs at time t, a Bayesian network is automatically constructed, correctly representing the probability space at t according to the semantics defined in Section 5.1. Probabilities of new events are then computed using standard Bayesian network methods.

Two important properties that must be maintained in any algorithm for event derivation (both in the deterministic and uncertain setting), are determinism and termination [40]. Determinism ensures that for the same set of explicit EIDs, the algorithm outputs the same set of derived EIDs. Termination ensures that the derivation algorithm terminates.

An algorithm for constructing a Bayesian network in this setting should take into account two main features. First, the Bayesian network is dynamically updated as information about events reaches the system. This is to ensure that the constructed network reflects, at each time point t, the probability space at t. Second, throughout the inference process additional information beyond the Bayesian network is stored. This additional information is used both to allow an efficient dynamic update of the network, and to make the inference process more efficient.

---

[3] For a binary operator f, idempotency is defined to be $f(x,x) = x$ (similar to [20], pp. 36).

The set of patterns is assumed to have no cycles, and a priority is assigned to each pattern so that determinism is guaranteed. By a cycle we mean a set of complex events that are both determined and participate in the decision making of each other. Priority determines a unique ordering of rule activation and can be set using rule quasi-topological ordering [1].

Recall that by definition, the occurrence time of each derived event is the time in which the pattern was applied. Therefore, the single possible occurrence time of each EID defines a full order on the EIDs (this single point in time for EID E is denoted by $E.occT$). In addition, according to Eq. 4, the uncertainty encoded with each EID is independent of all preceding EIDs, given the EIDs that may be selected by the selection expression. Therefore, a Bayesian network is constructed such that the nodes of the network consist of the set of random variables in the system event history, and an edge exists between EID E_1 and EID E_2 iff $E_1.occT \leq E_2.occT$ and E_2 is an EID corresponding to an event that may be inferred by pattern p, where the event corresponding to E_1 may be selected by sel_p^n. A network constructed by these principles encodes the probabilistic independence required by Eq. 4 (see [25]). This structure is now augmented with values based on Eq. 5-7. It is worth noting that this construction of the Bayesian network guarantees the probabilistic independence of EIDs. Any value dependency (*e.g.*, similar readings of close-by sensors) needs to be captured by pattern definition.

Based on the above principles, a Bayesian network is constructed and dynamically updated as events enter the system. At each point in time, nodes and edges may be added to the Bayesian network. The algorithm below describes this dynamic construction. The information regarding the new event is represented by some EID E, the system event history is represented by EH, and the constructed Bayesian network by BN. The algorithm follows:

1. $EH \leftarrow EH \cup \{E\}$
2. Add a node for E to BN.
3. For each such pattern p in a decreasing priority order:

    a. Denote by $sel_p^n(EH)$ the subset of EIDs in EH that may be selected by sel_p^n (these are all EIDs whose *type* attribute is of one of the types specified by sel_p^n).
    b. If there is a subset of events in $sel_p^n(EH)$ that may be selected by sel_p^n such that $pred_p^n$ is true, add a vertex for the derived event's EID E_p. In addition, add edges from all events in $sel_p^n(EH)$ to the event E_p.
    c. For E_p, fill in the quantities for the conditional probabilities according to Eq. 5-7.

4. Calculate the required occurrence probabilities in the probability space defined by the constructed Bayesian network.

The algorithm describes at a high level the calculation of the required probabilities, omitting the details of several steps. The omitted details include the mechanism for selection of events as indicated by sel_p^n, the evaluation of the predicates defined by $pred_p^n$, and the exact algorithm used to infer the required probabilities from the Bayesian network. In all of these cases, standard algorithms from the domains of deterministic event composition and Bayesian networks may be used and extended. The specific algorithms used for these tasks will determine the complexity of our algorithm. However, the dominant factor will be the calculation of the required probabilities from the Bayesian network, which is known to be computationally expensive. Therefore, ways to speed up this step, including reduction in network size and approximate computation, are topics that warrants future research (see discussion in Section 7).

6.1 Inference Example

This section illustrates the above algorithm using a specific example. Assume that in the system there exists a pattern p_1 designed to recognize an illegal stock trading operation, and which is defined as follows: $sel_{p_1}^n$ is

$$\langle \varepsilon_1 \in stockSell, \varepsilon_2 \in stockPurchase \rangle$$

$pred_{p_1}^n$ is

$$(\varepsilon_1.occT \leq \varepsilon_2.occT \leq \varepsilon_1.occT + 5) \wedge (\varepsilon_1.stockTicker = \varepsilon_2.stockTicker) \wedge (\varepsilon_1.customerID = \varepsilon_2.customerID)$$

$eventType_{p_1}$ is $illegalStockTrading$, and $mappingExpression_{p_1}$ consists of two functions: The first maps $\varepsilon_1.stockTicker$ to the $stockTicker$ attribute of the inferred event, and the second maps $\varepsilon_1.customerID$ to the $customerID$ attribute of the inferred event. Finally, $prob_p$ is 0.7. The intuition underlying such a definition is that the sale of a stock, followed closely by a purchase of the same stock, is an indication of suspicious activity.

Consider now the following information about the possible occurrence of an event e_1 such that

$$e_1 \in stockSell, e_1.occT = 5, e_1.stockTicker = \text{``IBM''}, e_1.customerID = \text{``C1''}$$

In addition, the probability that this event occurred is 0.6. This information is represented in the system by an EID E_1 with two possible states: $\{notOccurred\}$ and $\{occurred, stockSell, 5, IBM, Customer1\}$ (we will abbreviate the second state by $\{occurred\}$). The constructed Bayesian network will consist of a single node E_1 with $\Pr(E_1 = \{notOccurred\}) = 0.4$ and $\Pr(E_1 = \{occurred\}) = 0.6$.

Another information is also received regarding the possible occurrence of an additional event

$$e_2 \in stockSell, e_2.occT = 9, e_2.stockTicker = \text{``IBM''}, e_2.customerID = \text{``C1''}$$

This is represented in the system by the EID E_2 with two states as above, which is added to the Bayesian network. At this stage, the Bayesian network consists of two disconnected nodes, E_1 and E_2.

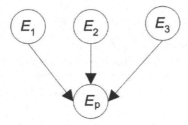

Fig. 2 Inference Example, Part I

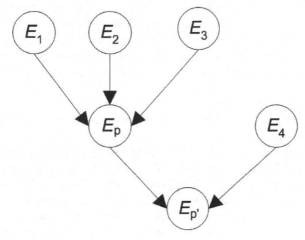

Fig. 3 Inference Example, Part II

Note that although two possible events have occurred, there is no possible world in which two events are selected by $sel^n_{p_1}$, and, therefore, the pattern p_1 is not recognized. Now, assume that information regarding a third event e_3 reaches the system, such that

$$e_3 \in stockPurchase, e_3.occT = 5, e_3.stockTicker = \text{``IBM''}, e_3.customerID = \text{``C1''}$$

This is represented in the system by the EID E_3. Now there is one possible world in which there is a non-zero probability that E_p occurs - this is the world in which the event history is e_2, e_3. Therefore, a node E_p is added to the network, and edges will be added from E_1, E_2, E_3 to E_p. This will result in the Bayesian network depicted in Figure 2.

In addition, the event corresponding to the EID E_p occurs only if e_1 did not occur, and e_2 and e_3 both occurred. Therefore, according to Eq. 5-7,

$$\Pr(E_p = \{occurred\}|E_1 = \{notOccurred\}, E_2 = \{occurred\}, E_3 = \{occurred\}) = 0.7$$

and
$$\Pr(E_p = \{occurred\}|E_1, E_2, E_3) = 0$$
for all other value combinations of E_1, E_2 and E_3.

Finally, if we define an additional pattern p' which states that an event has a non-zero occurrence probability whenever e_p and an additional event of type e_4 occurs, and e_4 is signaled, this will result in the network depicted in Figure 3.

7 Conclusions

In this chapter we have provided the basics of uncertain event processing. We have provided a basic model of uncertain events, demonstrated the use of probability theory, fuzzy set theory, and possibility theory in measuring uncertainty of events and the inferencing of such uncertainty in complex events. A specific simple event language is presented, highlighting the role of uncertainty management in event-based systems.

The challenges, associated with the use of uncertainty in event processing, may be classified into three categories, namely model, usability, and implementation issues, as detailed next.

In the modeling area, we have shown that probability-based models are suitable for some cases, but for other cases, there are more suitable models such as possibility theory or fuzzy set theory. The challenge is to construct a flexible generalized model that can match the appropriate model for a specific implementation.

In the usability area, a major difficulty is the practicality of obtaining the required rules and probabilities. As in many cases, it may be difficult even for domain experts to correctly specify the various cases as well as the probabilities associate with them. Machine learning techniques may apply for automatic generation of rules and probabilities, but the state-of-the-art in this area supports mining only simple patterns; furthermore, in some cases, the history is not a good predictor of the future. Initial work regarding automatic derivation of such rules appears in [34].

In the implementation area, event processing systems may be required to comply with scalability and performance requirements. Therefore, there is a need to develop algorithmic performance improvements to the current models such as Bayesian networks, which are known to be computationally intensive. See [37] for some initial steps in this direction. Possible additional directions for future work include performance improvements of existing derivation algorithms, either by general algorithmic improvements, or by developing domain and application specific efficient algorithms.

References

1. Adi, A.: A Language and an Execution Model for the Detection of Reactive Situations. PhD thesis, Technion – Israel Institute of Technology (2003)
2. Adi, A., Etzion, O.: Amit - the situation manager. The International Journal on Very Large Data Bases 13(2), 177–203 (2004)

3. The Apama home page,
   http://web.progress.com/en/apama/index.html
4. The AutoPilot home page,
   http://www.nastel.com/autopilot-m693.80.html
5. Bacchus, F.: Representing and Reasoning with Probabilistic Knowledge. MIT Press, Cambridge (1990)
6. Balazinska, M., Khoussainova, N., Suciu, D.: PEEX: Extracting probabilistic events from rfid data. In: Proceedings of the IEEE CS International Conference on Data Engineering, Cancun, Mexico (2008)
7. Breese, J.S., Goldman, R.P., Wellman, M.P.: Introduction to the special section on knowledge-based construction of probabilistic and decision models. IEEE Transactions on Systems, Man and Cybernatics 24(11), 1577 (1994)
8. Constantin, A., Gregory, C.: A structurally and temporally extended bayesian belief network model: Definitions, properties, and modelling techniques. In: Proceedings of the 12th Annual Conference on Uncertainty in Artificial Intelligence (UAI 1996), pp. 28–39. Morgan Kaufmann, San Francisco (1996)
9. Drakopoulos, J.: Probabilities, possibilities and fuzzy sets. International Journal of Fuzzy Sets and Systems 75(1), 1–15 (1995)
10. Dubois, D., Prade, H.: Possibility Theory: An Approach to Computerized Processing of Uncertainty. Plenum Press, New York (1988)
11. Etzion, O., Niblett, P.: Event Processing in Action. Manning publications (2010)
12. Gal, A., Anaby-Tavor, A., Trombetta, A., Montesi, D.: A framework for modeling and evaluating automatic semantic reconciliation. VLDB Journal 14(1), 50–67 (2005)
13. Green, T., Tannen, V.: Models for incomplete and probabilistic information. IEEE Data Eng. Bull. 29(1), 17–24 (2006)
14. Gustavo, A.-F., Luis, S.: A temporal bayesian network for diagnosis and prediction. In: Proceedings of the 15th Annual Conference on Uncertainty in Artificial Intelligence (UAI 1999), pp. 13–20. Morgan Kaufmann, San Francisco (1999)
15. Hajek, P.: The Metamathematics of Fuzzy Logic. Kluwer Acad. Publ., Dordrecht (1998)
16. Halpern, J.Y.: An analysis of first-order logics of probability. Artificial Intelligence 46(3), 311–350 (1990)
17. Halpern, J.Y.: Reasoning About Uncertainty. MIT Press, Cambridge (2003)
18. Kanazawa, K.: A logic and time nets for probabilistic inference. In: AAAI, pp. 360–365 (1991)
19. Kjaerul, U.: A computational scheme for reasoning in dynamic probabilistic networks. In: Proceedings of the Eighth Conference on Uncertainty in Artificial Intelligence, pp. 121–129 (1992)
20. Klir, G.J., Yuan, B. (eds.): Fuzzy Sets and Fuzzy Logic. Prentice-Hall, Englewood Cliffs (1995)
21. Liebig, C., Cilia, M., Buchman, A.: Event composition in time-dependant distributed systems. In: Proceedings of the Fourth IFCIS International Conference on Cooperative Information Systems, pp. 70–78. IEEE Computer Society Press, Los Alamitos (1999)
22. Liu, H., Jacobsen, H.-A.: Modeling uncertainties in publish/subscribe systems. In: Proceedings of the IEEE CS International Conference on Data Engineering, pp. 510–522 (2004)
23. The Netcool impact policy language home page, http://www-01.ibm.com/software/tivoli/products/netcool-impact/
24. The Oracle cep home page, http://www.oracle.com/technologies/soa/complex-eventprocessing.html

25. Pearl, J.: Probabilistic Reasoning in Intelligent Systems: Networks of Plausible Infer-
    ence. Morgan Kaufmann, San Francisco (1988)
26. Raiffa, H.: Decision Analysis: Introductory Lectures on Choices under Uncertainty.
    Addison-Wesley, Reading (1968)
27. The RTI data distribution service, http://www.rti.com/products/dds/
28. The RuleCore home page, http://www.rulecore.com/
29. Soulé, R., Hirzel, M., Grimm, R., Gedik, B., Andrade, H., Kumar, V., Wu, K.-L.: A
    universal calculus for stream processing languages. In: Proceedings of the 19th European
    Symposium on Programming, pp. 507–528 (2010)
30. The StreamBase home page, http://www.streambase.com/
31. The Sybase cep home page, http://www.sybase.com/products/
    financialservicessolutions/sybasecep
32. The Tibco cep home page, http://www.tibco.com/products/soa/
    messaging/rendezvous/default.jsp
33. The Tibco rendezvous home page, http://www.tibco.com/software/
    complex-eventprocessing/businessevents/default.jsp
34. Turchin, Y., Wasserkrug, S., Gal, A.: Rule parameter tuning using the prediction-
    correction paradigm. In: Proceedings of the third International Conference on Distributed
    Event-Based Systems (DEBS 2009), Nashville, TN, USA (July 2009)
35. Wasserkrug, S., Gal, A., Etzion, O.: A model for reasoning with uncertain rules in event
    composition systems. In: Proceedings of the 21st Conference in Uncertainty in Artificial
    Intelligence (UAI 2005), Edinburgh, Scotland, pp. 599–608 (July 2005)
36. Wasserkrug, S., Gal, A., Etzion, O., Turchin, Y.: Complex event processing over un-
    certain data. In: Proceedings of the 2nd International Conference on Distributed Event-
    Based Systems (DEBS 2008), Rome, Italy (2008)
37. Wasserkrug, S., Gal, A., Etzion, O., Turchin, Y.: Efficient processing of uncertain events
    in rule-based systems. IEEE Transactions on Knowledge and Data Engineering, TKDE
    (2010) (accepted for publication)
38. The WebSphere business events home page,
    http://www-01.ibm.com/software/integration/wbe/
39. The WebSphere MQ home page,
    http://www-01.ibm.com/software/integration/wmq/
40. Widom, J., Ceri, S. (eds.): Active Database Systems: Triggers and Rules for Advanced
    Database Processing. Morgan Kaufmann, San Francisco (1996)
41. Zadeh, L.A.: Fuzzy sets. Information and Control 8, 338–353 (1965)

Index

Author Index